# Two-Dimensional NMR Methods for Establishing Molecular Connectivity

## A Chemist's Guide to Experiment Selection, Performance, and Interpretation

# Methods in Stereochemical Analysis

**Series Editor: Alan P. Marchand**
Department of Chemistry
North Texas State University
Denton, Texas 76203

## Advisory Board

# Two-Dimensional NMR Methods for Establishing Molecular Connectivity

## A Chemist's Guide to Experiment Selection, Performance, and Interpretation

By

**Gary E. Martin**
**Andrew S. Zektzer**

**VCH**

Gary E. Martin and Andrew S. Zektzer
Department of Medicinal Chemistry
  and Pharmacognosy
College of Pharmacy
University of Houston-University Park
4800 Calhoun Road
Houston, TX 77204-5515

**Library of Congress Cataloging-in-Publication Data**

Martin, Gary E., 1949–

Two-dimensional NMR methods for establishing molecular connectivity : a chemist's guide to experiment selection, performance, and interpretation / Gary E. Martin and Andrew S. Zektzer.
    p. cm.
    Bibliography: p.
    Includes index.
    ISBN 0-89573-703-5
    1. Nuclear magnetic resonance spectroscopy. I. Zektzer, Andrew S., 1960–  . II.
Title. III. Title: Molecular connectivity.
QD96.N8M37    1988                                                     88-27705
543'.0877—dc 19                                                          CIP

**British Library Cataloguing in Publication Data**

Martin, Gary E., 1949–

    Two-dimensional NMR methods for establishing molecular connectivity
    1. Molecules. Structure. Determination. Nuclear resonance magnetic spectroscopy
    I. Title II. Zektzer, Andrew S., 1960–
    541.2'8
    ISBN 0-89573-703-5

Printed in the United States of America.

ISBN 0-89573-703-5 VCH Publishers
ISBN 3-527-26858-8 VCH Verlagsgesellschaft

Distributed in North America by:        Distributed Worldwide by:

VCH Publishers, Inc.                     VCH Verlagsgesellschaft mbH
220 East 23rd Street                     P.O. Box 1260/1280
Suite 909                                D-6940 Weinheim
New York, New York 10010                 Federal Republic of Germany

# Preface

Two-dimensional NMR spectroscopy has assumed a position of continually increasing importance in the assignment of complex NMR spectra and the determination of new molecular structures. Presently, it is possible to routinely tackle problems that were unthinkable even a scant few years ago. The number of techniques available has grown exponentially and there is now, quite literally, an alphabet soup of acronyms to describe the myriad of experiments possible. To the novice, the assortment of acronyms and their respective pulse sequences can be quite bewildering, especially when it is necessary to choose between them to accomplish a particular task. It is the specific intent of this monograph to provide a "road map" to guide people interested in using these techniques through this bewildering array of techniques.

The book is subdivided by objective. Chapter 1 introduces the techniques and some of the terminology. The discussion goes on to address the conceptual problem of what a two-dimensional NMR experiment is and how the information ends up where it does. Two-dimensional J-resolved experiments are used as a medium to demonstrate the function of the various basic intervals common to all two-dimensional NMR experiments and the data processing necessary to process the information. Alternative means of presenting the data are also considered. The chapter concludes with the often neglected subject of pulse calibration and the operation of specialized pulse clusters -- BIRD and TANGO pulses -- now beginning to play an important role in the modification of existing experiments.

Chapter 2 deals with the important goal of establishing proton-proton connectivity networks. The discussion begins with classical homonuclear decoupling and proceeds quickly to the now standard COSY experiment first proposed by Jeener. The pulse sequence, phase cycling, and parameter selection are considered in turn, after which the processing and presentation of the resultant data are considered. Next, the interpretation of COSY spectra is addressed using examples that range from natural products to complex polynuclear aromatic heterocycles. Each of the examples selected also serves as a medium through which additional points about the COSY experiment are addressed, culminating with molecules of sufficient complexity that the spectra become intractable to analysis with a COSY spectrum. The chapter then proceeds to variants of the basic COSY experiment, which include phase sensitive data collection and processing, long range optimized COSY (LRCOSY), broadband homonuclear decoupled COSY (HDCOSY), super COSY, edited COSY sequences (MIS/T and MIS/I) and the spin echo correlated (SECSY) experiment. A group of variants treated as a separate category are the multiple quantum filtered experiments. This segment of the chapter deals principally with the double quantum filtered technique, which is the only member of this group now in common usage. Finally, approximately the last half of the chapter is devoted to "alternative" quantum methods for establishing proton-proton connectivity networks. This section begins with proton double quantum coherence (DQCOSY) including a brief consideration of the creation of multiple quantum coherence. The location and types of responses in proton double quantum coherence spectra are considered, and examples are presented to illustrate considerations in the optimization and interpretation of the resulting data. The discussion then continues on to zero quantum coherence in its various forms (ZQCOSY and ZECSY). Results obtainable with multiple quantum experiments are compared to the COSY family of experiments. The overall objective of the chapter, in addition to providing a basic understanding of this vast group of experiments, is to provide the reader with some insight into the rationale for selecting one experiment over another.

The third chapter considers the powerful heteronuclear chemical shift correlation experiment. Experiments such as spin population inversion, which was one of the forerunners to the modern two-dimensional heteronuclear shift correlation experiment, are considered first to lay the groundwork for the development of the two-dimensional techniques. The development of two-dimensional heteronuclear correlation is historical, beginning with the simplest experiment, which suffered from heteronuclear spin coupling in both frequency domains. Heteronuclear decoupling is then introduced stepwise on first the $F_1$ and then the $F_2$ frequency axes to give the contemporary Freeman-Morris sequence now in wide usage. Aspects of phase cycling, parameter selection, data processing and presentation are considered in turn. The simple, readily available and highly soluble alkaloid norharmane ($\beta$-carboline) is used as an example to provide the reader new to the techniques a compound that can be used to implement the experiments discussed quickly and easily on his own spectrometer. The concern here is to use a compound that can be run "in-house" and the results obtained compared to those presented herein. Artifacts arising in heteronuclear correlation experiments are considered next, followed by alternative correlation sequences such as the DEPT based heteronuclear correlation technique. The chapter then considers heteronuclear correlation sequences using a BIRD pulse in incremented and constant evolution experiments to provide broadband homonuclear proton decoupling. Contemporary proton detected heteronuclear correlation sequences are considered next, and their advantages in sensitivity are contrasted with difficulties in performing the experiments and obtaining adequate digital resolution in the carbon ($F_1$) frequency domain. Finally, the chapter addresses the area of long-range heteronuclear chemical shift correlation techniques, which have been rapidly growing in importance since the first reports in 1984. Long range optimization of the Freeman-Morris sequences and the attendant problems, specifically, one bond modulation of long range response intensity, are considered in detail, as are modifications of the basic sequence which are capable of "decoupling" these modulations. Long range optimized DEPT, fully coupled (FUCOUP), J-filtered techniques and others are considered before turning attention to the constant evolution experiments, COLOC and XCORFE. Finally, the most recent arrival on the scene, proton detected long range heteronuclear multiple quantum coherence (HMBC) experiments, are treated in detail and contrasted to their heteronucleus detected counterparts.

Transfer of magnetization between non-directly coupled homo- and heteronuclear spins provides the central theme of Chapter 4. The relayed coherence transfer experiments (RCOSY and 2RCOSY) pioneered by Eich, Bodenhausen and Ernst are considered first and provide the point of departure for the discussion of isotropic mixing based HOHAHA and related experiments found later in the chapter. Several natural products are used as the basis for comparing the COSY, RCOSY and 2RCOSY techniques. Here again, one of the model compounds, strychnine, was chosen for its wide availability as a model system for new users interested in having data to compare with results generated in their own laboratories. Relayed coherence transfer in multiple quantum coherence experiments (zero and double quantum) is discussed next. Finally, the last of the proton techniques considered are the isotropic mixing experiments, specifically HOHAHA. The development of the concepts underlying these experiments is presented first and leads to a discussion of the MLEV and related sequences, which form the basis for the isotropic mixing intervals on which these experiments depend. Once again, strychnine is used as an example to allow direct comparison to the conventional relay experiments and to provide the reader with data for comparison. Chapter 4 next directs the attention of the reader to the heteronuclear relayed coherence transfer experiments beginning with the heteronucleus detected variants and then turning to the proton detected analogs. A number of examples are presented, here again using the simple alkaloid norharmane to allow the reader to quickly generate data for comparison to that presented in the figures contained in the chapter. Modified heteronuclear relay experiments are considered next including low pass J-filtered and isotropic mixing experiments followed, finally, by proton detected heteronuclear relay experiments.

The last techniques chapter, Chapter 5, is devoted to the topic of $^{13}C$-$^{13}C$ double quantum INADEQUATE and related experiments. The fundamentals of the INADEQUATE experiment are considered in conjunction with parameter selection and the spectral result. Intentional folding in the second frequency domain as a means of improving digital resolution in this highly insensitive and demanding experiment is considered in detail using menthol as an example which allows the generation of usable INADEQUATE spectra in about 6 hours, in a 5 mm probehead. The establishment of the sesquiterpene cedrol is next considered in detail, followed by a discussion of potential breaks in the INADEQUATE spectrum of benzo[3,4]phenanthro[1,2-b]thiophene. Applications of the INADEQUATE experiment are briefly reviewed and alternatives such as the proton detected INSIPID experiment and the carbon relayed proton-carbon heteronuclear correlation experiment are discussed.

Chapters 6 and 7 are best treated in tandem. Problems are presented in Chapter 6, which begins with the application of a single technique in an assignment problem followed by more complex concerted applications of techniques: first to assigning the spectra of compounds whose structures are given, then the elucidation of unknown structures. Experiments utilized run the gamut from simple COSY through proton detected long range heteronuclear multiple quantum (HMBC). Chapter 7 provides detailed solutions to the problems presented in the preceding chapter. Individual spectra provided with the problems are annotated and information that can be gleaned from them considered in detail. It is the specific intent of these chapters, when utilized in tandem, to reinforce the material presented in the preceding five chapters. This is not to say that the limited number of problems contained in these chapters is sufficient to cover all or even a fraction of the possible applications of two-dimensional NMR spectroscopy.

Two-dimensional NMR is a dynamic field, and numerous new experiments have been added since the completion of this monograph. It is the authors' hope that the treatment provided herein will afford the reader a sufficient background in the area that they can begin to tackle new experiments contained in the primary literature on their own.

**Gary E. Martin**
**Andrew S. Zektzer**

# Acknowledgments

There are numerous individuals who must be acknowledged for providing encouragement, assistance, compounds for use as examples, preprints, etc. These individuals include Professors M. Alam, R. N. Castle, R. Freeman, M. D. Johnston, Jr., H. J. Kohn, W. F. Reynolds, M. G. Richmond, A. J. Weinheimer and S. C. Welch, Drs. A. Bax, G. S. Linz, V. V. Krishnamurthy, G. A. Morris, M. J. Musmar, J. N. Shoolery, M. J. Quast, D. Sanduja, R. Sanduja, Mr. M. Salazar, Mr. L. R. Soltero and Mr. D. Barnekow.

We especially want to express our appreciation to Mr. L. D. Sims whose superb help in maintaining the spectrometer, and in designing and building new circuitry made it possible to perform the reverse detected experiments and HOHAHA experiments discussed in Chapters 3 and 4. The assistance of Mr. R. S. Masters in producing the photographs used in illustrating this monograph was likewise invaluable.

A number of publishers have very generously granted permission for the use of published materials which we would like to acknowledge. These include: Academic Press for permission to use Figures. 3-39, 3-49, 6-19 and 7-19; American Chemical Society for permission to use Figures. 2-34, 7-3, 7-4 and 7-5; American Society of Pharmacognosy for permission to use Figures 2-32, 2-35, 2-36, 2-37, 2-47 4-25 and 5-8; Heterocorp for permission to use Figures 1-89, 1-9, 1-19, 2-8, 2-13, 2-41, 2-42, 2-43, 2-44, 2-45, 2-46, 3-1, 3-20, 3-27, 4-24, 5-10, 5-11, 6-1, 6-10, 6-11, 6-12, 6-13, 7-1, 7-2, 7-11, 7-12, 7-13 and 7-14; Pergammon Press for permission to use Figure 3-3; and finally, John Wiley & Sons, Ltd. for permission to use Figures 3-37, 3-40, 3-41, 3-42, 3-43, 3-51, 3-52, 3-53, 3-61, 3-62, 3-63, 6-7, 6-8, 6-9, 6-24, 6-25, 6-26, 6-27, 6-28, 7-9, 7-10, 7-24, 7-25, 7-26 and 7-27.

Finally, we are especially indebted to our wives for their support, immense patience and understanding during the writing of this monograph.

Gary E. Martin
Andrew S. Zektzer

**Dedicated to**

Josh, Casey, Linda and Judy

# Table of Contents

# Chapter 3  Heteronuclear Chemical Shift Correlation                    162

# Chapter 4  Relayed Coherence Transfer and Related 2D-NMR Experiments                    280

# Chapter 5 $^{13}$C-$^{13}$C Double Quantum Coherence 2D-NMR The INADEQUATE Experiment          348

# Chapter 6 Applications Problems                          381

## Chapter 7 Solutions to Problems                                        **418**

## CHAPTER 1

## INTRODUCTION AND PRACTICAL CONSIDERATIONS OF TWO-DIMENSIONAL NMR SPECTROSCOPY

## INTRODUCTION

While there have been numerous developments in the area of nuclear magnetic resonance (NMR) during the 40-plus years since the first observation and report of what was then described as "nuclear induction,"[1-3] the past decade has produced what are probably the most exciting advances. One such advance is magnetic resonance imaging (MRI), which provides the advantages of X-ray computed axial tomography (CAT) as an imaging modality without the danger associated with the radiation. MRI has essentially grown into a discipline of its own, spawning several journals and numerous monographs. No further consideration is given to MRI here; rather, we focus on the other major development that has occurred in NMR over the past decade: two-dimensional NMR (2D-NMR) spectroscopy.

Two-dimensional NMR spectroscopy was first suggested conceptually by Jeener at an Ampere International Summer School held in Yugoslavia in 1971. Unfortunately, Jeener's first presentation was never published. Indeed, it was not until 1975 that the concept was experimentally realized.[4,5] Basically Jeener's idea of 2D-NMR differed from conventional NMR spectroscopy in that response intensity would be a function of two frequencies rather than a single frequency as in a conventional NMR experiment. We may denote such experiments as $S(F_1,F_2)$, where F denotes the separate frequencies. Experimentally, data is collected as a function of two independent time domains giving a time domain data set, which we may denote $S(t_1,t_2)$. By subjecting the data matrix to double Fourier transformation,[6,7] the two independent time domains are converted to the frequency domain as shown by Eqn. [1-1].

$$S(t_1,t_2) \rightarrow S(F_1,F_2) \qquad [1-1]$$

1

It cannot be stressed strongly enough that 2D-NMR does not represent merely an alternative display strategy. The information contained in a 2D-NMR data matrix frequently cannot even be accessed via a conventional (one-dimensional) NMR experiment. We will develop many aspects of this theme, but for now it is sufficient to note that it is possible to detect multiple quantum (zero, double, triple, etc.) transitions that have no analogy in any one-dimensional NMR experiment.

One-dimensional NMR experiments are limited to the portrayal of response intensity as a function of the observation frequency. In contrast, 2D-NMR spectra have available a second frequency domain, which opens vast new horizons in regard to information content of the spectrum. For example, both axes may correspond to a chemical shift frequency. In the case where both axes are the same chemical shift, for example proton, we refer to the spectrum as an autocorrelated spectrum. Historically, the autocorrelated proton 2D-NMR spectrum, which has come to be known as a COSY spectrum (COrrelated SpectroscopY), was the exact experiment initially proposed by Jeener. Alternately, the two frequency axes may correspond to different nuclides, for example proton and carbon, in which case we have a heteronuclear chemical shift correlation spectrum, which is extremely useful. Alternately, the second frequency axis may correspond to a multiple quantum frequency such as zero or double quantum. Here again useful information may be obtained. Importantly, the second frequency domain provides the means of establishing correlations and hence connectivity information, which can be very useful in determining molecular structure. We will discuss the performance of these various experiments and describe the interpretation of the data they contain in the chapters that follow. For now, however, it is sufficient to note that these exciting possibilities exist and can be exploited to provide useful chemical structure information.

After the preamble above, it is appropriate to consider the workings of 2D-NMR experiments. We shall tackle this problem in several stages. First, it is necessary to examine the parts of the experiment and how the second frequency domain is introduced. Second, we must consider how the doubly subscripted data sets common to all 2D-NMR experiments are processed to provide usable information in the frequency domain. Third, we will address the problem of presenting spectral information as a function of two frequencies. Usefully we may tackle these problems using a very simple 2D-NMR experiment, the one pulse, two-dimensional J-resolved (2DJ) experiment. Finally, before continuing to discuss specific 2D-NMR experiments of more general interest than the 2DJ experiment, we will also consider the very necessary evil of calibration of pulse widths from the observe transmitter and the decoupler. While this undertaking is mandatory if 2D-NMR experiments are to be performed successfully, the manufacturers' instruction manuals are generally woefully inadequate in describing how these calibrations can best be performed.

# FUNDAMENTAL SEGMENTS OF
# TWO-DIMENSIONAL NMR EXPERIMENTS

While seemingly highly diverse when judged on the basis of the data they generate, two-dimensional NMR experiments are, in contrast, all fundamentally similar. All two-dimensional NMR experiments may be broken down into four basic intervals. Three of these intervals are mandatory: preparation, evolution and detection. The fourth interval, generally referred to as a mixing period, is optional in the sense that it is not used in all two-dimensional NMR experiments. Before continuing into the specifics of encoding data in the second frequency domain and the first two-dimensional NMR experiments, it is worth briefly reviewing the function of the "building blocks."

## Preparation

Preparation periods are employed in every Fourier transform NMR experiment although we may not think of them as such. In a simple one-pulse experiment (see Fig. 1-1 below), the preparation period will be more commonly referred to as the "interpulse delay." Regardless of how we refer to this interval, it perhaps most frequently serves the purpose either of initially bringing the nuclear spins to thermal eqilibrium or of returning the spins to eqilibrium following an acquisition. Alternatively, the preparation period may be used to saturate one nucleus while another is brought to thermal equilibrium.

Generally, the preparation period ends with the application of a pulse that perturbs the equilibrated spins to initiate the NMR experiment. In most two-dimensional NMR experiments, the pulse applied to end the preparation period is a 90° pulse applied to one of the nuclei in the sample. Here we refer to pulse application to one of the nuclei because, as we shall see in subsequent chapters, we shall frequently be applying pulses to more than one nuclide.

## Evolution

Following disruption of the equilibrium state created during the preparation period, nuclear spins will typically be allowed to precess in the xy-plane of the rotating frame. Precessional events are typically referred to as evolution. As we shall see in the following major section of this chapter, it is the evolution period that is responsible for providing the information encoded into the second frequency domain. The information we choose to encode in the second time domain of the two-dimensional NMR experiment will be a function of homo- or heteronuclear scalar coupling, cross relaxation through space, multiple quantum coherence or some other event. We shall consider evolution periods and how they operate in quite some detail in a later section.

## Mixing

At the completion of the evolution period, we have the option of either detecting the evolved magnetization or somehow causing a further distribution of the evolved magnetization (eg see Relayed Coherence Transfer experiments in Chapter 4; another type of redistribution would be via dipolar relaxation). As noted above, not all two-dimensional NMR experiments employ mixing periods. The choice of whether to utilize a mixing period depends on the information sought.

## Detection

The detection period is familiar to anyone already acquainted with Fourier transform NMR spectroscopy. The duration of the detection period follows well defined rules, which derive from sampling theory and are dependent on the width of the spectral region being observed and the level of digitization to be employed. In two-dimensional NMR experiments, the duration of the detection period is defined by the parameters utilized in digitizing the spectrum of the nuclide being observed (spectral width and level of digitization).

Although the detection period in each two-dimensional NMR experiment will be identical, the information detected in each will generally differ as a function of the changes introduced by the evolution period. Most commonly, the spectrum that results from the first Fourier transformation will bear some resemblance to the conventional or one-dimensional spectrum of the nuclide being observed. It should be noted that while the spectrum will be similar, eg responses will occur in the same location, the intensity and/or phase of the responses contained in the a given "block" of two-dimensional NMR data will frequently have little similarity to the conventional spectrum. This difference is a result of what has transpired during the evolution period and corresponds to the frequency and phase information encoded in the data, which will ultimately provide the information obtained following the second Fourier transformation. The range of information that may be encoded during the detection period is amazingly diverse. Examples include homo- or heteronuclear scalar coupling information, homo- or heteronuclear chemical shfit frequencies, information about multiple quantum coherences that are detectable only by indirect observation as in a two-dimensional NMR experiment and information about homo- or heteronuclear dipolar relaxation among others.

The preceding discussion of detection periods leads us conveniently into the consideration of NMR data collection in two domains and the function of the evolution period in this process. This is also a convenient point to introduce some of the bookkeeping notations which are employed in two-dimensional NMR, which are used in the following sections.

## NMR DATA IN TWO TIME DOMAINS:
## HOW THE INFORMATION GETS THERE

Before we may even consider collecting a 2D-NMR data set, we must consider how we will record data from two presumably simultaneously running time variables. Conventional NMR experiments record data in the time domain. A pulse is applied to excite the system and the free induction decay (FID) is recorded in the form of an exponential decay as the system returns toward equilibrium. Fourier transformation of this data affords response intensity as a function of frequency. It is convenient to denote this operation as

$$S(t_2) \rightarrow S(F_2) \qquad\qquad [1\text{-}2]$$

where the time domain designator, $t_2$, has no relationship to the transverse or spin-spin relaxation time, $T_2$. The pulse sequence for the conventional "one pulse" NMR experiment just described is shown in Fig. 1-1. You will note that there is a short, fixed span of time between pulse application and the initiation of data acquisition in the receiver. This delay is placed here to prevent transmitter pulse breakthrough. Put another way, it allows time for the output of the power amplifier used to generate the excitation pulse to fall completely to zero before the receiver is enabled. In this fashion, we ensure that the observation pulse will not be recorded by the receiver. Since this fixed interval is constant in every experiment we perform, the delay becomes transparent. During the delay, magnetization generated by the pulse is not static but rather has undergone some evolution. However, since the delay is constant in every replicate of the experiment, the evolution is constant and thus may be ignored by most novice users. The concept of evolving magnetization during a delay does, however, provide the mechanism of introducing our second frequency domain of the two-dimensional NMR experiment.

### Evolution Times

If the duration of an evolution delay is varied from one experiment to the next and the events transpiring during the varied interval can be "read. out" in some fashion, we have the means to encode additional information into an NMR data set. As a useful point of departure, let us consider the simple inversion-recovery experiment that is used to measure spin-lattice ($T_1$) relaxation times before turning our attention to the encoding of other spectral parameters such as coupling constants and chemical shift in the second frequency domain.

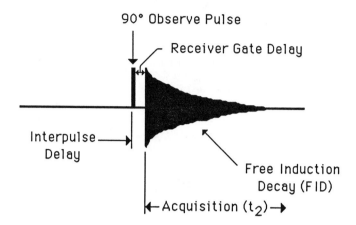

Fig. 1-1.      Pulse sequence schematic for a one pulse Fourier transform NMR experiment. Following an interpulse delay, which is designed to allow the sample to come to equilibrium, magnetic or otherwise, an observe pulse is applied, 90° in this case. Following the application of the pulse, a short, fixed and hence transparent delay is allowed to elapse before the receiver is gated on. After the completion of the receiver gate delay, the receiver is enabled and the free induction decay or FID is recorded.

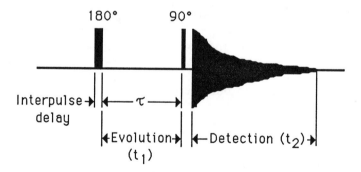

Fig. 1-2.      Inversion-recovery pulse sequence for the measurement of spin-lattice $(T_1)$ relaxation times. The initial 180° pulse inverts magnetization, after which relaxation occurs along the z'-axis during the interval $\tau$ , which is identical to the evolution time, $t_1$, which will be developed in the context of the two-dimensional NMR experiment. The evolved magnetization is rotated into the xy-plane by the 90° pulse, at which point it is sampled and the FID recorded.

# Encoding Spin-Lattice ($T_1$) Relaxation Information via an Evolution Period

The inversion-recovery pulse sequence is shown in Fig. 1-2 and initially employs a 180° pulse to invert magnetization. According to the Bloch equations, the recovery of magnetization following a 180° pulse will be along the z'-axis with no x'y'-component. If we were to follow the 180° pulse immediately by a 90° pulse, we would produce an effective net rotation of 270° and would thus have a response exhibiting unit, negative intensity. In this case, the interval between the two pulses, $\tau$, will be essentially zero and there will be virtually no recovery of equilibrium magnetization along the z'-axis. We may represent the sequence of events just described using the vector diagrams shown in Fig. 1-3.

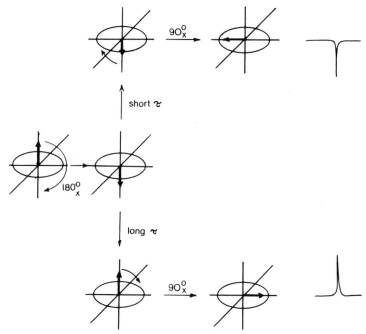

Fig. 1-3.     Vector diagrams illustrating the effect of the duration of the evolution time on spin-lattice relaxation in an inversion recovery experiment. When the duration of the evolution time, $\tau$, is short, only minimal relaxation along the z'-axis will occur. When the evolved magnetization is sampled by the 90° pulse, the -z' magnetization will be rotated to the -y'-axis, ultimately giving rise to an inverted peak in the spectrum. When the duration of $\tau$ is longer, the other condition shown occurs. The +z' magnetization will be rotated to the +y'-axis by the 90° pulse, resulting in the positive signal shown.

When the interval between the two pulses is longer, appreciable relaxation along the z'-axis will occur during the evolution period. Sampling the magnetization after a longer delay is also illustrated in Fig. 1-3. By progressively lengthening the delay during which relaxation is allowed to transpire between the 180° pulse and the 90° "read" pulse, we will observe a continuum of response intensities ranging from essentially -1 to +1 when each file is Fourier transformed and plotted in a vertical series as shown in Fig. 1-4.

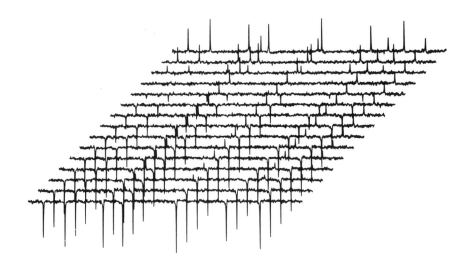

Fig. 1-4    Stack plotted series of spectra illustrating progressively longer evolution periods with correspondingly greater $T_1$ relaxation. The data has been subjected to only a single Fourier transformation as represented in Eqn. [1-3]. All spectra are phased with respect to the first spectrum, which exhibits negative peak intensity because of the 100 μsec evolution period allowed to elapse between the 180° and 90° pulses.

The process that we have used to this point is to collect a series of frequency domain spectra in the time domain, $t_2$, in which a variable duration evolution time, $t_1$ or $\tau$, was inserted to allow spin-lattice or $T_1$ relaxation to occur. $T_1$ is not to be in any way confused with $t_1$. In this fashion, we have just seen how the evolution period, $t_1$, may be used to encode additional information, in this case spin-lattice relaxation information, into our data set. Hence, we have produced a doubly subscripted data set which we may denote $S(t_1,t_2)$, which is analogous to the doubly subscripted data sets necessary for 2D-NMR spectra. Furthermore, the procedure that we use to prepare the data set for presentation, single Fourier transformation with respect to $t_2$, is also exactly analogous to the first step in the processing of a two-dimensional NMR data set. This process is shown in Eqn. [1-3].

$$S(t_1,t_2) \rightarrow S(t_1,F_2) \qquad\qquad [1-3]$$

## Encoding Spin Coupling Information
## via Evolution Periods

In the preceding section we have demonstrated how spin-lattice relaxation information may be encoded into a doubly subscripted data set via an evolution period. Next let us examine how heteronuclear spin coupling information may also be encoded using an evolution period of systematically incremented duration.

To accomplish the task of encoding heteronuclear scalar coupling information, we must first consider the design of a very simple modification of the familiar one pulse experiment. Specifically, we will employ the pulse sequence shown in Fig. 1-5, which employs a single 90° pulse applied to the heteronucleus, $^{13}C$, followed by a variable duration delay. Acquisition will follow the delay in the normal fashion. Now, superimposed over the pulse scheme just described, let us also introduce a decoupler gating scheme. Specifically, broadband proton decoupling will be gated off with the completion of pulse application and will be gated back on when acquisition begins. Thus, during the evolution period, $t_1$, $^{13}C$ magnetization, in addition to precession, will also evolve under the influence of heteronuclear scalar coupling.

Before examining the experimental result of this simple experiment, let us consider it from a vectorial standpoint. In the first acquisition, when the duration of $t_1$ will be zero or minimal, the vector ensemble for a given carbon will be rotated to the y'-axis by the application of the pulse. To simplify matters further, let us also assume that we will be on resonance so that precession can be ignored. Next, when the decoupler is gated off, the vector ensemble will begin to dephase under the influence of the heteronuclear coupling(s) experienced by the carbon. Since during the first experiment the duration of $t_1$ will be minimal, only scant defocusing will take place. Hence, when the decoupling is gated back on, the vector resultant will still be near unity and coaligned with the +y'-axis as shown by Fig. 1-6a.

Next let us consider the effect of a longer evolution time where $t_1$ = J/2. This case is shown in Fig. 1-6b. Here, during the evolution period the vector components will assume an antiphase relationship colinear with the ±x'-axis by the end of the evolution period. Thus, when the decoupler is gated on, the vector components will cancel giving no net signal in the receiver.

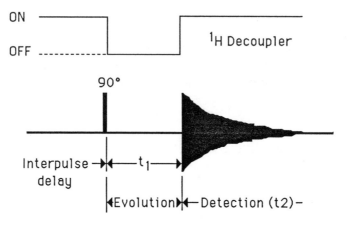

*Fig. 1-5*    Pulse sequence for one pulse, two-dimensional J-resolved NMR spectroscopy. The decoupler is gated off for the evolution period, $t_1$, allowing magnetization to evolve under the influence of heteronuclear spin coupling. When the decoupler is gated on coincidentally with the acquisition, the signal intensity will be amplitude modulated by $^nJ_{CH}$ as a function of the incrementation of the evolution period. In this fashion, scalar heteronuclear spin coupling information may be encoded during the evolution time, $t_1$.

Finally, consider the case where the evolution period $t_1 \sim J$. In this case the vector components will have essentially refocused on the -y'-axis, giving a signal 180° out of phase when the decoupler is gated on and the data acquired (Fig. 1-6c).

**1-1**

From the three cases just described, we begin to see that it is possible to encode heteronuclear scalar coupling information into our doubly subscripted data set by systematic variation of the evolution period. Experimentally, the results obtained with the experiment just described are demonstrated for one carbon resonance of the simple alkaloid norharmane (**1-1**) in Fig. 1-7; we will employ norharmane as a model compound in many instances in this monograph.  As shown in Fig. 1-7, the amplitude of the $^{13}$C resonance modulates from one experimental trace to the next in the series of spectra in the "stack plotted" presentation.

The experiment just described represents only one type of two-dimensional NMR experiment now relatively infrequently used, of what has grown to be a formidable array.  Other uses of evolution will be described later in this chapter and in the following chapters.

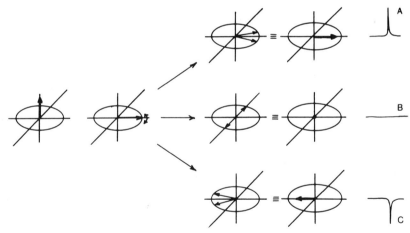

*Fig. 1-6.*      Vector diagrams illustrating the effect of variation of the evolution time on response intensity.  (A) When the evolution time duration is quite short, only minimal defocusing of the vector ensemble will occur due to $J_{CH}$.  The resultant, when the decoupler is gated back on, will be essentially net +y magnetization, the Fourier transformed result of which will be a signal with full positive intensity.  (B) When the evolution time is proportional to J/2, magnetization will be antiphase, colinear with the ±x'-axis.  When the decoupler is gated back on, the antiphase components of magnetization will cancel and no net signal will be recorded.  (C) When the evolution time is proportional to J, the components of magnetization will have refocused on the -y'-axis. When the decoupler is gated back on the resultant will have essentially full net -y magnetization, resulting in a 180° phase shift relative to A.

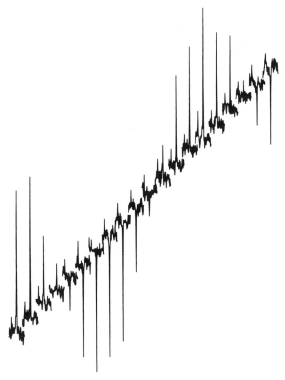

*Fig. 1-7.*        Amplitude modulated series of spectra obtained following the first
                  Fourier transformation (Eqn. [1-3]) of data acquired using the one
                  pulse, 2D J-resolved pulse sequence shown in Fig. 1-5.


## Dwell Times: Control of Spectral
## Width in the Second Dimension

Before we may continue on with the development of the one
pulse 2DJ experiment we have been considering, we must address a
fundamental issue of how we control the spectral width in the second
frequency domain, $F_1$, which is derived by Fourier transformation with
respect to $t_1$.  In a normal one-dimensional NMR experiment we must
decide upon the spectral width we wish to examine and the number of
data points with which it is to be digitized.  Once these decisions have
been made, a number of others will have been made for us
transparently by the operating software of the spectrometer we are
using.  If we wish to discriminate spectral features separated by $\Delta v$ Hz,
then we are required to sample for at least the reciprocal of $\Delta v$ sec.
Based on this, we may define a time known as the dwell time (DW),
which is the reciprocal of the spectral width (SW) to be observed.The
number of points we have selected to digitize the spectral width in
question multiplied by the dwell time gives the acquisition time (AT).

$$DW = 1/SW \qquad\qquad [1\text{-}4]$$

Returning to the simple one pulse 2DJ experiment we are in the process of developing, we may now consider the impact of the dwell time in controlling the spectral width in the second frequency domain. If we initially let the evolution time, $t_1$, equal some minimal value, typically 100 μsec or less, we then may systematically increment the duration of the evolution period by a dwell time chosen to control the spectral width that will result by Fourier transformation with respect to $t_1$. If we wish to have the spectral width in both frequency domains equal, which is convenient for reasons that will become apparent in the present case shortly, we may increment the evolution period using the dwell time from the first frequency domain. In a more practical case, we will typically wish to generate autocorrelated proton or COSY spectra that have identical spectral widths in both frequency domains, since the spectrum is correlated with itself (see Chapter 2). Other spectral widths may be defined by appropriately redefining the dwell time necessary for a given case. We will consider further requirements with individual experiments.

## Data Processing Beyond the
## First Fourier Transformation

In the development of our prototypical 2DJ experiment before we interrupted the discussion to consider the importance of dwell times, we had processed our data matrix as far as shown by Eqn. [1-3]. Assuming that we have incremented the evolution period in successive experiments by the dwell time from the first frequency domain ($t_2 \rightarrow F_2$) to ultimately produce a square data matrix, we are ready to consider how to complete the data processing.

## Transposition

Older dedicated spectrometer computers generally had insufficient memory to load large two-dimensional NMR data matrices into memory in a single large piece. After the first Fourier transformation, the data was arrayed in such a way that it would be necessary for the second Fourier transformation to be performed in a direction orthogonal to the first. Obviously the data files were not formatted to allow this in a file by file fashion. To circumvent this problem it is necessary to relabel or "transpose" the axes of the data matrix resulting from the first Fourier transformation.

Transposition represents nothing more than the restructuring of the data files generated by the first Fourier transformation. Simply, we take the first point of each file, 1 through n, and write them into a new file. The second through $m^{th}$ points are then manipulated in turn. When we are finished, we will have m new files, each containing n time domain points, $t_1$, which vary as a function of frequency, $F_2$, from one file to the next. The transposition process may be denoted as shown in

Eqn. [1-5] and produces files that are amenable to the performance of the second Fourier transformation.

$$S(t_1,F_2) \rightarrow S(F_2,t_1) \qquad\qquad [1\text{-}5]$$

Considering the data points flanking our single amplitude modulated resonance, which was shown in Fig. 1-7, the transposition process produces the set of files shown in Fig. 1-8, which strongly resemble free induction decays (FIDs). However, as suggested by Freeman and colleagues[5] we shall denote these files as interferograms to prevent their confusion with FIDs. In the series of traces plotted in Fig. 1-8, the most intense trace represents the data point closest to the top of the resonance, while those of lower intensity correspond to data points lower on the sides of the resonance. Those files with virtually no information content represent data points that digitized regions of baseline in the spectral traces shown in Fig. 1-7.

Before continuing with a discussion of the second Fourier transformation, it is worth noting that some newer spectrometers have sufficient computer memory to permit entire data matrices to be loaded into memory intact. In such cases, transposition is obviated and even the relabeling of the data matrix axes may become another transparent operation for the disinterested user. However, whether transparent or not, either transposition or relabeling of the axes must occur prior to the second Fourier transformation, which is discussed next.

## The Second Fourier Transformation

Having the data files in a form amenable to further processing, by whatever means, we may next consider subjecting the data matrix to the second Fourier transformation. Returning to our prototypical 2DJ experiment, the first Fourier transformation afforded us proton decoupled carbon resonances whose intensity was amplitude modulated in successive data files. The information encoded as amplitude modulation was the frequency of the heteronuclear scalar coupling(s). Hence, after transposition, Fourier transformation with respect to $t_1$ will provide us with two frequency domains as shown by Eqn. [1-6].

$$S(F_2,t_1) \rightarrow S(F_2,F_1) \qquad\qquad [1\text{-}6]$$

Returning to the one pulse 2DJ experiment with which we began this discussion, we have thus far examined the results obtained following the first Fourier transformation. $^{13}$C chemical shift information is partitioned in $F_2$, while heteronuclear scalar coupling was encoded as an amplitude modulation as a function of $t_1$. Following transposition,

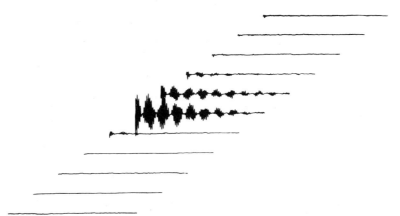

*Fig. 1-8.*        Interferograms obtained by transposition of the amplitude modulated
                  series of traces shown in Fig. 1-7.

Fourier transformation with respect to $t_1$ gives an $S(F_2,F_1)$ data set as
defined by Eqn. [1-6] with heteronuclear scalar coupling information
now available as frequency information in $F_1$. The experimental result
obtained by the Fourier transformation of the single amplitude
modulated resonance we have been following is presented in Fig. 1-8.
The modulation encoded in the time domain, $t_1$, represented a direct
heteronuclear coupling, $^1J_{CH}$, as well as a three bond heteronuclear
coupling, $^3J_{CH}$.
    The entire data matrix collected using this experiment for
norharmane (**1-1**) and processed as just described is shown in Fig. 1-
9 as a whitewashed stack plot. Since chemical shift terms in addition to
heteronuclear scalar coupling terms were allowed to evolve during $t_1$,
there is also dispersion in $F_1$ due to the $^{13}C$ chemical shift.

## Data Presentation

    Continuing to use the data from the one pulse 2DJ experiment for
the moment, we next must consider alternatives in data presentation.
Several are available. The whitewashed stack plot shown in Fig. 1-9 is
by no means the most common form of presentation and certainly not
the most readily interpreted.

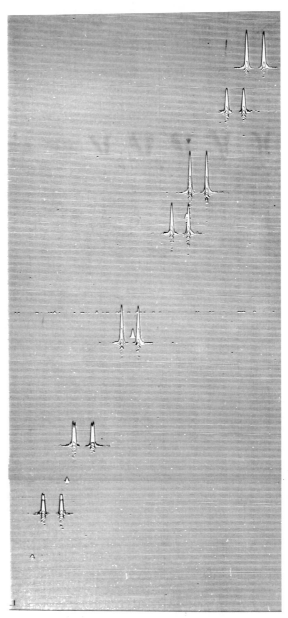

*Fig. 1-9 .*      Whitewashed stack plot of the one pulse 2DJ spectrum of norharmane
(**1-1**) acquired using the pulse sequence shown in Fig. 1-5.

## Whitewashed Stack Plots

The complete spectrum of norharmane (**1-1**) is presented in Fig. 1-9 as a whitewashed stack plot.  Presentations of this form derive readily from simple plot algorithms.  In their earliest form, each successive spectral trace was plotted in its entireity without benefit of the whitewash routine.  Quite obviously, such a presentation would contain numerous overlaps and would be very difficult to interpret.  The whitewash routine improves matters somewhat.  Quite simply, the whitewash algorithm keeps track of the vertical deflection of each point in previous traces.  Whenever the condition arises where the preceding trace contains a point of higher intensity, the computer lifts the pen, interrupting the plot and thereby preventing overlap.

Despite the improvement offered by the whitewashed stack plot, the interpretation of data in such presentations is difficult and peaks are frequently obscured by other responses in the spectrum.  Hence, it would be highly preferable to have an alternative means of presenting two-dimensional NMR data, especially in cases where highly congested spectra must be interpreted.  This situation is solved by the preparation of contour plots.

## Contour Plots

A two-dimensional NMR spectrum is nothing more than a three-dimensional surface.  Generally, the spectroscopist will be more interested in the relationship of the frequencies along the two axes, not in the absolute intensity of the individual responses.  Given this provision, it is possible and preferable to prepare contour plots of the data.

Contour plots are exactly analogous to a topographic map of the earth's surface.  Contour lines are drawn at points of equal elevation.  In the case of a 2D-NMR spectrum, the contour lines will obviously define the shape of responses contained in the data matrix.  Intensity information is indirectly portrayed in the numbers of contour levels that define a given response.  Weak responses will be defined by one or a few contour levels, while intense responses will require numerous contour levels for definition.

The contour plot of the one pulse 2DJ spectrum of norharmane (**1-1**) is shown in Fig. 1-10.  As noted above, both axes represent carbon chemical shift.  Contour plots have a further advantage over stack plots in the ease with which the relation of  frequency information and relationships can be examined. The $F_2$ axis running vertically contains only chemical shift information.  Individual carbon resonances are sorted by chemical shift along this axis as shown in the 90° projection of the data matirx.  The orthogonal $F_1$ axis, in addition to chemical shift, contains heteronuclear scalar coupling information. The 0° projection flanking this axis of the contour plot in Fig. 1-10 thus presents a proton coupled carbon spectrum.  Overlapping multiplet

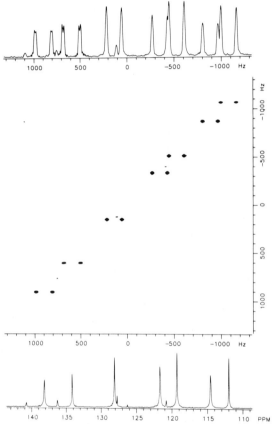

Fig. 1-10.    Contour plot equivalent of the whitewashed stack plotted spectrum of norharmane (1-1) presented in Fig. 1-9 (the matrix has been subjected to an additional transpose to facilitate comparison of the 0° projection and the normal spectrum). The 0° projection of the data matrix is shown horizontally above the contour plot and recovers, in this case, the proton decoupled carbon NMR spectrum of the compound. The 90° projection of the data matrix is plotted vertically and recovers the proton coupled carbon NMR spectrum in this example. Finally, the normal, high resolution carbon spectrum is plotted horizontally below the contour plot for comparison.

structures, which would hamper the extraction of spin coupling information from a conventional proton coupled carbon spectrum, are clearly simplified in the contour plot, since the spin coupling information for each carbon is segregated at the chemical shift of that carbon in $F_2$. Furthermore, information contained in any of the final spectra (slices) that comprise the contour plot shown in Fig. 1-10 may be extracted and examined individually.

Whitewashed stack plots, while esthetically pleasing, are cumbersome and difficult to interpret. Hence, contour plots are in virtually every case the preferred method of two-dimensional NMR data presentation. Probably the sole exception is the comparison of data from multiple two-dimensional NMR experiments. In such cases, it will be most convenient to prepare "slice summaries" of the appropriate traces of each spectrum.

## Projections

Projections provide a useful means of portraying the information content of a contour plot in a familiar form. Generally, you will find contour plots flanked either by projections or by one-dimensional high resolution reference spectra.

Projections are prepared by creating a computer file, which becomes a histogram of the contour plot. Each data point is monitored for intensity in each successive file of the $S(F_2, F_1)$ data matrix. By means of a simple bookkeeping operation, the computer, in examining the entire data matrix, records the highest intensity level of each data point along a given axis. The resulting histogram is the analog of a normal spectrum. In actual practice, there are only three useful angles for projection of a data matrix. The applications of 0° and 90° projections through the two principal frequency axes are obvious. These projections are shown flanking the contour plot in Fig. 1-10. The third angle of projection that has utility is a 45° projection of proton 2DJ spectra. The computation of a 45° projection of a proton 2DJ spectrum provides the spectral equivalent of a broadband proton decoupled proton spectrum that has no analog in conventional NMR spectroscopy.

## Two-Dimensional J-Resolved
## NMR Spectroscopy

Experimentally, the two-dimensional J-resolved experiments are probably the simplest to perform and understand using classical vector descriptions. Practically, the 2DJ experiment may be performed using the one pulse technique employed above as an example. Generally though, it is preferable to restrict the width of the $F_1$ frequency domain to contain only scalar coupling information rather than scalar coupling and chemical shift terms as in the one pulse method. Chemical shift terms may be refocused in J-resolved experiments by resorting to the classical Hahn spin echo.[8] In the heteronuclear variant of the experiment we have also have several further choices. Data may be collected in either an amplitude or a phase modulated form. It is also possible to J-scale the heteronuclear coupling constants or to observe them at their true size using the so-called proton spin-flip technique. In the homonuclear 2DJ experiment we are restricted to phase modulation.

Each of these variants will be discussed briefly in the following sections. However, since the 2D J-resolved experiments are now generally considered to be of limited applicability and are almost uniformly accepted to represent a somewhat poor investment in time for the amount of information they provide, the treatment afforded these first two-dimensional NMR experiments will be somewhat cursory relative to the treatment accorded the more powerful experiments described in the following chapters.

## Heteronuclear Two-Dimensional J-Resolved Spectroscopy

Heteronuclear 2DJ experiments may be broadly subdivided into amplitude and phase modulated variants. Experimentally, the differences in the pulse sequence are minimal. Data processing is largely similar and the extraction of information from the spectrum is comparable.

**Amplitude modulated heteronuclear two-dimensional J-resolved spectroscopy**. The amplitude modulated heteronuclear 2DJ experiment differs only slightly from the one pulse experiment described in detail above.[9] The pulse sequence is shown in Fig. 1-11. The difference is the insertion of a 180° pulse to refocus chemical shift terms during the evolution period. In addition, the decoupler is gated off only during the second half of the evolution period which leads to a down scaling of the size of the heteronuclear coupling constants so that they appear as J/2 rather than J.

Since chemical shift terms are eliminated from the second ($F_1$) frequency domain, the spectral width in $F_1$ may be significantly reduced. The spectral width needs to be only slightly greater than $\pm J/2$, where J corresponds to the largest coupling in the molecule, if we are restricted to a purely methine containing molecule. In the case of methylene and methyl containing systems, spectral width may be limited to $\pm J$ or $\pm 3J/2$, respectively. Regardless of the case, the dwell interval for the 2DJ experiment is defined as a function of:

$$DW = 1/2(SW \text{ in } F_1) \qquad [1\text{-}7]$$

As an example, if a spectral width of 100 Hz ($\pm 50$ Hz) is needed in $F_1$ to accommodate typical aromatic/heteroaromatic one bond couplings (remember that they will appear as J/2), then each half of the evolution period (see Fig. 1-11) would be incremented by 50 msec in successive experiments.

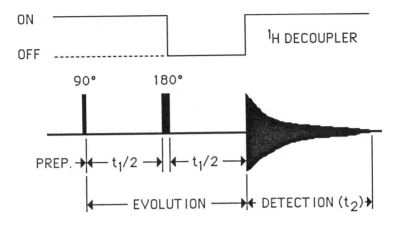

*Fig. 1-11.*    Amplitude modulated heteronuclear 2D J-resolved pulse sequence. The decoupler gated off for half of the evolution time ($t_1/2$) results in a downscaling of the width of the multiplet by a factor of 2 in the $F_1$ dimension.

The operation of the amplitude modulated heteronuclear 2DJ experiment is easily rationalized using a vector formalism. For the sake of simplicity, let us consider a heteronuclear AX spin system such as chloroform. To simplify matters even further, let us eliminate the need to consider precession by performing the experiment "on resonance." Thus, the initial $90°_x$ carbon pulse will rotate magnetization onto the y'-axis, where it will remain stationary during the first evolution period, $t_1/2$. The vector ensemble will also remain collapsed to the resultant during this interval since decoupling is applied. Next, the 180° $^{13}$C pulse will rotate the vector ensemble to the -y'-axis and the decoupler will be gated off. When the evolution period is short, minimal defocusing will occur, giving the result shown in the top trace of Fig. 1-12. When the evolution time is proportional to J/2, the vector components will assume an antiphase relationship and will cancel when the decoupler is gated back on as shown in the center trace. Finally, when the evolution period is proportional to J, the vector components will begin to refocus on the +y'-axis, giving the bottom trace shown in Fig. 1-12. In this fashion we again achieve amplitude modulation of $^{13}$C resonance intensity, which provides the means of encoding scalar coupling information.

The heteronuclear 2D J-resolved spectrum of norharmane (**1-1**) recorded using the amplitude modulation pulse sequence shown in Fig. 1-11 is presented in Fig. 1-13. Note that the information content is the same as contained in the spectrum shown in Fig. 1-9, but we now have a much more efficient use of available digital resolution in $F_1$, since areas of the spectrum that cannot possibly contain any responses are no longer being digitized.

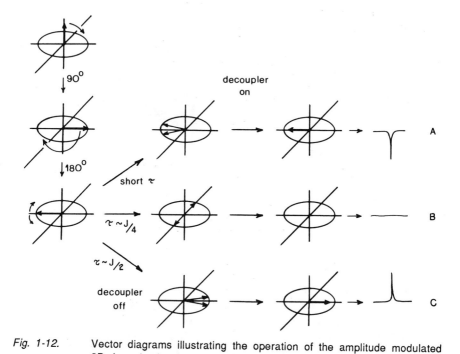

*Fig. 1-12.*    Vector diagrams illustrating the operation of the amplitude modulated 2D J-resolved pulse sequence shown in Fig. 1-11. Assuming an AX heteronuclear spin system in an "on-resonance" condition, precessional effects may be ignored. Hence, during the first half of the evolution period following the 90° pulse, the vector resultant will remain stationary. The $180°_x$ pulse will rotate the resultant to the -y'-axis, at which point the decoupler will be gated off and defocusing due to $J_{CH}$ will begin. When the evolution interval is short (A), minimal defocusing will occur and the resultant will be collapsed when decoupling is resumed on the -y'-axis. The Fourier transformed result is the resonance with negative intensity shown. When the evolution time is proportional to J/2 (B), the vector components will defocus to an orientation colinear with the ±x'-axis (antiphase). Resumption of decoupling will cause cancellation and result in no net signal being observed. Finally, when the evolution time is proportional to J (C), the vector components will refocus on the +y'-axis, which will result in a positive signal following collapse to the resultant and Fourier transformation.

## Phase modulated heteronuclear two-dimensional J-resolved spectroscopy. To proceed from amplitude modulation to phase modulation, the change in the pulse sequence is trivial. Rather than gating the decoupler back on for data acquisition ($t_2$), the decoupler is left off (Fig. 1-14) and the spectrum is recorded with full heteronuclear spin coupling in $F_2$. Obviously, this approach to

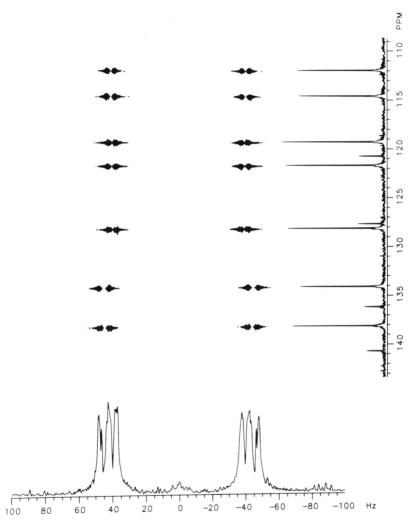

*Fig. 1-13.*　　Amplitude modulated 2D J-resolved spectrum of norharmane (**1-1**). The contour plot, as shown, was obtained directly from the second Fourier transformation. The normal proton decoupled carbon spectrum is plotted vertically beside the contour plot and could also be recovered by 90° projection of the data matrix. Spin multiplets are symmetric about the axis $F_1$ = 0 Hz. The J-spectrum is shown horizontally below the contour plot and is obtained by 0° projection of the data matrix. The J-spectrum contains all of the spin coupling information but no chemical shift dispersion, and its computation has little utility other than for illustration.

acquiring heteronuclear 2DJ data has some stringent disadvantages. First, since data is recorded with full heteronuclear spin coupling during $t_2$, the inherent sensitivity of the experiment will be considerably lower than the amplitude modulated variant. Second, it will no longer be possible to directly extract heteronuclear coupling information from the experiment. Heteronuclear coupling information will be spread in $F_2$ by virtue of the data being recorded with full proton decoupling during $t_2$, as shown in Fig. 1-15. Hence it will be necessary to "tilt" the data matrix to reorth-ogonalize chemical shifts and spin coupling information. While it is possible to eventually retrieve heteronuclear coupling information following the additional data manipulation, this is clearly undesirable when a more sensitive alternative that does not require the manipulation is available. Hence, phase modulated 2D J-resolved heteronuclear spectroscopy is no longer employed.

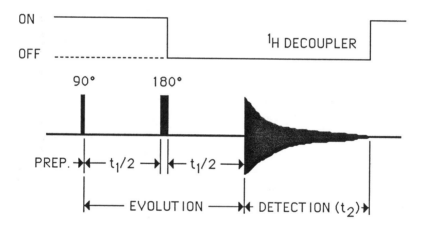

*Fig. 1-14.*    Phase modulated heteronuclear 2D J-resolved pulse sequence. The decoupler is gated off for the second half of the evolution period and the acquisition period ($t_2$). As a result of the decoupler being off during acquisition, carbon spectral information is recorded as the proton coupled multiplet rather than as the decoupled resonance with a significant penalty in sensitivity. Multiplet structures exhibit chemical shift effects in both frequency domains as shown in Fig. 1-15 and the data matrix must hence be tilted before spin multiplet information may be extracted from the spectrum.

**Proton spin-flip heteronuclear two-dimensional J-resolved spectroscopy**. Scaling heteronuclear couplings by J/2 has some advantages in terms of the efficiency of data point utilization in $F_1$. Conversely, it becomes more difficult to observe very small heteronuclear spin couplings when they have been halved in size. To allow the direct obervation of heteronuclear couplings at full size, the proton spin-flip experiment was developed.[7,10] Quite simply, if the decoupler is gated off for the entire evolution period in the amplitude modulated 2DJ

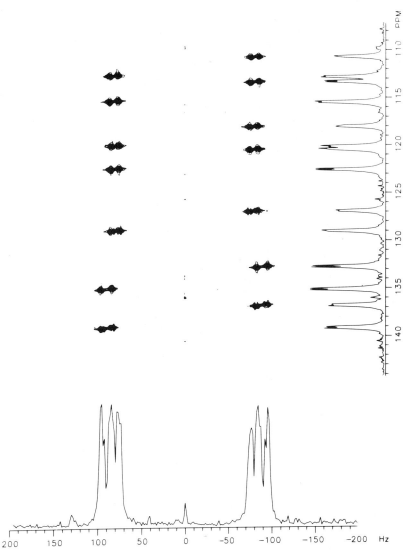

Fig. 1-15.    Phase modulated heteronuclear 2D J-resolved spectrum of norhar-
              mane (1-1) recorded using the pulse sequence shown in Fig. 1-14.
              The 90° projection plotted vertically along the contour plot recovers
              the proton coupled carbon spectrum while the 0° projection plotted
              horizontally beneath the contour plot affords the J-spectrum.   Note
              that spin multiplet structures are now no longer orthogonal to the $F_2$
              axis but rather have a chemical shift component in $F_1$.   Extraction of
              spin multiplet information can be completed only after "tilting" of the
              data matrix.

experiment, heteronuclear couplings will be refocused along with $^{13}C$ chemical shift terms by the application of the 180° $^{13}C$ pulse midway through the evolution period. However, if a proton 180° pulse is applied followed immediately by the 180° $^{13}C$ pulse (essentially simultaneous application), the evolution of heteronuclear couplings will continue for the entire evolution period while chemical shifts are refocused. By gating the decoupler back on during acquisition (see Fig. 1-16), the data will be amplitude rather than phase modulated, with corresponding sensitivity and processing advantages as defined in the preceding section. The sole difference between this experiment and the amplitude modulated experiment described above is that coupling constants are no longer scaled as J/2.

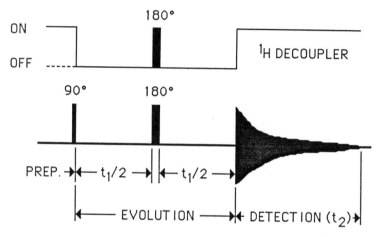

Fig. 1-16.   Proton spin-flip heteronuclear 2D J-resolved pulse sequence. The decoupler may be gated off for the entire evolution period as a result of the inclusion of the 180° proton pulse applied simultaneously with the carbon 180° pulse. The experiment has the beneficial effect of circumventing the J-scaling, which arises when the decoupler is off for only half of the evolution period (see Figs. 1-11 and 1-14) and is thus useful when smaller heteronuclear couplings must be observed.

## Homonuclear Two-Dimensional J-Resolved Spectroscopy

The homonuclear J-resolved 2D-NMR experiment is identical to the heteronuclear analogs except that the decoupler plays no role in the experiment at all. The basic concept of the homonuclear experiment was described by Ernst and coworkers[11] and the pulse sequence is shown in Fig. 1-17. There have been a few modifications.[12-15]
    As with the phase modulated heteronuclear experiment, the homonuclear 2DJ experiment does not have the capacity to eliminate chemical shift from the second frequency domain, $F_1$. Hence, as shown in the schematic presentation in Fig. 1-18, the responses of individual

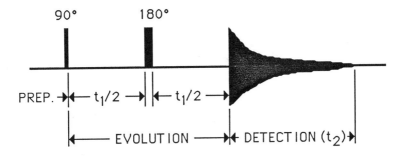

*Fig. 1-17 .*    Homonuclear 2D J-resolved pulse sequence. For homonuclear
applications, since the decoupler is not used at all, spin coupling
constants are observed at full size as in the proton spin-flip
heteronuclear experiment (see Fig. 1-16); there is no ability to choose
between amplitude and phase modulated data -- phase modulated data
must be acquired.

protons are located in a cross section that proscribes a 45° angle
relative to the axis $F_1 = 0$ Hz . Hence, as alluded to above, a 45°
projection of a homonuclear 2D J-resolved spectrum will recover the
equivalent of a proton broadband decoupled proton NMR spectrum that
has no analogy in conventional NMR spectroscopy. Response locations
prior to tilting are denoted by open squares. It should also be noted that
responses on the axis $F_1 = 0$ Hz are unaffected by the tilt algorithm.

After tilting the data matrix as shown in Fig. 1-18 (solid squares),
the matrix is reorthogonalized and individual proton spin multiplets are
now contained in a single data file and may be examined conveniently.
The proton 2DJ NMR spectrum of the simple heterocycle, 1,4,9-triaza-
phenoxathiin (**1-2**) is shown in Fig. 1-19a and 1-19b at observation
frequencies of 100 and 200 MHz, respectively. As will be quickly noted
from the structure, the one dimensional NMR spectrum would be
expected to contain an AB and an ABX pattern for the two isolated spin
systems contained in the molecule. This expectation is indeed realized,
as shown in the contour plot. There are in addition, however, eight
"extra" responses located between the A and B spins of the ABX system,
which arise due to strong coupling. In nearly simultaneous papers,
Kumar[16] and Freeman and coworkers[17] reported expressions for the
analysis of simple homonuclear 2DJ spectra to which the spectrum
shown in Fig. 1-19 has been subjected.[18] It should also be noted that
the position of the strong coupling responses relative to the $F_1 = 0$ Hz
axis and the intensity of these responses are functions of observation
frequency.[19] Compare the location of the strong coupling responses at
100 MHz shown in Fig. 1-19a to their location in the 200 MHz spectrum
shown in Fig. 1-19b. Finally, strong coupling responses, while in many
senses a nuisance, can be used productively to correlate coupled
spins.[20,21]

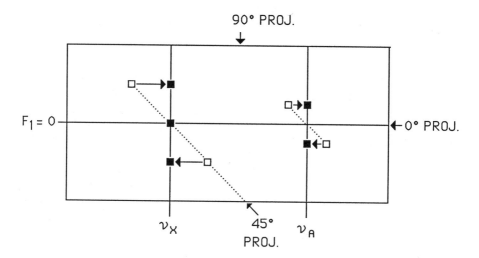

**1-2**

**Fig. 1-18.**    Schematic representation of the data location of a homonuclear 2D J-resolved experiment. Data distribution following the second Fourier transformation is shown. Multiplets contain chemical shift informtion in both frequency domains. Responses observed in the tilted spectrum are shown as solid squares, while their location prior to manipulation of the data matrix is represented by open squares. Responses on the $F_1$ = 0 Hz axis are unaffected by tilting. Prior to tilting, 0° projection would recover the J-spectrum, 45° projection would recover the equivalent of a proton broadband decoupled proton spectrum that has no analog in conventional NMR spectroscopy and 90° projection would recover the normal proton NMR spectrum. Following tilting, the 0° projection would again recover the J-spectrum and the 90° projection would recover the same spectrum as the 45° projection prior to tilting. $F_1$ slices of the data matrix taken after tilting will provide individual proton spin multiplet structures (see Fig. 1-19).

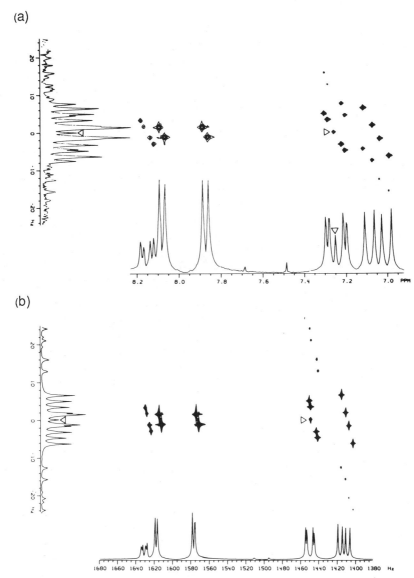

*Fig. 1-19.*     Contour plots of the proton 2D J-resolved NMR spectrum of 1,4,9-
triazaphenoxathiin (**1-2**) recorded at 100 and 200 MHz. (a)  The
contour plot of **1-2** recorded at 100 MHz, in addition to the expected
responses for the AB and ABX spin system, contains eight responses
arising due to strong coupling between the A and B spins of the ABX
spin system. (b) At 200 MHz, the responses due to strong coupling in
the ABX spin system have been shifted away from the $F_1$ = 0 Hz axis
and exhibit diminished intensity as predicted theoretically.[18,19]

## PULSE WIDTH CALIBRATION
## AND SPECIALIZED PULSES

Introductory chapters in treatments of two-dimensional NMR, in addition to some descriptive examples such as those just presented, normally treat the effects of various weighting techniques on line shape. The sensitivity of the techniques or alternately the amounts of sample required for performing various experiments are another favorite topic for discussion. We will make a departure from this protocol. The choice of weighting factors, in our opinion, is more meaningful when considered with regard to processing the data for a given experiment. Additionally, sensitivity and/or sample size requirements are also more appropriately considered with each experiment.   One topic that frequently doesn't receive much attention, however, is pulse calibration.

Pulse calibration is important in that without accurately calibrated pulses, even the simplest NMR experiments requiring a specific flip angle cannot be performed properly.   Further, two-dimensional NMR experiments frequently make use of either a proton or X-band decoupler as a pulse transmitter, something not routinely done in conventional NMR experiments. Hence, it also becomes necessary to calibrate pulse lengths from decoupler amplifiers to perform some of the heteronuclear 2D NMR experiments.

Next, pulses may be strung together in quite useful and ingenious ways to achieve specific tasks.  Examples include composite pulses, BIRD pulses and TANGO pulses.  Pulse clusters of the types just alluded to are also generally unfamiliar to newcomers to two-dimensional NMR spectroscopy.  After we complete the treatment of pulse calibration, we will describe some of the pulse clusters now commonly used in 2D NMR.

### Pulse Width Calibration

One of the most important parameters to be determined before starting an  NMR experiment is the relationship between pulse duration and  spin-flip angle.  Two-dimensional NMR experiments, especially those involving multiple quantum coherence (MQC),   are critically dependent on accurately calibrated pulse durations (widths).  Several methods available for pulse calibration are compared in the following sections. This treatment is not intended to be exhaustively comprehensive but rather to provide a collection of useful methods for the convenience of the novice.  Unfortunately, most manfacturers do not provide adequate descriptions of the calibration of pulse lengths for their spectrometers.

### Pulse Width Versus Spectral Width

Before discussing the specific methods used for pulse calibrations, we must be consider one important factor: power outputs of the transmitter and decoupler are subject to variation over the course of time, and for this reason pulse calibration should be performed on a regular basis. It is important to note that pulse widths are power dependent, if the power output drifts significantly over time it may be necessary to adjust the power level of either the transmitter or the decoupler to produce pulses with 90° pulse widths in the range of 15-25 μsec. It is desirable to obtain pulse widths in the range of 15-25 μsec to maintain uniform excitation over an adequate frequency range.

The importance of pulse length may be dramatically illustrated by using the equations contained in a treatment of pulse width by Martin, Delpuech and Martin.[22] Using the dimensionless parameter $\phi$, which is defined in Eqn. [1-8]:

$$\phi = 4P_W(90°)\Delta v = 2\pi\Delta v/\gamma B_1 \qquad [1-8]$$

Consider the net $M_z^+$ magnetization produced with pulses of 5, 10, and 15 μsec correspond to a 90° flip, where the resonance is 10 kHz from the carrier frequency. Magnetization components $M_x^+$, $M_y^+$ and $M_z^+$ are defined by Eqns. [1-9]-[1-11]:

$$M_x^+ = M_o \sin(\phi_{eff}) \cos(\phi_{eff}) (1 - \cos(\Omega)) \qquad [1-9]$$

$$M_y^+ = M_o \cos(\phi_{eff}) \sin(\Omega) \qquad [1-10]$$

$$M_z^+ = M_o \{\sin^2(\phi_{eff}) + \cos^2(\phi_{eff}) \cos(\Omega)\} \qquad [1-11]$$

where $\phi_{eff}$ is defined by Eqn. [1-12]

$$\phi_{eff} = \arctan(\phi) \qquad [1-12]$$

and $\Omega$ is given by Eqn. [1-13] :

$$\Omega = \mu / \cos(\phi_{eff}) \qquad [1-13]$$

Using the equations above we note that for 90° pulse widths of 5, 10 and 15 μsec, a 180° pulse applied to a signal 10 kHz from the transmitter gives $M_z^+/M_o$ = -0.71, -0.64 and +0.05. These values come from solving Eqns. [1-8]- [1-13] with $\Delta v$ = 10 kHz and $\mu$ = 180°.

The calculations just described clearly illustrate the decreasing effectiveness of an inversion pulse applied to a signal 10 kHz from the transmitter. If a pulse length becomes too long, it will be impossible to produce uniform excitation over the desired spectral width. To remedy the problem of nonuniform excitation, an increase in the power output of the amplifier would decrease the pulse duration for a given tip angle, and, in turn, would increase the effective excitation bandwidth.

## Relaxation Times and
## Pulse Calibration

Relaxation times of the sample used to calibrate pulse widths represents another consideration. Saturation of the signal must be avoided during calibration to ensure accurate pulse length determination. From the Bloch equations, it is easily demonstrated that complete relaxation will be obtained when a delay of $5{*}T_1$ is used between pulses. Quite obviously, accurate determination of $T_1$ relaxation times requires accurately calibrated pulses and is unduly time consuming. The approximate $T_1$ relaxation time of a signal being used for calibration purposes can be determined using the inversion-recovery pulse sequence. To minimize the time required for calibration of pulse lengths, it is thus prudent to choose, if possible, a compound with a relaxation time in the range of one second or less. When small molecules, which tend to have longer relaxation times, are used to calibrate pulse widths, it may be convenient to add a small amount of a relaxation reagent, such as $Cr(acac)_3$ or $Gd(FOD)_3$, to decrease the interval necessary between acquisitions.

Given some approximation of the duration of a 180° pulse, usually determined by increasing the pulse length until the signal disappears (null point, assuming sufficiently long interpulse delays) we may employ half that duration to give us a sufficiently accurate 90° pulse to determine $T_1$ relaxation times of the molecule being used as a calibration standard. For convenience, we normally find it useful to employ a sample of *p*-dioxane in deuterochloroform or deuterobenzene. Using a sufficient sample concentration to afford usable signal-to-noise ($S{:}N$) ratios in one acquisition and allowing 60 sec between acquisitions, an inversion-recovery experiment may next be performed with delays between the 180° and 90° pulses (see Fig. 1-2) incremented in steps of a few seconds beginning with a delay of approximately 250 μsec to provide a spectrum that may be used for phasing. After processing and phasing the spectra in the series, the first spectrum exhibiting any positive signal intensity may be used as the basis for calculating an approximate $T_1$ relaxation time for the sample. By dividing the duration of the interpulse delay used to produce the trace by 0.693, we obtain a relaxation time that may be used as a guide in establishing the interpulse delays used during pulse calibration.

## Observation (Transmitter) Pulses

There are several pulse widths that have utility in two-dimensional NMR experiments. These pulse lengths should thus be calibrated periodically using one of the methods described below.

Basic two-dimensional NMR experiments most frequently utilize 90° and 180° pulses from both the observe transmitter and the decoupler amplifer. Multiple quantum experiments may utilize 45° and 135° pulses in addition to the 90° and 180° pulses. Finally, some composite pulses will require other pulse flip angles, the most common being 240°.

**Determination of the 180° pulse width: The null point method.** Determination of the "null point," that corresponds to a 180° rotation of magnetization from equilibrium is probably the simplest pulse calibration technique.[23] A second null point will be observed when ø = 360°. By assuming a linear relationship between pulse duration and effective flip angle, once the 180° and/or 360° pulse is determined, other requisite flip angles may be approximated.

Generally, the "null point" is determined by recording a series of spectra in which the pulse duration is successively incremented in small steps (usually 5 μsec) beginning with a duration short enough to ensure a flip angle < 90°. In this fashion, response intensity will increase in successively recorded spectra, passing through a maximum at a 90° flip angle and then declining as the pulse width approaches a 180° flip angle. At pulse widths near 180°, responses will be observed ranging from a weak dispersion signal to, ideally, no signal at all. The first null point located in this fashion corresponds to a 180° flip angle. Once an approximation of the 180° pulse is known, a more reliable determination may be made by incrementing the pulse width in smaller steps, eg, 1 μsec.

Accuracy of the null point technique depends on several factors including sufficient time for relaxation and the increment used to locate the null point. Utilization of this method is generally limited to occasions when exact pulse widths are not mandatory and the duration determined by this method provides an acceptable compromise between time spent and calibration accuracy. If the null method is used to determine only the duration of a 180° pulse or only the duration of a 360° pulse, then nothing about the linearity of flip angle versus duration can be stated. A simple modification of the technique just described provides a more comprehensive method of pulse calibration and is discussed next.

**Intensity interpolation.** Automation of the procedure used in the null method is easily introduced and results in greater reliability of the calibration technique. Automation of the method discussed above involves the acquisition and storage of a succession of one pulse experiments. The pulse duration is uniformly incremented with each spectrum acquired over a range of values from <90° to >360°. Fig. 1-20 shows the result obtained using this experiment to calibrate $^{13}C$ pulse widths with a sample consisting of 80% chloroform in hexadeutero dimethyl sulfoxide ($d_6$-DMSO) with a trace of Cr(acac)$_3$ added to minimize relaxation delays.

The first pulse width duration should be chosen such that a flip angle of less than 90° is produced; this ensures that proper phase adjustments can be made for spectra obtained with longer pulse lengths. Incrementation of the pulse width in Fig. 1-20 proceeded with

64 values from 2 to 128 µsec in steps of 2 µsec. This procedure used a preset list of values that were input into the pulse experiment

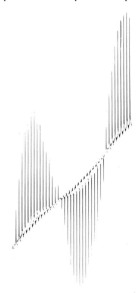

*Fig. 1-20 .*    Series of spectra acquired for pulse calibration purposes using a sample consisting of 80% v/v chloroform in $d_6$-DMSO with a trace of $Cr(acac)_3$ added to minimize interpulse delay duration. A total of 64 increments of the pulse width were taken in 2 µsec steps ranging from 2 to 128 µsec.

under computer control for operator convenience.    Plotting peak intensities allows the interpolation to the 180° pulse width, which theoretically has zero intensity.    Duration of the 360° pulse should be approximately double the length of the 180° pulse. One can thus establish that there is a linear relationship between pulse duration and flip angle.    Correspondingly, the 90° pulse can be assumed to be half the length of the 180° pulse.    Since there is a delay from the time that the pulse is turned on to the time it comes up to full power, the 90° and shorter pulses will inherently have the most uncertainty in their measurement. Where linearity can be assumed, interpolation is effective for calculating pulse durations of any conceivable flip angle. In cases where linearity is questionable or the accuracy of the length of the 90° pulse must be better than ±0.5 µsec, there are two options. First it may be preferable to calibrate the 90° pulse by a more accurate method, such as the method of Lawn and Jones[24] discussed in the next section. Alternately, it may be advisable to use a selfcompensating pulse cluster that is designed to produce accurate flip angles even when there are small (< 10%) errors in the calibrated pulse width.    Compensating pulses will be discussed later in this chapter.

### Calibration of 90° pulses using the double Fourier transformation method: The Lawn and Jones method.

Recognizing that it is difficult to accurately determine the pulse duration necessary to give the most intense response (a 90° pulse), Lawn and Jones[24] devised a simple method for accurately determining the 90° pulse duration. Their technique assumes that there is only a slight deviation in the linearity of flip angleversus pulse width, if any.

Essentially, the method of Lawn and Jones employs the same procedure as above with two additional steps. A stacked set of spectra is collected with the duration of the pulse width stepped in 2-5 μsec increments from < 20° to > 360° over 64 or more values. The accuracy of the 90° pulse measured in this manner is inversely proportional to the value of the pulse increment and proportional to the number of spectra acquired. Once collected, the data is processed in a manner analogous to that used in performing the first Fourier transformation on a set of 2D NMR data (see Eqn. [1-3]). The resultant spectra, once correctly phased, will exhibit a modulation of response intensity induced by the incrementation of the pulse width used to generate one spectrum relative to the next. The data points in the region of the response to be used for calibration are transposed (Eqn. [1-5]) and may be digitally filtered, if necessary. The file containing the most intense response should next be selected and zero filled to between 4 and 16K points. The file is then Fourier transformed and magnitude calculated. The resultant spectra will contain only two signals symmetrically disposed about the zero frequency as shown in Fig. 1-21.

Using Eqn. [1-14], the duration of a 90° pulse may be determined by measuring the offset of the peaks relative to zero after setting the "spectral width" of the plot to ±180° (Hz):

$$t_{90} = 90(\Delta t/\Delta\emptyset) \qquad\qquad [1-14]$$

The accuracy of pulse widths determined by this method is presumably superior to that obtained using interpolation at the expense of a small cost in time for transposition and the second Fourier transformation, not to mention the convenience of not having to perform a regression analysis of the data. This additional accuracy is only true however, if the assumption of linearity is valid.

The methods described above are applicable to the calibration of both proton and carbon pulse widths. In practice, for the calibration of carbon pulses a $^{13}C$ labeled compound sufficiently high in enrichment (>50%) can be used to allow the experiment to be carried out with a single scan per duration increment rather than the use of a sample at very high concentrations doped with a relaxation reagent as used in the example.

### The Wesener and Gunther method.

A second double Fourier transformation method has been described by Wesener and

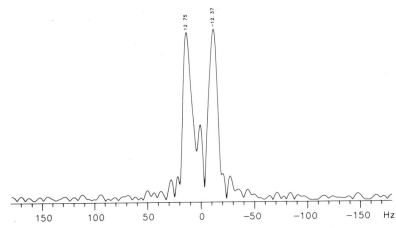

*Fig. 1-21.*     Pulse calibration using the method of Lawn and Jones.[24] The data shown in Fig. 1-20 was transposed. The interferogram corresponding to the highest point of the resonance was zero-filled to 4K points and subjected to the second Fourier transformation. After setting the "plot width" to $\pm 180°$, the peak position may be measured and the duration of the 90° pulse determined according to Eqn. [1-14].

Gunther.[25] Generally, pulse calibration techniques require a delay between sampling on the order of $5*T_1$ to allow reestablishment of the Boltzmann distribution. Wesener and Gunther demonstrated that the procedure described by Lawn and Jones could be used for the calibration of pulse widths under nonequilibrium conditions. Two simple modifications were necessary however; first quadrature phase cycling was necessary to avoid distortions due to incomplete relaxation and, second, four dummy scans were required before actual data acquisition began. By this method insensitive nuclei and/or dilute concentrations can be used to calibrate pulse widths.

## Calibration of Proton Decoupler Pulses for Heteronuclear Experiments

Carbon detected heteronuclear two-dimensional NMR experiments rely not only on accurate pulses from the observe transmitter but also on the accuracy of the pulses generated from the proton decoupler. Power output for the transmitter and the decoupler invariably will cause pulses generated by the two sources to have different durations. Therefore, it cannot be assumed that the 90° proton pulse width calibrated for the transmitter can be used for a 90° proton pulse generated from the decoupler. If proton pulses are to be generated with the proton decoupler, they must be calibrated to assure an accurate flip angle. In addition there are usually two seperate modes in which the decoupler can act: nonmodulated (continuous wave) or modulated. The nonmodulated mode produces a signal that is centered at one frequency and is the mode, used in pulse applications, as opposed to the modulated mode which is used to broadband decouple through modulation of its frequency and phase. Calibration techniques discussed here will assume that the spectrometer is producing a nonmodulated signal for pulsing and a modulated signal for decoupling.

**Calibration of the decoupler pulse through off-resonance decoupling.** One of the earliest methods for calibration of proton pulses from the decoupler relies on the principle of off-resonance decoupling. In the absence of proton decoupling, a methyl carbon would be coupled to its directly attached (via $^1J_{CH}$) protons and the carbon signal would appear as a quartet in the absence of any long range heteronuclear couplings. When a nonmodulating continuous wave field is applied at the frequency of the methyl protons, the carbon signal is decoupled from the protons, collapsing from a quartet to a singlet. This is the basic principle of single frequency on-resonance decoupling. If the proton decoupling field were applied at a point 1 kHz away from the proton resonance frequency, partial decoupling would occur, producing a quartet for the carbon signal with a residual coupling constant, $^RJ_{CH}$, which is smaller than the coupling constant measured in the complete absence of a decoupling field. Furthermore, if this decoupling field were incrementally shifted in frequency toward the frequency of the proton, then the apparent coupling proton-carbon constant would become successively smaller until it collapsed to a singlet when the frequency of the decoupler matched the frequency of the directly attached proton.[26]

Decoupler calibration is carried out by first determining the frequency of the proton signal in megahertz by acquiring a proton spectrum. The resonant frequency of the proton signal is then entered as the proton decoupler frequency. Next the proton-carbon coupling constant ($^1J_{CH}$) is measured by acquisition of a carbon spectrum without proton decoupling. To ensure that the correct proton frequency has been input as the decoupler frequency, the on-resonance decoupled carbon NMR spectrum is next acquired. If the frequency has been entered properly, a singlet will result for the $^{13}C$ resonance. Once confirmation of the proton resonance frequency has been completed, the decoupler should be offset 5-7 KHz from the on-resonance location, a carbon spectrum acquired and the residual coupling constant recorded. The frequency of the decoupling field is then moved back toward the proton resonant frequency in 500-1000 Hz steps, the residual coupling constant ($^RJ_{CH}$) recorded after each carbon spectrum is acquired. Once 4 - 8 values have been obtained, the 90° pulse from the decoupler can be calculated using the following equations:

$$\gamma H_2/2\pi = [(\text{offset in Hz})^2 * ((^1J_{CH} / {}^RJ_{CH})^2 - 1)]^{1/2} \qquad [1\text{-}15]$$

where the 90° proton pulse from the decoupler coils is determined as a function of Eqn. [1-16].

$$90° \text{ pulse} = 1/4(\gamma \, H_2/2\pi) \qquad\qquad [1\text{-}16]$$

Actual values from an experiment using 60 mg 99% $^{13}$C enriched sodium acetate in $D_2O$, are shown, along with the calculated $\gamma H_2/2\pi$ and 90° pulse width are shown in Table 1-1.

Table 1-1.    Measured residual heteronuclear couplings ($^RJ_{CH}$) at various offsets from on-resonance and the resulting calculated values of $\gamma H_2/2$ and 90° pulse widths.

| OFFSET (Hz) | $^RJ_{CH}$ | $\gamma H_2/2\pi$ | 90° pulse width ($\mu$sec) |
|---|---|---|---|
| 2000 | 29.1 | 8480 | 29.5 |
| 2500 | 36.0 | 8460 | 29.6 |
| 3000 | 43.1 | 8330 | 30.0 |
| 3500 | 49.1 | 8360 | 29.9 |
| 4000 | 55.5 | 8240 | 30.3 |
| 4500 | 61.0 | 8230 | 30.4 |
| 5000 | 65.8 | 8260 | 30.3 |
| 5500 | 70.8 | 8190 | 30.5 |

The values contained in Table 1-1 average to give 30 $\mu$sec and vary by no more than ±0.5 $\mu$sec, which is quite reasonable for the 90° pulse from the decoupler. Keep in mind that spectral width is less of a problem than with $^{13}$C spectral acquisition and that consequently longer 90° pulse durations may be experimentally tolerated. In addition the value of $\gamma H_2/2\pi$ gives 8000 Hz as the relative effective sweep width that this pulse will uniformly excite (assuming a perfect pulse), since this measurement was carried out at a proton observation frequency of 300 MHz there is more than sufficient pulse power to excite the normal proton spectral width of 0 to 10 ppm (±1.5 KHz).

One drawback to the method just described is the dependence of the technique on the dielectric constant of the solvent. Therefore, the same solvent used for the calibration should be used for the actual experiment using the value obtained. The alternative technique discussed next is less sensitive to the problem of the solvent dielectric constant. One further note: in many instances the proton decoupler will be used to generate 180° pulses. When 180° decoupler pulses are to be used, it is advisable to employ composite 180° pulses rather than a single 180°pulse. Composite pulses and the rational for their use are discussed later in this chapter.

**Pulse techniques for decoupler pulse calibration**. The use of off-resonance decoupling as a method of calibrating the pulse width from the decoupler coils is somewhat tedious and solvent dependent. Normally an average value from a series of measurements needs be calculated. Several pulsed methods have been proposed that provide both greater sensivity and accuracy than the off-resonance method. Thomas et al.[27] devised a two pulse method for the calibration of the decoupler pulse that was some time latter reproposed by Bax.[28] The pulse sequence referred to as TBPDF is shown in Fig. 1-22.

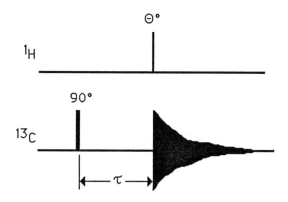

Fig. 1-22.          The TBPDF sequence for the calibration of decoupler pulses.[27] The duration of the interval $\tau$ is fixed as a function of $1/2(^1J_{CH})$. The duration of the pulse applied from the decoupler, $\theta$, is then varied. When $\theta \neq 90°$, a doublet is observed for the resonance being used for calibration with one limb positive and the other negative (see Fig. 1-23) as in an SPI spectrum (see Chapter 3). When $\theta = 90°$ the components of the doublet null. A virtually identical sequence has also been described by Bax.[28]

The sequence requires an $X_nY_n$ (typically $^1H/^{13}C$ will be used in this example and the chapters that follow, but $^1H/^{15}N$ and others also certainly satisfy the requirement equally well) heteronuclear spin system where n = 1 with previous calibration of the 90° X ($^{13}C$) pulse by one of the transmitter calibration methods presented above, as well as measurement of the heteronuclear coupling constant. The duration $\tau$ is equal to $1/2(^1J_{XY})$ and $\theta$ is varied from < 90° to > 90° in several steps. For an XY pair, a doublet with one positive and one negative component analogous to a spin population inversion or SPI spectrum (see Chapter 3 for a discussion of the SPI experiment) results when $\theta$ is $\neq$ 90°. As the value of $\theta$ is increased from ~45° to 90°, the two limbs of the doublet approach zero intensity with an absolute slope of 1. When $\theta$ is exactly a 90° pulse, a null signal results for both limbs of the doublet. For calibration of the pulse width from the decoupler with the TBPDF sequence, it is important to note that broadband decoupling should not be applied during detection or the two vectors will collapse into a singlet and all information will be lost.

### Calibration of decoupler pulses using the pulsed methods of Nielsen and coworkers: SEMUT and SINEPT.

Problems with the TBPDF pulse sequence inspired Nielsen et al. to propose two methods for calibration of the decoupler pulse.[29] Becaue the TBPDF pulse sequence requires the acquisition of a coupled spectrum, the sensitivity of this method is obviously lower than if a decoupled spectrum could be employed.   Newer sequences allow decoupling during acquisition, thereby increasing sensitivity.

Differences in the two proposed methods are significant and will be discussed with regard to the mechanism each sequence uses.  The SEMUT sequence can be regarded as a refocused version of the TBPDF sequence to allow the use of homonuclear decoupling during the acquisition period.  When observing $^{13}C$ this allows the buildup of a nuclear Overhauser enhancement. The SINEPT sequence relies on polarization transfer, as the INEPT experiment does.  This sequence is useful for nuclei that have negative nOe's such as $^{15}N$; this sequence has the added advantage of allowing the use of long range ($^{n}J_{XY}$ n > 1) couplings to transfer magnetization. This is useful for $^{15}N$ when the directly coupled proton is exchangeable. The SINEPT technique can be used for $^{13}C$ if accurate pulse calibration is necessary or if sensitivity is a problem.  Since the intensity of the responses with the SINEPT pulse sequence is a function of $\sin(\theta)\sin(2\theta)$ instead of $\cos(\theta)$, the slope of the response intensity is -2 as $\theta$ approaches 90° compared to -1 for the two methods.

## Calibration of X-Band
## Decoupler Pulses

The TBPDF method discussed above in the context of proton decoupler pulse calibration can also be used for the calibration of a 90° pulse generated by an "X-band" decoupler.  In the case of proton detected experiments, for example, long range multiple bond heteronuclear multiple quantum chemical shift correlation (see Chapter 3), where the decoupler would be employed to pulse carbon, then a $^{13}C$ labeled compound should be used.  This method is also applicable to the calibration of decoupler pulses  for other nuclei  such as $^{31}P$, $^{15}N$ or $^{19}F$.

Data generated with the TBPDF pulse sequence to calibrate the 90° $^{13}C$ from an "X" band decoupler is shown in Fig. 1-23. In this particular application, the proton 90° pulse was previously determined by the method of Lawn and Jones.[24]  A solution of 99% $^{13}C$ enriched sodium acetate (60 mg) in 0.4 ml of $D_2O$ was employed. The spectrum shown in Fig. 1-23 consists of  positive and negative components of the proton doublet arising from the 99% enriched $^{13}C$ to which it is attached, where the components of the doublet are separated by the proton-carbon coupling constant.  As can be seen, the signals decrease in absolute intensity as $\theta$ is increased until at $\theta$ = 90° a null is reached.

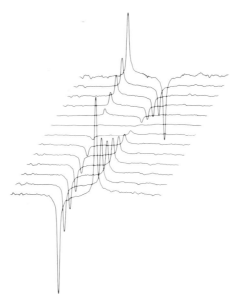

*Fig. 1-23.*     Calibration of X-band decoupler pulses using a sample of 99% $^{13}C$
enriched sodium acetate in $D_2O$. The proton doublet arising due to $^{1}J_{CH}$
coupling to $^{13}C$-C2 is shown. The spectra have the appearance of an
SPI or SPT spectrum when the X-band pulse $\neq 90°$. The two limbs of
the doublet are canceled when the X-band pulse = 90°. The same
technique can be used for the calibration of proton decoupler pulses by
recording the signal for $^{13}C$.

## Composite Pulses and
## Specialized Pulse Clusters

Having considered the various means available for the
calibration of pulse lengths from both the observe transmitter and the
decoupler, it is appropriate to direct our attention to composite pulses
and specialized clusters of pulses (pulse operators). Composite pulses
will be considered first. In general, composite pulses are specialized
pulse clusters that have the specific task of producing very accurate spin
flips, which are compensated for offsets from the transmitter and/or
miscalibration of some pulse, typically the 180° pulse. Specialized
pulse clusters are comprised of bilinear rotational decoupling or BIRD
pulses and a pulse cluster known as TANGO, which is short for Testing
for Adjacent Nuclei with a Gyration Operator. Both BIRD and TANGO
pulses operate on the differential between the one bond and longer
range heteronuclear spin coupling constants, thereby providing the
opportunity to selectively manipulate components of magnetization
associated through directly attached or remote protons. These pulse
clusters will be the last topic discussed in this chapter.

## Composite Pulses

Radiofrequency pulses used to cause excitation in two-dimensional nuclear magnetic resonance are often required to produce accurate spin-flip angles. The two most widely utilized flip angles in 2D NMR experiments are pulses that produce net rotations of 90° or 180°. Especially in the case of a 180° pulse used for inversion of magnetization, deviations in flip angle and phase imparted to signals at various offsets from the transmitter represent a major concern. Hence, multipulse sequences, referred to as composite pulses, were developed to produce accurate 180° pulses over a broad range of frequencies and/or miscalibration of the 180° pulse. Collectively, composite pulses are pulse schemes that would produce the same $\theta°$ flip that a single "perfect" $\theta°$ pulse produces. More recently, methods have been devised to allow the construction of composite pulse sequences for any filp angle. Consideration of when it is appropriate to use composite pulses and the form of these composite pulses will be undertaken in the following sections.

**Compensating pulses**. In some applications, as shown by the example associated with Eqns. [1-8] - [1-13], a single pulse is capable of producing a desired flip angle over a specified spectral width. As a result of this shortcoming, compensating pulses were devised. Compensating pulses are composed of a number of radiofrequency pulses of specific duration and phase applied in succession. The purpose of a compensating pulse cluster is to produce a "more accurate" pulse. Composite pulses have the ability to compensate for deviations in the actual flip angle from the desired flip angle and for offset effects in flip angle imparted at varied offsets from the transmitter.

***Sandwich pulses as a replacement for single pulses.***
The initial composite pulse reported by Levitt and Freeman[30,31] was referred to as a "sandwich" since this pulse cluster consisted of three pulses; a 180°(y) pulse flanked by two 90°(x) pulses, resembling a sandwich. This protypical composite pulse replaced a single 180°(y) pulse to produce the same overall effect on flip angle. The results of Levitt and Freeman's initial report[30] concluded that a composite pulse of the form:

$$90°(x)\ 180°(y)\ 90°(x) \qquad\qquad [1\text{-}17]$$

was more effective than a single 180°(+y) pulse with respect to offset compensation and unintentional miscalibration of pulse widths.

The vector trajectories of the 180° composite pulse shown by Eqn. [1-17] are compared to the single 180°(y) pulse in Fig. 1-24. From this presentation, it can easily be seen that both the single pulse and the composite pulse produce the same net rotation of the vector about the y'-axis. Although it is beyond the scope of this chapter, it can further be demonstrated that the composite pulse performs the rotation more accurately, particularly in the case where the 180° pulse has been inadvertently miscalibrated.

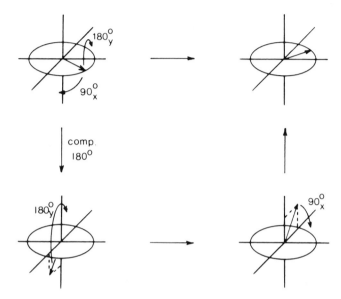

*Fig. 1-24 .*      Vector diagrams showing the vector trajectories of a 180°(y) composite pulse compared to a normal 180°(y) pulse.

Measurement of the spin-lattice relaxation time ($T_1$) of the high field signal of butane-1,3-diol was performed using the inversion-recovery sequence (see Fig. 1-2) and was then compared to the null-point method of $T_1$ measurement using a composite 180° pulse. Comparison of the two techniques is reasonable since the determination of $T_1$ using the inversion-recovery method is insensitive to missetting the 180° pulse used to invert magnetization. Therefore, the inversion-recovery method was used to determine the actual $T_1$ time. Next, the much faster yet, less forgiving, null-point method was used. The null method requires a 180° pulse to invert magnetization followed by acquisition of the signal. In the report of Levitt and Freeman[30,31] the inversion-recovery and null-point methods of measuring spin-lattice relaxation times were compared to ascertain the ability of composite pulses to alleviate problems of transmitter offset and radiofrequency field strength (or missetting of flip angle).

Following the initial communication,[30] a more comprehensive treatment of composite pulses was reported by Freeman, Kempsell and Levitt.[31] In the latter report, an alternative composite 180° pulse Eqn. [1-18] was found to be less sensitive than that shown in Eqn. [1-17] to missetting of pulse angle and offset effects.  This result was demonstrated by use of the second pulse scheme to measure the $T_1$ relaxation time of the high field signal of butane-1,3-diol by the null-point method. A most striking feature of this composite pulse, as compared to the original shown in Eqn. [1-17], is the flip angle of the center pulse. This composite takes the form:

$$90°(x) \ 240°(y) \ 90°(x) \hspace{3cm} [1-18]$$

This composite pulse uses a 240° pulse, the value of which must be interpolated from data generated by the interpolation technique discussed above in the section on pulse calibration.

*Higher order compensated pulses*. Following the initial papers of Freeman and coworkers, Levitt published an exhaustive treatment of the theory and experimental verification of the improvement that composite pulses having three or more components offered over single pulses with respect to   radiofrequency field inhomogeneity.[32] Compen-sation of offset effects was treated in a second paper.[33] Levitt reported that the composite pulse shown in Eqn. [1-19]

$$90°(x) \ 200°(y) \ 80°(-y) \ 200°(y) \ 90°(x) \hspace{2cm} [1-19]$$

was more effective than the composite pulses presented in Eqns. [1-17] and [1-18]. A composite sequence of the type shown in Eqn. [1-19] is the most effective for both offset effects and radiofrequency field strength inhomogeneity when compared to the simpler three pulse composites. Once again, however, it is necessary to determine the duration of the 80° and 200° pulses by interpolation.

The next step in composite pulse research was the derivation of a procedure to construct composite pulses of arbitrary flip angle described by Levitt and Ernst.[34] The effectiveness of composite pulses with up to sixteen 90° pulses used to produce a 180° rotation was considered. All pulse schemes reported by Levitt and Ernst[34] utilized routine phase shifts of 0°, 90°, 180° and 270°, making these composite pulse simple to implement on most commercial NMR spectrometers.

Problems with the sequence shown in Eqn. [1-19] and composite pulses with greater numbers of pulses arise from the total time required for sequence execution. Most applications discussed in this book will require nothing more complex than the sequences shown in either Eqn. [1-17] or [1-18] in place of a conventional 180° pulse. Furthermore, none of the experiments described in subsequent chapters require a compensated pulse in lieu of 90° pulses. This posture is consistent with the opinion of Tycko and co-workers,[35] who stated that in most instances only 180° pulses need be replaced by their equivalent composite pulse. The interested reader is referred to the papers of several groups for a more comprehensive discussion covering the conditions that dictate the utilization of composite pulses and which composite pulses are suitable in particular applications.[35-38]

*Composite pulses intended to remove phase distortions*. In addition to imperfections in the flip angle produced by an isolated 180° pulse, it is possible for single pulses to lead to phase distortions that are intolerable in some experiments such as those creating multiple quantum coherences. Consequently, composite pulses have also been developed that remove the effects of phase distortion better than sequences shown in Eqns. [1-17] - [1-19]. In order to generate pulses without phase distortions, Tycko, Schnider and Pines[35] employed pulse schemes that employ non-90° phase shifts as well as sequences that contain routine 90° phase shifts. Although not the first composite pulse schemes to contain non-90° phase shifts,[32,39] these composite pulses were the first to address the problem of compensation for phase distortions. Tycko et al.[35] have described in detail the theory behind these compensating pulses. Once again, a discussion of the theory behind these compensated pulses is beyond the scope of this chapter and the interested reader is referred to the work of Tycko and coworkers for a rigorous treatment. It is, however, useful to present a few examples of phase compensated pulses that utilize both 90° phase shifts and non-90° phase shifts. Phase compensated 90° pulses are shown in Eqns. [1-20] and [1-21]

$$385°(x) \; 320°(-x) \; 25°(x) \hspace{3cm} [1\text{-}20]$$

$$90°(0) \; 180°(105) \; 180°(315) \hspace{3cm} [1\text{-}21]$$

while their 180° counterparts are shown in Eqns. [1-22] and [1-23].

$$336°(x) \; 246°(-x) \; 10°(y) \; 74°(-y) \; 10°(y) \; 246°(-x) \; 360°(x) \hspace{1cm} [1\text{-}22]$$

$$180°(0) \; 180°(105) \; 180°(210) \; 360°(59) \hspace{2cm} [1\text{-}23]$$

The sequences shown in Eqns. [1-20] and [1-22] are examples of composite pulses that employ routine 90° phase shifts. Sequences shown in Eqns. [1-21] and [1-23] use non-90° phase shifts, where the shift in degrees is given parenthetically. Most commercially available spectrometers are not capable of the non-90° phase shifts used in these examples without modification of hardware and/or software. In general, the sequences shown in Eqns. [1-20] and [1-22] are appropriate for removing the problems associated with phase distortions and offset effects and may be directly implemented without spectrometer modification.

As discussed later in Chapter 2, double quantum filtered COSY spectra may be quite useful in the elucidation of chemical structures that involve complex proton NMR spectra. The use of a double quantum filter requires accurate rotation of magnetization and a minimum of phase distortion if the filter is to work optimally. The composite pulse shown in Eqn. [1-21] was successfully employed to replace the two 90° pulses used to provide for double quantum filtering in a one-dimensional spectrum.[35] This composite pulse compensates for both miscalibration of pulse widths and phase distortion. In principle this same scheme could be used to replace the final two pulses in double quantum filtered

COSY (DQF-COSY). In the example described, non-90° phase shifts were used. There should, however, be no reason for not using the composite pulse shown in Eqn. [1-20] in place of the sequence shown in Eqn. [1-21] with instruments that do not have a non-90° phase shifting capability.[35,38]

**Concluding remarks about composite pulses**. The use of composite pulses that either are self compensating or are compensating when used in combination with other composite pulses has been well documented. The interested reader is urged to consult the literature for a complete discussion. Applications of composite pulses in homonuclear and heteronuclear chemical shift correlation are discussed by Bax.[40] The use of composite pulses in multiple quantum coherence is considered by Weitekamp.[41] Further information can be also be obtained by consulting references 31, 35, 37, 39, 42.

## Pulse Operators: BIRD and TANGO Pulses

Pulse operators are groups of pulses that permit discrimination on the basis of the size of a spin-spin coupling constant. Differentiation is based on the fact that the direct or one bond heteronuclear coupling constant, $^1J_{CH}$, typically is in the range from 125 - 225 Hz. In contrast, longer range heteronuclear coupling constants, $^nJ_{CH}$ where n = 2-4, are generally less than 20 Hz. This approach lends itself to a wide variety of applications. Examples include but are not limited to decoupling of homonuclear scalar interactions, selective observation of direct versus long range heteronuclear coupling constants and "decoupling" modulations due to direct proton-carbon coupling constants in long range heteronuclear chemical shift correlations.

### BIRD Pulses

The first pulse cluster was introduced by Pines and coworkers.[43] The sequence was termed a Bilinear Rotational Decoupling operator, and the acronym "BIRD pulse" is now in general usage. Originally, the BIRD pulse was designed as a method of providing broadband homonuclear decoupling for organic compounds as demonstrated by Pines and coworkers.[43] The sequence has the form:

$^1$H          90°(+x)  - τ-  180°(+x)  - τ- 90°(+x)

$^{13}$C                          180°(+x)                          [1-24]

where $\tau = 1/2(^1J_{CH})$. The BIRD pulse cluster, shown in Eqn. [1-24], utilizes the difference in the size of the direct versus long range heteronuclear coupling constant to invert all protons directly bound to

[13]C, while leaving protons long-range coupled to a given carbon unaffected. The motion of direct and long range couplings is shown by the vector diagram of the BIRD operator in Fig. 1-25.

To better understand the differential actions of the BIRD pulse, a step-by-step treatment of the effect of the BIRD pulse shown in Eqn. [1-24] and illustrated in Fig. 1-25 follows. Initially, Fig. 1-25a assumes that the magnetization vector ensemble is aligned on the $+z'$-axis. After a nonselective $90°(x)$ [1]H pulse is applied, the vector ensemble containing components from both $^1J_{CH}$ and $^nJ_{CH}$ is rotated to the y-axis as shown in Fig. 1-25b. During the first evolution period of duration $\tau = 1/2(^1J_{CH})$, vectors due to $^1J_{CH}$ have each precessed by $90°$, establishing a $180°$ phase error, leaving them antiphase as shown in Fig. 1-25c. During the same time interval, the vectors due to $^nJ_{CH}$ will have diverged only $[(^nJ_{CH}/^1J_{CH}) \times 90°]$ degrees. After the first evolution delay, a $180°(+x)$ pulse is applied to [1]H, which serves to rotate the $^nJ_{CH}$ components to the $-y'$-axis while leaving the $^1J_{CH}$ components coaligned with the $\pm x$-axis and unaffected (Fig. 1-25d). In the absence of any other influencing pulse, all of the vector components would refocus on the $-y'$-axis after a second delay interval equal in length to the first, forming a classical spin echo.[44] In contrast, when a $180°$ [13]C pulse is applied immediately after the $180°(+x)$ [1]H pulse, a very different situation develops (Fig. 1-25e). The $180°$ [13]C pulse serves to exchange the [1]H spin labels and, as a result, components due to $^nJ_{CH}$ continue to defocus while those due to $^1J_{CH}$ completely refocus along the $+y'$-axis (Fig. 1-25f). The final $90°(x)$ [1]H pulse rotates the long range components of magnetization back to the $+z'$-axis, essentially returning them to equilibrium. Direct components of magnetization are placed on the $-z'$-axis, in a polarized state (Fig. 1-25g). The manipulated magnetization components are now in a position analogous to what we would observe in a selective population transfer (SPT) experiment.[45,46] Importantly, while the SPT experiment operates selectively on single transitions, the process we have just described operates nonselectively on all $^1J_{CH}$ magnetization components while simultaneously excluding $^nJ_{CH}$ components. Continuing, if the magnetization were to be sampled with a $90°$ "read" pulse applied to carbon, only those components due to $^1J_{CH}$ would affect the observed [13]C signal.

A second form of the BIRD pulse that selects for long range couplings is described as follows in Eqn. [1-25].

[1]H        $90°(+x) - \tau - 180°(+x) - \tau - 90°(-x)$

[13]C                        $180°(+x)$                        [1-25]

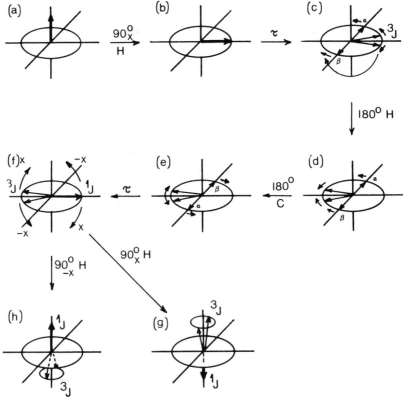

*Fig. 1-25.*     (a) Proton magnetization at equilibrium coaligned with the +z'-axis.
(b) Vector alignment following an initial 90°(x) proton pulse (see Eqn.
[1-24]). (c) Dephasing during the initial interval, $\tau = 1/2(^{1}J_{CH})$:
components of magnetization from coupling to the directly attached
proton develop a net 180° phase error while those for longer range
couplings dephase to a much smaller extent. (d) Magnetization
following the 180°(x) proton pulse. (e) Magnetization following the
180° $^{13}$C pulse (note that the labeles for α and β have been
exchanged). (f) Vector orientation following the completion of the
second interval. (g) Vector orientation following a final 90°(x) proton
pulse. (h) Vector orientation following a final 90° (-x) proton pulse
(see Eqn. [1-25]).

The difference between the two versions of the BIRD pulse lies in the phase of the final proton pulse. Since all prior events are equivalent in the two versions up to the application of the final proton pulse, we may continue the discussion from this point (Fig. 1-25f). As described above, before the final pulse is applied, the direct and long range components of magnetization are essentially 180° out of phase. The final 90°(-x) $^1$H pulse rotates the direct components of magnetization back to the +z'-axis, returning them to equilibrium, while the long range components of magnetization are placed on the the -z'-axis, in a polarized state (Fig. 1-25h). At this point, if the magnetization were to be sampled with a 90° "read" pulse applied to carbon, only those components due to $^nJ_{CH}$ would affect the observed $^{13}$C signal.

**Applications of BIRD pulses.** Following the initial report of Pines and coworkers[43], several applications of BIRD pulses appeared quite quickly. The use of the BIRD pulse in semiselective 2D J-resolved spectroscopy was reported first,[47] followed by an application to broadband homonuclear decoupling in heteronuclear chemical shift correlation.[48] These and other applications are treated in detail in the following sections.

*Semiselective 2D J-spectroscopy*. Although two-dimensional J-resolved (2DJ) experiments have rather limited utility, one paper by Bax[47] is interesting in that it represents a practical application of the BIRD pulse. Bax utilized the BIRD pulse to selectively measure direct ($^1J_{CH}$) or long range ($^1J_{CH}$) heteronuclear coupling constants. The pulse sequence devised for this application is a modification of a pulse sequence initially reported by Ernst and coworkers as a method of obtaining heteronuclear J-resolved spectra.[49] Modification of the original pulse sequence required the replacement of the 180° proton pulse with a BIRD operator as shown below in Fig. 1-26.

Normally, the 2D J-resolved experiment employs either a single 180° $^{13}$C pulse midway through the evolution period[49] or a simultaneous $^1$H/$^{13}$C 180° flip at the same point in the so-called proton flip experiment (see Fig. 1-16).[50] Using the conventional experiment as a point of departure, Bax replaced the 180° pulse(s) with a BIRD pulse. Discrimination between long range and direct couplings is accomplished with the phase of the final 90° proton pulse of the BIRD pulse. When the phase is +x as shown in Eqn. [1-24], only responses due to direct couplings will be observed. Alternately, when the phase is -x as shown in Eqn. [1-25], only responses due to long range couplings will be observed.

In work quite similar to that reported by Bax, Wong and Clark have also reported the use of BIRD pulses for selectively measuring geminal proton-proton coupling constants in steroids.[51]

**Other applications of BIRD pulses.** Several other interesting applications of the BIRD pulse in heteronuclear chemical shift correlation have appeared. The first application was reported by

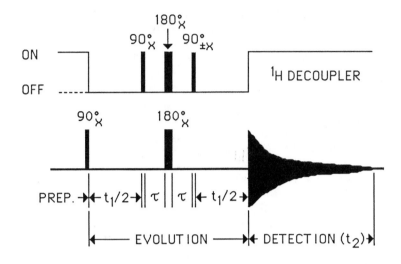

*Fig. 1-26.*      Semiselective 2D J-resolved pulse sequence devised by Bax.[47] The 180° proton pulse normally located midway through the evolution period is replaced by the BIRD pulse in which the phase of the second 90° pulse is chosen to select for either direct couplings (+x, as in Eqn. [1-24]) or long range couplings (-x, as in Eqn. [1-25]).

Bax,[48] who used a BIRD pulse (Eqn. [1-24]) midway through the evolution period of the conventional heteronuclear chemical shift correlation experiment to provide broadband homonuclear decoupling in the $F_1$ dimension. Similar applications have been reported by Wong and his colleagues for the DEPT based heteronuclear correlation experiment.[52] These applications of the BIRD pulse are treated in detail in Chapter 3. More recently, several groups have utilized BIRD pulses of the type shown in Eqn. [1-25] to provide  one bond "modulation decoupling" in long range heteronuclear chemical shift correlation experiments.[53-56]  Once again, these applications are covered in considerable detail in Chapter 3.

**_Compensated BIRD pulses_**. No experimental confirmation of the range of coupling over which the BIRD pulse effectively operates has appeared.  Bax has suggested that the $\tau$ interval be optimized to within 10% of the actual coupling constant for the sequence to operate properly.[48] Wimperis and Freeman[57] have reported the dependence of the BIRD sequence on the accuracy of the pulse calibration.  Pines's original paper suggested that compensated BIRD sequences might improve the tolerance of the pulse operator to variation in $^1J_{CH}$.[43]  One such compensated pulse related to the principle of compensated inversion pulses of Levitt and Freeman,[58] which reflects a suggestion by Pines and coworkers, is given by Eqn. [1-26]:

$^1H$  90°(+X) - τ - 180°(+X) - τ - 90°(+X) - τ - 180°(+X) - τ - 90°(-X)

$^{13}C$              180°(+X)      180°(+X)      180°(+X)              [1-26]

This type of compensated pulse operator, which increases the range of coupling constants affected, has recently been examined by Krishnamurthy and Casida[59] as a replacement for a conventional BIRD pulse of the type shown in Eqn. [1-25] to provide one bond modulation decoupling (see Chapter 3). Their study reports that the compensated BIRD pulse shown in Eqn. [1-26] gave rise to an unacceptable number of artifacts. Consequently, the BIRD has remained the operator of choice.

## TANGO Pulses

A BIRD pulse is one example of a pulse operator that relies on the difference in magnitude betweeen direct ($^1J_{CH}$) and long range ($^nJ_{CH}$, n = 2-4) heteronuclear coupling constants. A second example is the TANGO pulse cluster (Testing for Adjacent Nuclei with a Gyration Operator) devised by Wimperis and Freeman.[58] Differences in these two pulse operators lie in the flip angles of the first and last proton pulses resulting in a difference of the final orientation of the long range and direct components of magnetization. As illustrated above the BIRD pulse inverts one set of vectors relative to the other. Selection is dependent on the phase of the last proton pulse. In contrast, the TANGO pulse, shown in Eqn. [1-27], acts to excite directly bound protons to the extent $M_o sin^2(45°)$. Remote protons are returned to an equilibrium orientation on the +z'-axis. The TANGO takes the form:

$^1H$          135°(+x)  - τ - 180°(+x) - τ - 45°(+x)

$^{13}C$                      180°(+x)                      [1-27]

where $τ = 1/2(^1J_{CH})$. As described for the BIRD pulse above, a step-by-step explanation of the effects of the TANGO pulse, shown in Fig. 1-27, is useful. Initially, all components of magnetization are aligned on the +z'-axis (Fig. 1-27a). After a nonselective 135°(+x) $^1H$ pulse is applied, the resultant vector ensemble containing components from both $^1J_{CH}$ and $^nJ_{CH}$ is 45° below the +y'-axis in the z'y'-plane (Fig. 1-27b). During the operations that follow, vector components will be rotated on the conical surfaces as shown in Fig. 1-27b. During the first evolution period of duration $τ = 1/2(^1J_{CH})$, $^1J_{CH}$ vectors components have each diverged 90°, leaving these vectors 180° apart or antiphase on the surface of the conic (Fig. 1-27c). During the same time interval, the $^nJ_{CH}$ vector components have diverged only $[(^nJ_{CH}/^1J_{CH}) \times 90°]$ degrees. After the

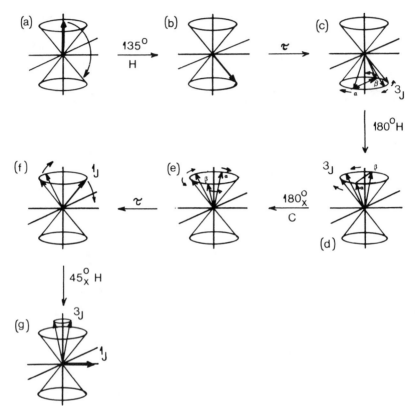

*Fig. 1-27.*        Vector motion observed with the TANGO pulse shown in Eqn. [1-27].
(a) Initial equilibrium proton magnetization. (b) Vector rotation caused
by the application of a 135°(x) proton pulse. (c) Dephasing on the
conical surface during the interval $\tau = 1/2(^1J_{CH})$, long range
components of magnetization defocus minimally while direct
components of magnetization are antiphase. (d) Effect of the 180°
proton pulse; (e) effect of the 180° $^{13}C$ pulse (note that the $\alpha$ and $\beta$
spin labels have been exchanged and that they are beginning to refocus
on the upper conical surface in an orientation colinear with the +y'-
axis). (f) Vector orientation following evolution during the second $\tau = 1/2(^1J_{CH})$ interval. (g) Direct components ($^1J_{CH}$) of magnetization
have been selectively rotated 90° and are coaligned with the +y'-axis
while long range components of magnetization have been rotated
through 360° and are returned to the +z'-axis.

first evolution period, $\tau$, the 180°(+x) pulse applied to $^1$H rotates vectors to the upper conical surface (Fig. 1-27d). Components arising due to $^nJ_{CH}$, if not acted upon by any other force, would refocus along the surface of the conic above the -y'-axis, while those arising due to $^1J_{CH}$ would refocus 180° out of phase above the +y'-axis after time = 2$\tau$. The $^{13}$C 180°(+x) pulse exchanges the populations of the higher and lower energy levels and once again allows the continued defocusing of vectors due to $^nJ_{CH}$ (Fig. 1-27e). After the second evolution period, vector components due to $^1J_{CH}$ will refocus on the conical surface above the +y'-axis, 45° from the +z'-axis (Fig. 1-27f). The final 45°(x) $^1$H pulse rotates the long range components of magnetization back onto the +z-axis, essentially in equilibrium, while the direct components of magnetization are rotated to the the +y'-axis (Fig. 1-27g). In summary, a selective 90° pulse has been applied to the directly bound protons.

A second form of the TANGO pulse that selects for long range couplings ($^nJ_{CH}$, n = 2-4), given by Eqn. [1-28].

$^1$H          45°(+x) - $\tau$ - 180°(+y) - $\tau$ - 45°(-x)

$^{13}$C                          180°(+x)                          [1-28]

The difference between the two versions of the TANGO pulse, shown in Eqns. [1-27] and [1-28], lies in the manipulation of the direct components relative to the long range components of magnetization. A discussion of the vector manipulation described by Eqn. [1-28] is shown in Fig. 1-28 and may be described as follows. Initially magnetization is aligned on the +z-axis. After a nonselective 45°(x) $^1$H pulse is applied (Fig. 1-28a), the vector ensemble containing components from both $^1J_{CH}$ and $^nJ_{CH}$ is rotated 45° from the +z'-axis onto the upper conical surface (Fig. 1-28b). During the first evolution period of duration $\tau = 1/2(^1J_{CH})$ that follows, the vector components due to $^1J_{CH}$ each diverge 90°, leaving these vectors antiphase in the z'x'-plane, on the upper conical surface (Fig. 1-28c). During the same time interval, the vector components due to $^nJ_{CH}$ have diverged only $[(^nJ_{CH}/^1J_{CH}) \times 90°]$ degrees. Next a 180°(+y) $^1$H pulse is applied, which serves to refocus the vector components in addition to rotating the vectors 180° onto the lower conical surface (Fig. 1-28d). The $^{13}$C 180°(+x) pulse exchanges the populations of the higher and lower energy levels; vector components due to $^nJ_{CH}$ continue to defocus while those due to $^1J_{CH}$ refocus after time = 2$\tau$ on the lower conical surface below the +y'-axis (Fig. 1-28e). The final 45°(-x) $^1$H pulse rotates the direct components of magnetization back on the -z-axis (Fig. 1-28f). Thus, a selective 180° pulse for the $^1J_{CH}$ components of magnetization is the result. This same 45°(-x) $^1$H pulse rotates the long range components of magnetization onto the +y'-axis. Thus, a selective 90° pulse is the result for the long range components of magnetization.

These two versions of the TANGO pulse allow the selection of which type, $^1J_{CH}$ or $^nJ_{CH}$, responses will be excited by a 90° pulse. Several applications of TANGO pulses will next be discussed to demonstrate the utility of being able to selectively excite one type of response over another.

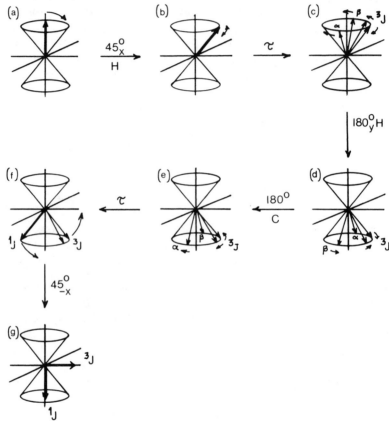

*Fig. 1-28.*        Vector motion observed with the TANGO pulse shown in Eqn. [1-28]. (a) Initial equilibrium proton magnetization. (b) Vector rotation caused by the application of a 45°(x) proton pulse. (c) Dephasing on the conical surface during the interval $\tau = 1/2(^1J_{CH})$, long range components of magnetization defocus minimally while direct components of magnetization are antiphase. (d) Effect of the 180° proton pulse; (e) effect of the 180° $^{13}C$ pulse (note that the $\alpha$ and $\beta$ spin labels have been exchanged and that they are beginning to refocus on the upper conical surface in an orientation colinear with the -y'-axis). (f) Vector orientation following evolution during the second $\tau = 1/2(^1J_{CH})$ interval. (g) Direct components ($^1J_{CH}$) of magnetization have been selectively inverted, while long range components of magnetization have been rotated through 90° and are coaligned with the +y'-axis (selective 90° pulse).

**Applications of TANGO Pulses**. Practical application of the TANGO pulse is somewhat different from the BIRD operator. Due to the ability of this operator to selectively apply a 90° pulse to remote protons, its use has been limited to applications where suppression of direct responses is desirable. Examples are found in the long range optimized heteronuclear chemical shift correlation experiments (see Chapter 3). The TANGO pulse is used to selectively promote magnetization transfer from remote protons while suppressing the transfer from direct protons.

*Long range heteronuclear chemical shift correlation.* The only applications of the TANGO pulse have been as a replacement of a nonselective 90° pulse in long range heteronuclear chemical shift correlation experiments. A relatively small number of applications of this pulse operator have been reported. Two papers from Freeman's group reported the use of this operator. The first paper dealt with the theoretical aspects, including the range of $^1J_{CH}$ values over which the operator was effective and the comparative merit of using compensated TANGO sequences.[57] The second paper demonstrated the utility of this operator in suppressing responses due to direct couplings in long range heteronuclear chemical shift correlation experiments.[60] Since these initial publications, several applications in the area of long range heteronuclear correlation have appeared.[55,59,61-63] These are described in detail in Chapter 3.

# REFERENCES

1.　　F. Bloch, W. W. Hansen and M. Packard, *Phys. Rev.*, **69**, 127 (1946).
2.　　F. Bloch, *Phys. Rev.*, **70**, 460 (1946).
3.　　E. M. Purcell, H. C. Torrey and R. V. Pound, *Phys. Rev.*, **69**, 37 (1946).
4.　　R. R. Ernst, *Chimia*, **29**, 179 (1975).
5.　　L. Müller, A. Kumar and R. R. Ernst, *J. Chem. Phys.*, **63**, 5490 (1975).
6.　　W. P. Aue. E. Bartholdi and R. R. Ernst, *J. Chem. Phys.*, **64**, 2229 (1976).
7.　　G. Bodenhausen, R. Freeman, R. Niedermeyer and D. L. Turner, *J. Magn. Reson.*, **26**, 133 (1977).
8.　　E. L. Hahn and D. E. Maxwell, *Phys. Rev.*, **88**, 1070 (1952).
9.　　G. Bodenhausen, R. Freeman and D. L. Turner, *J. Chem. Phys.*, **65**, 839 (1976).
10.　　A. Kumar, W. P. Aue, P. Bachmann, J. Karhan, L. Müller and R. R. Ernst, *"Proc. XIXth Congress Ampere,"* Heidelberg, 1976.
11.　　W. P. Aue, E. Bartholdi and R. R. Ernst, *J. Chem. Phys.*, **65**, 4226 (1976).
12.　　K. Nagayama, P. Bachmann, R. R. Ernst and K. Wüthrich, *Biochem. Biophys. Res. Commun.*, **86**, 218 (1977).
13.　　A. Bax, A. F. Mehlkopf and J. Smidt, *J. Magn. Reson.*, **35**, 167 (1979).
14.　　A. Bax, A. F. Mehlkopf and J. Smidt, *J. Magn. Reson.*, **40**, 213 (1980).
15.　　K. Nagayama, *J. Chem. Phys.*, **71**, 4404 (1979).
16.　　A. Kumar, *J. Magn. Reson.*, **30**, 227 (1978).
17.　　G. Bodenhausen, R. Freeman, G. A. Morris and D. L. Turner, *J. Magn. Reson.*, **31**, 75 (1978).

18.   G. E. Martin, R. T. Gampe, Jr., J. J. Ford, M. R. Willcott, III, M. Morgan, A. L. Ternay, Jr., C. O. Okafor and K. Smith, *J. Heterocyclic Chem.*, **20**, 1063 (1983).
19.   S. Puig-Torres, R. T. Gampe, Jr., G. E. Martin, M. R. Willcott, III and K. Smith, *J. Heterocyclic Chem.*, **20**, 253 (1983).
20.   M. J. Musmar, R. T. Gampe, Jr., G. E. Martin, W. J. Layton, S. L. Smith, R. D. Thompson, M. Iwao, M. L. Lee and R. N. Castle, *J. Heterocyclic Chem.*, **21**, 225 (1984).
21.   E. L. Ezell, R. P. Thummel and G. E. Martin, *J. Heterocyclic Chem.*, **21**, 817 (1984).
22.   M. L. Martin, J. J. Delpuech, and G. J. Martin, *Practical NMR Spectroscopy*, Heyden, Philadelphia, 1980.
23.   E. Fukushima and S. B. Roeder, *Experimental NMR: A Nuts and Bolts Approach*, Addison-Wesley, London, 1981.
24.   D. B. Lawn and A. J. Jones, *Aust. J. Chem.*, **35**, 1717 (1982).
25.   J. R. Wesener and H. Gunther, *J. Magn. Reson.*, **62**, 158 (1985).
26.   F. W. Wehrli and T. Wirthlin, *Interpretation of Carbon-13 NMR Spectra*, Heyden, New York, pp. 66-72, 1976.
27.   D. M. Thomas, M. R. Bendall, D. T. Pegg, D. M. Doddrell and J. Field, *J. Magn. Reson.*, **42**, 298 (1981).
28.   A. Bax, *J. Magn. Reson.*, **52**, 76 (1983).
29.   N. C. Nielsen, H. Bildsen, H. J. Jakobsen, and O. W. Sørensen, *J. Magn. Reson.*, **66**, 456 (1986).
30.   M. Levitt, and R. Freeman, *J. Magn. Reson.*, **33**, 473 (1979).
31.   R. Freeman, S. P. Kempsell, and M. Levitt, *J. Magn. Reson.*, **38**, 473 (1980).
32.   M. Levitt, *J. Magn. Reson.*, **48**, 473 (1982).
33.   M. Levitt, *J. Magn. Reson.*, **50**, 95 (1982).
34.   M. Levitt, and R. R. Ernst, *J. Magn. Reson.*, **55**, 247 (1983).
35.   R. Tycko, H. M. Cho, E. Schnider, and A. Pines, *J. Magn. Reson.*, **61**, 90 (1985).
36.   M. Levitt, *Mol. Phys.*, **50**, 1109 (1983).
37.   M. H. Levitt, *Prog. NMR Spectrosc.*, **18**, 61 (1986).
38.   S. Wimperis and G. Bodenhausen, *J. Magn. Reson.*, **71**, 355 (1987).
39.   R. Tycko, *Phys. Rev. Lett.*, **51**, 775 (1983).
40.   A. Bax, *Two Dimensaional Nuclear Magnetic Resonance in Liquids*, Delft Univ. Press, Delft, 1982.
41.   D. P. Weitekamp, in *Advances in Magnetic Resonance*, J.S. Waugh, Ed., Vol. 11, Academic Press, New York, 1983.
42.   A. J. Shaka, 8th European Experimental NMR Conference, Spa, 1986.
43.   J. R. Garbow, D.P. Weitekamp and A. Pines, *Chem. Phys Lett.*, **93**, 504 (1982).
44.   E. L. Hahn and D. E. Maxwell, *Phys. Rev.*, **84**, 1246 (1952).
45.   D. W. Nagel, K. G. R. Pachler, P. S. Steyn, R. Vleggaar and P. L. Wessels, *Tetrahedron*, **32**, 2625 (1976).
46.   K. G. R. Pachler, P. S. Steyn, R. Vleggaar, P. L. Wessls and D. B. Scott, *J. Chem. Soc., Perkin Trans I*, 1182 (1976).
47.   A. Bax, *J. Magn. Reson.*, **53**, 517 (1983).
48.   A. Bax, *J. Magn. Reson.*, **52**, 330 (1983).
49.   L. Müller, A. Kumar, and R.R. Ernst, *J. Magn. Reson.*, **25**, 383 (1977).
50.   G. Bodenhausen, R. Freeman, and C. Turner, *J. Chem. Phys.*, **65**, 839 (1976).
51.   T. C. Wong and G. R. Clark, *J. Chem. Soc., Chem. Commun.*, 1518 (1984).
52.   T. C. Wong, and V. Rutar, *J. Magn. Reson.*, **63**, 524 (1985).
53.   C. Bauer, R. Freeman, and S. Wimperis, *J. Magn. Reson.*, **58**, 526 (1984).

54.  W. F. Reynolds, D. W. Hughes, M. Perpick-Dumont and, R. G. Enriquez, *J. Magn. Reson.*, **63**, 413 (1985).

55.  A. S. Zektzer, B. K. John, R N. Castle, and G. E. Martin, *J. Magn. Reson.* **72**, 556 (1987).

56.  A. S. Zektzer, B. K. John, and G. E. Martin, *Magn. Reson. Chem.*, **25**, 752 (1987).

57.  S. Wimperis and R. Freeman, *J. Magn. Reson.*, **58**, 348 (1984).

58.  M. H. Levitt and R. Freeman, *J. Magn. Reson.*, **33**, 473 (1979).

59.  V. V. Krishnamurthy and J. E. Casida, *Magn. Reson. Chem.*, **26**, 362 (1988).

60.  C. Bauer, R. Freeman and, S. Wimperis, *J. Magn. Reson.*, **58**, 526 (1984).

61.  A. S. Zektzer, M. J. Quast, G. S. Linz, G. E. Martin, J. D. McKenney, M. D. Johnston, Jr. and R. N. Castle, *Magn. Reson. Chem.*, **24**, 1083 (1986).

62.  V. V. Krishnamurthy and J.E. Casida, *Magn. Reson. Chem.* **26**,367 (1988).

63.  M. J. Quast, A. S. Zektzer, G. E. Martin and R. N. Castle, *J. Magn. Reson.*, **71**, 554 (1987).

# CHAPTER 2

## ESTABLISHING PROTON-PROTON CONNECTIVITIES

### CLASSICAL PROTON HOMONUCLEAR DECOUPLING

Any discussion of natural products structure elucidation by NMR spectroscopythat touches on classical techniques assuredly includes homonuclear proton decoupling techniques for relating protons in the molecular structure in proximity to one another.[1] When homonuclear decoupling was introduced, the technique had a substantial impact on structure elucidation. Indeed, there are still numerous instances when a few well selected homonuclear decouplings are capable of providing sufficient structural information to completely solve a problem.

Homonuclear proton decoupling experiments provide useful structural information through their ability to selectively disrupt couplings between a given pair of protons in a molecule. In cases where the molecules are relatively simple or the individual resonances are well resolved, homonuclear decoupling can provide unambiguous information on molecular connectivities. Unfortunately, molecules eventually become intractable to homonuclear decoupling techniques whenever they are of sufficient complexity that numerous resonances either are in close proximity to one another or are partially and/or completely overlapped. In such cases, homonuclear decoupling experiments may lead to ambiguous or mistaken conclusions regarding molecular structures. A convenient alternative to conventional homonuclear decoupling is, however, provided by the first of the proton two-dimensional NMR techniques to be discussed below, homonuclear autocorrelated (COSY) proton spectroscopy.

## AUTOCORRELATED HOMONUCLEAR PROTON COSY EXPERIMENTS

### Introduction

Autocorrelated homonuclear proton two-dimensional NMR experiments were, so to speak, responsible for the eventual development of the now massive field of two-dimensional NMR spectroscopy. Jeener in 1971 proposed the autocorrelated proton two-dimensional NMR experiment in a paper presented at the Ampere International Summer School but was unfortunately never published.[2] Jeener's idea was visionary and it took fully five years for technology to catch up with theory in this case. It wasn't until 1976 when Aue, Bartholdi and Ernst[3] published their very comprehensive first work on two-dimensional NMR spectroscopy that Jeener's idea came to fruition. In the intervening years, Jeener's experiment, which is now almost universally referred to using the acronym COSY which stands for COrrelated SpectroscopY, has become a true workhorse. In conjunction with other experiments, the COSY experiment has dramatically expanded our ability to probe the structure of large molecules such as proteins and nucleic acids.[4,5] In this chapter we will be primarily concerned with applications of the COSY experiment, its variants and related experiments to small molecules with molecular weights < 2000 daltons.

### The Basic COSY Experiment

Pulse sequencing for the basic COSY experiment is quite simple, requiring the application of only two 90° proton pulses separated by a systematically incremented evolution period, $t_1$, and an acquisition period.[2,3] Modern variants of the experiment also employ phase cycling, which will be discussed below to provide the equivalent of quadrature phase detection in both frequency domains.[6] The pulse sequence for the COSY experiment is shown in Fig. 2-1. Although the pulse sequence is quite simple, the fundamental workings of the COSY experiment are considerably more complex and have been the focus of several density matrix treatments[2,7] as well as a somewhat more readily understood product operator treatment by Ernst and coworkers in their recent review on product operator formalisms.[8] The interested and capable reader is referred to any of these more rigorous treatments for a detailed discussion of the workings of the COSY experiment.

Limitations associated with classical homonuclear proton decoupling techniques are obvious and quickly are reached. Irradiations become nonselective as molecules increase in size and complexity, leading to multiple decoupling pathways arising from a single irradiation, which cannot necessarily be disentangled. Well resolved resonance multiplets are thus preferable for the use of conventional decoupling techniques. Likewise, it becomes rather cumbersome to use conventional decoupling techniques when numerous decouplings are required. In both of these instances, the COSY experiment serves as one of the preferred means of solving such problems. First, the COSY pulse sequence allows all connectivity pathways to be exploited largely without regard to overlap, a point that has made possible the study of complex biopolymers[4,5] that are intractable to conventional methods. While problems still exist in the interpretation of the COSY spectra of biopolymers, the utilization of COSY in concert with other techniques such as homonuclear single and double relay, proton double quantum INADEQUATE and NOESY makes it possible to probe the structures of these fascinating molecules. Second, since the COSY experiment probes all coupling pathways in a molecule simultaneously, the COSY experiment also becomes a preferred method when numerous connectivities must be established.

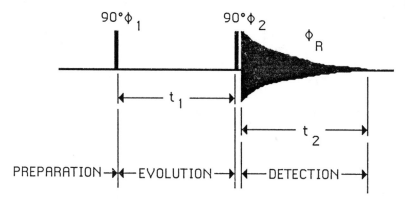

*Fig. 2-1.* The basic COSY or COSY90 pulse sequence consists of two 90° pulses separated by an evolution period, $t_1$, which is systematically incremented throughout the course of the experiment. The experiment produces an autocorrelated proton spectrum that correlates scalar coupled (J) spins via off-diagonal responses.

## Phase Cycling

Quadrature phase detection in normal spectral acquisition is achieved by a series of phase shifts. In the case of two-dimensional NMR experiments, it is desirable to have quadrature detection in the observed frequency domain as well as in the second frequency domain, $F_2$ and $F_1$, respectively, to maximize the digital resolution of the experiment while minimizing data storage requirements. Phase cycles for the COSY experiment have several levels of complexity. At a minimal level of complexity, it is possible to perform the COSY experiment using a very simple four step phase cycle as shown in Table 2-1, cycling the phase of the second 90° pulse in four steps. The next level of complexity is attained with a sixteen step phase cycle, which is also presented in Table 2-1. Here, the first 90° pulse phase, $\phi_1$, and the receiver phase, $\phi_R$, are cycled in four 90° steps. The phase of the second 90° pulse, $\phi_2$, is held constant during the first four steps and then is stepped by 90° for the next four acquisitions in a fashion analogous to the "cyclops" cycle proposed by Hoult.[9]

Generally speaking, it is advantageous to employ one of the sixteen step phase cycles for reasons of sample concentration and spectral quality. The four step cycle might be employed to produce a "quick and dirty" COSY spectrum and it will, in most cases, be just that! Spectra acquired using a four step phase cycle may contain any number of artifacts such as axial peaks and quadrature images, which can arise for a variety of reasons that include but are not limited to $T_1$ relaxation during evolution, imbalances due to imperfect phase shifts

Table 2-1.          Phase cycling possibilities for the basic COSY experiment.

| | Four Step | | 16 Step Coherence Transfer | | | | | |
| | | | Antiecho (P-Type) | | | Echo (N-Type) | | |
| Acq. | $\phi_2$ | $\phi_R$ | $\phi_1$ | $\phi_2$ | $\phi_R$ | $\phi_1$ | $\phi_2$ | $\phi_R$ |
|---|---|---|---|---|---|---|---|---|
| 1 | 0 | 0 | 0 | 0 | 0 | 0 | 0 | 0 |
| 2 | 1 | 2 | 1 | 0 | 1 | 3 | 0 | 1 |
| 3 | 2 | 0 | 2 | 0 | 2 | 2 | 0 | 2 |
| 4 | 3 | 2 | 3 | 0 | 3 | 1 | 0 | 3 |
| 5 | | | 0 | 1 | 0 | 0 | 1 | 0 |
| 6 | | | 1 | 1 | 1 | 3 | 1 | 1 |
| 7 | | | 2 | 1 | 2 | 2 | 1 | 2 |
| 8 | | | 3 | 1 | 3 | 1 | 1 | 3 |
| 9 | | | 0 | 2 | 0 | 0 | 2 | 0 |
| 10 | | | 1 | 2 | 1 | 3 | 2 | 1 |
| 11 | | | 2 | 2 | 2 | 2 | 2 | 2 |
| 12 | | | 3 | 2 | 3 | 1 | 2 | 3 |
| 13 | | | 0 | 3 | 0 | 0 | 3 | 0 |
| 14 | | | 1 | 3 | 1 | 3 | 3 | 1 |
| 15 | | | 2 | 3 | 2 | 2 | 3 | 2 |
| 16 | | | 3 | 3 | 3 | 1 | 3 | 3 |

and irreproducibility of the duration of the 90° pulses. In Table 2-1 and all other phase tables presented below, numeral designators 0, 1, 2, 3 are employed to designate phases x, y, -x and -y, respectively.

Selection of a phase cycle depends upon the ultimate disposition of the data. The four step phase cycle shown is occasionally employed for a preliminary or "quick and dirty" examination of molecular connectivity networks. Alternatively, the phase of the second 90° pulse,

$\phi_2$, could be cycled as 0, 3, 2, 1 to provide a second type of peak selection with a four step cycle. Better quality spectra with more complete suppression of quadrature images require the use of one of the two sixteen step phase cycling schemes shown in Table 2-1. The coherence transfer antiecho (P-type) phase cycle is now the most widely employed, and a brief explanation of the terminology is in order. Specifically, coherence transfer antiecho or P-type phase cycling selects for the components of magnetization that precess in the same sense during the evolution and detection periods, $t_1$ and $t_2$, respectively. In contrast, the coherence transfer echo or N-type phase cycle shown in Table 2-1 selects for the components of magnetization that precess with opposite senses during the evolution and detection periods.

## Parameter Selection

Selection of parameters for acquiring a COSY data set is relatively simple. There are only two parameters to be concerned with in most cases, the dwell time and the interpulse delay, assuming that the duration of the 90° pulse has been determined (see Chapter 1). Occasionally, it will also be necessary to choose between a 90° and a 45° read pulse. Finally, a decision must also be made regarding the digitization of the second or $F_1$ frequency domain.

The COSY experiment is autocorrelated and hence it is desirable to have the data matrix end up square with equivalent frequency ranges digitized in both frequency domains. It is also necessary for the matrix to be square if it is to be symmetrized[10] prior to plotting. Quite simply, it should be recalled that the dwell time is the reciprocal of the sweep width and that the dwell time multiplied by the number of data points being acquired determines the acquisition time. Thus, the dwell time should be noted when a high resolution reference spectrum is acquired to determine the appropriate spectral width prior to the acquisition of the COSY data set. The dwell interval will frequently be used as the initial evolution time duration and as the factor by which the evolution time will be incremented in successive experiments in the series.

Interpulse delays in the COSY experiment allow the system to return toward magnetic equilibrium. At a minimum, it is usually desirable for a period of 1.5 x $T_1$ to be employed for the sum of the acquisition time and the interpulse delay when only a routine examination of coupling pathways is desired. A useful starting point in the absence of any prejudices is to employ a 1 second interpulse delay plus the acquisition time. When it becomes necessary to obtain higher dynamic range, it is beneficial to have the sum of the acquisition time and the interpulse delay approach 5 x $T_1$. While individuals will frequently develop a "feel" for a usable interpulse delay, the $T_1$ relaxation time of the system can be estimated conveniently when there is any question from a simple inversion-recovery experiment in which the tau value is chosen to null the signals. With proton signals nulled using the inversion-recovery sequence, the approximate $T_1$ relaxation time of the protons in question is $\tau/0.693$.

Digitization of the second frequency domain is a third consideration in the selection of parameters, although it might more accurately be categorized as a factor in performing the experiment. In cases where the spectrum being examined is well resolved, it is entirely possible to use very coarse digitization of the $F_1$ frequency domain with as few as 64 increments of the evolution time. As spectra increase in complexity and congestion, finer digitization of the $F_1$ frequency domain becomes desirable, with 256 to 512 points being employed. It should be noted, however, that although it may be desirable to have 512 points digitizing the $F_1$ domain, signal decay due to $T_2$ processes will generally occur well prior to the completion of the experiment. Indeed, even if signal is still present in the file corresponding to the last $t_1$ increment, in many instances the weighting functions used in the data processing will obliterate any information contained in the last experiments in the series anyway. At this point, the user must thus make a conscious choice in determining the number of increments or blocks of data that the spectrometer will be allowed to acquire, and the choice frequently is governed by time constraints. Generally it is beneficial to instruct the computer to acquire either 256 or 512 increments of $t_1$ even though the experiment will be intentionally terminated prior to completion. In this sense, we have the flexibility of having essentially zero-filled the data matrix in the second frequency domain once or a fraction of a time, thereby improving resolution in $F_1$ somewhat. Whatever the case, the choice of the number of increments of $t_1$ that are to be performed is one that the user must make and it will doubtless vary from one sample to the next. Some of the illustrations shown below will address this choice further.

A further consideration enters into the decision process associated with the preceding discussion. Specifically, the user must decide whether it will be desirable to symmetrize[10] the final data matrix prior to plotting. Symmetrization routines, in order to operate, must have equal digitization in both the $F_1$ and $F_2$ frequency domains. Typically, if symmetrization will be performed, data matrices of the type $S(t_1,t_2)$ are acquired as 256 x 512 or 512 x 1K points which, after completion of the processing, give final matrices consisting of 256 x 256 or 512 x 512 points, respectively.

## Data Processing and Presentation

Having completed the selection of parameters and the acquisition of a COSY data set, it next becomes necessary to process the data and prepare it for presentation and interpretation. To illustrate the considerations implicit in the utilization of the COSY experiment, we shall first employ a structurally formidable but spectroscopically simple sesquiterpene, plumericin (**2-1**).[11] At an observation frequency of 300 MHz, the proton spectrum of plumericin (**2-1**) is essentially first order, resolved into its component multiplets as shown by the high resolution spectrum plotted for reference purposes below the COSY spectrum shown in Fig. 2-2. Thus, plumericin (2-1) represents an easy natural product with which to learn about the COSY experiment.

**2-1**

**Processing: selection of weighting factors, phasing and scaling.** Processing the COSY spectrum begins with the Fourier transformation of the first "block" or $t_1$ increment contained in the data set. After selecting how the data will be mathematically manipulated, eg, using an exponential versus sinusoidal multiplication of the free induction decay, the first block can be subjected to Fourier transformation. Although weighting factors must be selected early on, we find it convenient to discuss data processing before considering the consequences of the weighting factor(s) selected. The impact of weighting factors on the appearance of the data will be treated below. The phase of the spectrum that results from the Fourier transformation of the first block of data is completely arbitrary, and it is unnecessary to expend any time in attempting to phase the spectrum. It is mandatory, however, to apply some uniform phase correction to the data contained in this first spectrum. Following Fourier transformation, scaling parameters should also be set for application in all subsequent Fourier transformations of later blocks of data contained in the data set. At this point, with the initial parameters set, the balance of the raw COSY data set may be processed. The same mathematical treatment, phasing and scaling factors should be applied in each Fourier transformation.

After completing the first Fourier transformation of the COSY data set, the data has been manipulated as shown in Eq. [2-1].

$$S(t_1, t_2) \quad \rightarrow \quad S(t_1, F_2) \qquad\qquad [2-1]$$

At this stage, it is necessary to transpose the data prior to subjecting it to the second Fourier transformation. The transposition algorithm, as described in the preceding chapter (see Figs. 1-7 and 1-8), converts the phase modulated series of spectra prepared with the first Fourier transformation into a series of interferograms according to Eq. [2-2].

$$S(t_1, F_2) \quad \rightarrow \quad S(F_2, t_1) \qquad\qquad [2\text{-}2]$$

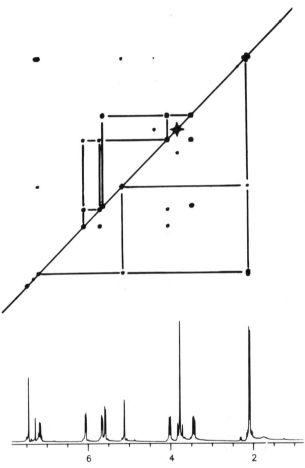

*Fig. 2-2.*    Symmetrized COSY spectrum of plumericin (**2-1**). The data was collected in 6 hours (overkill) using a solution containing 35 mg of **2-1** in 0.4 ml of deuterochloroform at a probe temperature of 20°C and an observation frequency of 300.068 MHz. The experiment was performed using coherence transfer antiecho phase cycling (P-type response selection). The initial $S(t_1, t_2)$ data matrix was collected as 512 x 1K complex points with 274 of the 512 $t_1$ increments actually completed. The spectral width in both $F_1$ and $F_2$ was 1110 Hz, requiring a dwell time of 901 μsec. Thus, the evolution time of the first experiment was set to 901 μsec and was incremented by this interval in each successive experiment performed, giving an evolution time in the final experiment of 246.85 msec.

At this stage, we are essentially ready to perform the second Fourier transformation. One consideration that does, however, enter into the second Fourier transformation is the scaling of the Fourier transformed result. There are basically two approaches to this problem. Older software required that the user visually locate the largest interferogram, Fourier transform it and then set the scaling parameters. Newer software generally incorporates a command that allows "after the fact" scaling. Specifically, the second Fourier transformation is done without regard to scaling and the computer, as it processes the data, keeps track of the address of the most intense Fourier transformed result. Following the Fourier transformation of the last block of data, a scaling factor is determined on the basis of the most intense data file which is then applied to all blocks of data. Obviously, the latter approach is far easier since it requires no user intervention. In any case, the step in the processing containing the second Fourier transformation is described by Eq. [2-3].

$$S(F_2,t_1) \quad \rightarrow \quad S(F_2,F_1) \qquad [2-3]$$

Following the completion of the second Fourier transformation, the COSY data set is ready for plotting or alternatively, for symmetrization.[10]

**Symmetrization of COSY NMR data sets.** Following the completion of data processing, COSY data can be either displayed directly or subjected to symmetrization[10] to cosmetically improve the appearance of the data prior to presentation. Symmetrization, to begin, requires that the data matrix be square. That is, the data matrix must consist of equal numbers of data points in both frequency domains, eg, 256 x 256 points for the final $F_2,F_1$ matrix. Thus, COSY spectra that are intentionally digitized coarsely in the $F_1$ frequency domain cannot be directly subjected to symmetrization. The operation of symmetrizing a COSY spectrum utilizes a relatively simple computer algorithm that examines locations in the COSY data file in a symmetrical fashion. Responses arising due to signals in a COSY spectrum are symmetrically located in the data matrix. In contrast, noise is a random process and will not occur symmetrically relative to the diagonal. To illustrate this point, consider Fig. 2-3 which schematically illustrates a simple AX spin system and the location of several potential types of noise in the data matrix. The symmetrization algorithm examines addresses in the data matrix that are symmetrically positioned about the diagonal. The algorithm examines the intensity of the signal stored at each address and takes the square root of the signal intensity stored at each address and then compares the square roots. Whichever value is smaller is then reinserted into the data matrix at both storage addresses. Noise, being a random event, will not occur symmetrically in the data matrix as is shown in Fig. 2-3. Actual responses, in contrast, are symmetrically disposed about the diagonal and are left intact. In summary, noise and any other unsymmetrically located artifacts can be cosmetically eliminated form the data matrix via the symmetrization operation.

**Consequences of weighting factor multiplication during data processing.** The COSY spectrum of plumericin (**2-1**) shown in Fig. 2-4 is presented as a contour plot consisting of 64 x 512 data points after final processing, leaving the digital resolution in the $F_1$ frequency domain rather coarse. The data was processed using sinusoidal multiplication[12] prior to both Fourier transformations. Under these conditions, although tailing that would result if exponential multiplication were employed has been curtailed, the shape of the responses is not symmetrical but rather is exaggerated in the $F_1$ dimension because of the low digital resolution. A third alternative, pseudo-echo weighting,[13] is also available. Pseudo-echo weighting can be produced by multiplying by a rising exponential function followed by a second multiplication by a falling Gaussian function. This process converts the decaying exponential signal to a Gaussian line shape, which is desirable since it eliminates the four pointed star shape characteristic of exponentially multiplied data. Sinusoidal multiplication and pseudo-echo weighting methods are largely comparable in terms of the final appearance of the data; either is definitely preferable to exponential

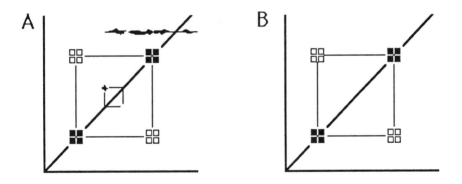

*Fig. 2-3.*    Schematic presentation of the symmetrization operation. (A) A simple schematic of an AX spin system is shown with two types of noise, the single spot of noise above the diagonal, which does not have a symmetrically positioned counterpart, and the noise "track." The symmetrization algorithm examines addresses in the data matrix that are symmetrically disposed about the diagonal and takes the square root of the signal intensity at each address. The square roots are compared, and whichever is smaller is reinserted at both data matrix addresses. Since noise is a random event, it will not generally have a symmetric pattern of occurrence and will thus be cosmetically eliminated as shown in (B).

Fig. 2-4. Contour plot of the COSY spectrum of plumericin (2-1) in deuterochloroform at 20° recorded at 300.068 MHz. The data was collected as 64 x 512 points and was processed using a sinusoidal multiplication prior to both Fourier transformations, resulting in the elongated responses due to the limited digital resolution in the $F_1$ frequency domain.

multiplication and should be employed routinely in the processing of COSY experimental data.

**Presentation of COSY spectra: stack plots versus contour plots.** Presentation of COSY data is almost invariably in the form of a contour plot simply because the numbers of responses encountered can easily lead to obscuration when stack plots are employed, thereby making interpretation much more difficult. Likewise, even if resonances are resolved in a stack plot, the precise location of responses relative to one another is more difficult to determine, thus making it harder to establish connectivity networks in a stack plot than in the corresponding contour plot.

## Interpretation of COSY Spectra

Prior to exploring the more subtle features of a COSY data matrix, it is necessary to familiarize individuals new to two-dimensional NMR spectroscopy with the fundamentals for interpreting COSY spectra. To begin this task, recall that the COSY experiment is an autocorrelated experiment and that the proton spectrum is thus correlated with itself via the scalar (J) couplings between resonances. In examining the schematic COSY spectrum of a hypothetical AX two spin system shown in Fig. 2-5, we will note that responses corresponding to the normal proton spectrum are contained along the diagonal $F_2 = F_1$. Beginning with the doublet downfield, responses on the diagonal correspond to the resonances of the normal spectrum. These responses have associated with them "near diagonal" responses, which correlate the individual lines of the multiplet with one another. Further removed are off-diagonal responses which serve to actually correlate multiplets that are associated with one another by scalar coupling.

A somewhat more complex example is shown schematically in Fig. 2-6. Here the example consists of two partially overlapped ABX spin systems. In the first case, the X spin furthest downfield is coupled to the B spin but not to the A spin. The B spin is coupled, in turn, to the A spin furthest upfield. The schematic of this particular ABX spin system also illustrates the nature of the diagonal and off-diagonal responses that would be expected given sufficient digital resolution in both $F_1$ and $F_2$ where $J_{AB} = J_{BX}$ and $J_{AX} = 0$. Such a situation is commonly encountered in the COSY spectra of aromatic and heteroaromatic molecules. The other ABX spin system shown schematically in Fig. 2-6 assumes all of the couplings to be nonzero. Further, the schematic of this system illustrates the problems that may be encountered when two coupled spins have very closely similar chemical shifts. Specifically in this case the A and B spins are close and thus the off-diagonal responses, which correlate them with one another, are very near the diagonal. In severe cases, the off-diagonal responses may be so close to the diagonal that they are virtually impossible to detect. This problem is one of the shortcomings of the COSY experiment that is circumvented

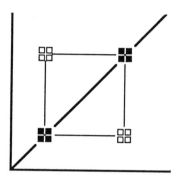

*Fig. 2-5.*    Schematic representation of a COSY spectrum of a simple two spin AX system. Normal responses for the two resonances are located along the diagonal and are represented by the solid squares. Off-diagonal responses which correlate the two resonances, are denoted by the open squares. Note that the off-diagonal responses are displaced vertically and horizontally from the diagonal.

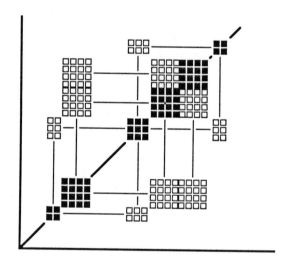

*Fig. 2-6.*    COSY contour plot schematic showing two ABX spin systems. Diagonal responses are shown as solid squares, off-diagonal responses are shown as open squares. In one case, the X spin is coupled only to the B spin, which is in turn coupled to the A spin. In the other case, each component spin is coupled to both others in the spin system. In the latter case note that the A and B spins are quite similar in chemical shift and that consequently their off-diagonal elements are quite close to the diagonal, which can make these connectivities difficult to observe in real COSY spectra.

by some of the other techniques available to establish proton-proton connectivities discussed below.

Having considered the two schematic cases shown in Figs. 2-5 and 2-6, it is appropriate to return to the plumericin (**2-1**) COSY spectrum (Fig. 2-2).[11] Spectroscopically, the simplest component of the COSY spectrum shown in Fig. 2-2 is the isolated response for H6, which resonates furthest downfield at 7.423 ppm. Increasing in complexity, we must next consider the ethylidene moiety, comprised of the H10 vinyl proton resonating at 7.226 ppm and appearing as a doubled quartet, the vinyl methyl protons, 11-Me, resonating at 2.034 ppm, which appears as a double doublet and the H3a bridgehead proton resonating at 5.093 ppm appearing as a complex multiplet due to couplings to both H10 and the ll-Me group in addition to other couplings. In the COSY spectrum shown in Fig. 2-2, these structural substituents constitute a second independent spin system whose connectivities are traced out below the diagonal. It should also be noted that while the H3a does have couplings ($^4J_{H3aH4a}$) to the third spin system, which is discussed below, there are no observable responses correlating these spin systems in the normal COSY spectrum shown in Fig. 2-2. The final spin system, which completes the analysis of the COSY spectrum of plumericin (**2-1**), is the five spin system comprised of H4a, H9b, H7a, H8 and H9 which resonate at 5.525, 3.403, 3.948, 5.618 and 6.013 ppm, respectively. Since these spins are spread across five separate carbon atoms, there is relatively limited coupling between them and they are, essentially, a linear system, which considerably simplifies the analysis of the COSY spectrum. Thus, H4a is readily seen to be coupled to H9b; H9b is in turn coupled to H7a and also to H8. Continuing, the H7a resonance is coupled to both the H8 and H9 resonances, completing the connectivity network traced above the diagonal in Fig. 2-2.

Structurally, the three connectivity networks derived from the COSY spectrum shown in Fig. 2-2 are oriented relative to one another in Fig. 2-7 (structural fragments **A**-**C**). Ideally, we would like to be able to assemble these fragments into the larger substructure represented by **D** in Fig. 2-7. Unfortunately, this is not possible using the simple COSY spectrum shown above. We will see below, however, that other two-dimensional NMR techniques are available that can accomplish this task.

## Application of the COSY experiment in the proton spectral assignment of the cembranoid diterpene jeunicin (2-2).

Since many applications of two-dimensional NMR spectroscopy are now directed toward the elucidation of natural product structures, we will next focus our attention on a more complex molecule, jeunicin (**2-2**), a cembranoid diterpene originally isolated from the marine coral *Eunicia mammosa* [14] and more recently from the mollusc *Planaxis sulcatus*.[15]

Application of the COSY experiment to molecules such as jeunicin will generally require good digital resolution in both frequency domains. The spectrum of **2-2** shown in Fig. 2-8 thus consists of 512 x 512 data points and has been symmetrized.[10] To begin to use the data contained in the contour plot, it is necessary to select a starting point. Convenient starting points are provided by the exomethylene protons and the H7 vinyl methine proton, which will be discussed in turn. Beginning with the exomethylene protons, H17a/b, which resonate furthest downfield, we note that one of these protons shows a coupling

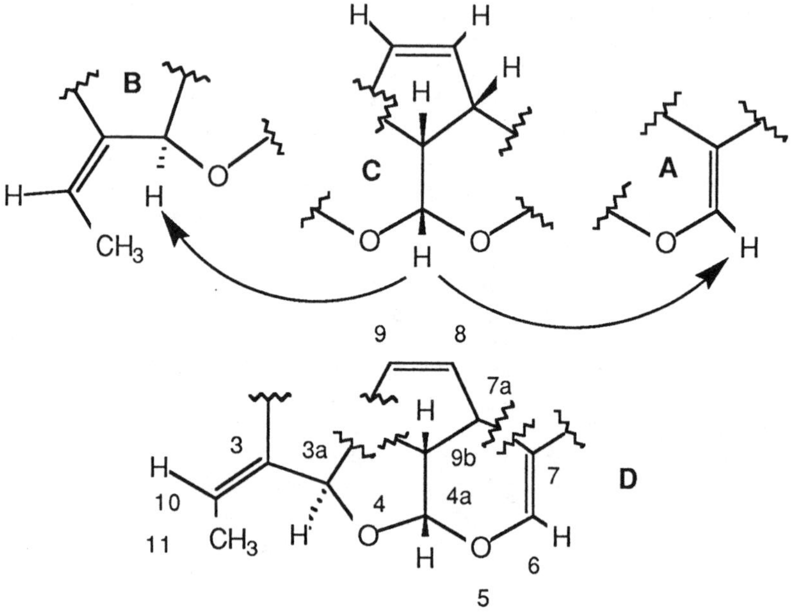

Fig. 2-7.    Structural fragments of plumericin (2-1) that can be assembled from the proton-proton connectivities elucidated using the COSY spectrum shown in Fig. 2-2. Fragments A-C can be assembled using the data contained in the spectrum. It would be desirable to be able to combine fragments A, B and C to provide the much more nearly complete fragment shown by D. Key couplings necessary to assemble D are $^4J_{H4aH3a}$ and $^4J_{H4aH6}$, which are represented by the arrows. Unfortunately, these connectivities were not visible either in the contour plot or in individual traces extracted from it.

to a proton resonating at 3.36 ppm, which can be assigned as H1. From H1, connectivities are also observed that correlate H1 with three other protons; H14 which resonates downfield at 4.42 ppm and the anisotropic H2 methylene protons, which resonate at 2.36 and 1.89 ppm. Proceeding outward from these assignments, the H14 resonance correlates with a proton resonating at 3.18 ppm, which is assigned as H13 which in turn correlates with a proton resonating upfield in the methylene envelope at 1.86 ppm, which is assigned as H12. Any futher connectivities along the "southern flank" of the molecule cannot be extracted from the COSY spectrum due to congestion in the "methylene envelope." Returning to the anisotropic H2 resonances, they correlate with a single proton resonating at 3.67 ppm, which can be assigned as the H3 resonance. There are no connectivities beyond this point in this direction either as expected from the structure. Thus, from this simple series of operations, it is possible to assemble a rather large piece of the structure of jeunicin as represented by **2-3.** In similar fashion, beginning with the H7 vinyl methine proton, correlations can be established to the H6 anisotropic methylene protons, which resonate at 2.27 and 2.08 ppm, and to the vinyl methyl protons, H19, resonating at 1.58 ppm. From the H6 protons, the anisotropic H5 protons, which resonate at 1.71 and 1.50 ppm, can also be assigned, reaching a termination point in the assignment process again as a consequence of the structure. The structural fragment thus assembled is shown by **2-4.**

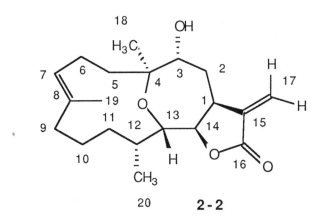

In summary, for jeunicin, the COSY experiment provides two substantial structural fragments, **2-3** and **2-4**, which would be a significant piece of information if we were dealing with a molecule of unknown structure. Successful usage of the COSY experiment requires that we have a starting point and that we can then track the off-diagonal responses to establish the connectivity from our starting resonance to its

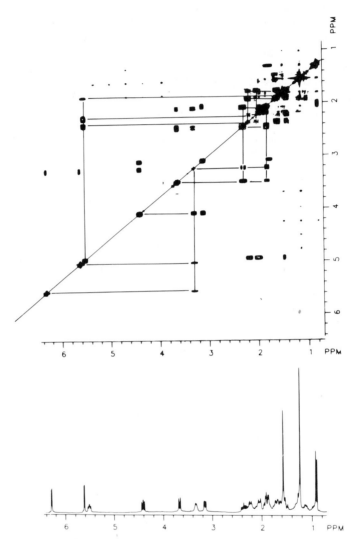

*Fig. 2-8.*    COSY spectrum (512 x 512 points) of the cembranoid diterpene jeunicin (**2-2**) recorded in deuterochloroform at 17°C and an observation frequency of 300.068 MHz. The connectivity network beginning from H17 is traced in below the diagonal; the connectivity network beginning from the H7 vinyl methine proton is traced in above the diagonal. The spectum shown was acquired on a 10 mg analytically pure sample in about 2 hours. Of the 512 blocks in the $F_1$ dimension, the spectrometer was allowed to complete 300 blocks, the balance were taken as zeros. The data was processed using sinusoidal multiplication before both Fourier transformations and was symmetrized[10] prior to plotting.

vicinal neighbor(s). Connectivity networks can, as shown, be pursued in some cases until there is a break caused by the absence of a protonated carbon. Alternatively, breaks in the connectivity network can arise because of congestion in the proton spectrum. For example, it is not possible to establish the connectivity between the H12 proton which resonates at 1.86 ppm and the anisotropic H11 vicinal neighbor protons which resonate at 1.72 and 1.14 ppm. There are, however, alternative experiments such as proton double quantum INADEQUATE which do facilitate the determination of these connectivities as described below.

**Phenanthro[3,4-*b*]thiophene: Benefits of reprocessing the COSY data matrix.** Occasionally, it may be beneficial to collect a COSY data set, process normally and plot it and then reprocess the data using a different set of weighting functions. Such cases arise with aromatics frequently because of the tendency of long range couplings to appear in the spectra of these molecules. Normal processing using sinusoidal multiplication prior to both Fourier transformations will allow long range couplings to appear in the COSY spectrum. To verify that these are indeed responses due to long range couplings, the data matrix can then be reprocessed using, for example, sinusoidal multiplication followed by a Gaussian multiplication with sufficient line broadening to eliminate the small long range couplings, which will generally be in the range $J < 1.5$ Hz.

A relatively simply polynuclear aromatic, phenanthro-[3,4-*b*]thiophene (**2-5**) can be used to illustrate the benefits of differential processing of the COSY data set. Assignments derived by two-dimensional NMR means have recently been reported for phenanthro[3,4-b]thiophene by Martin and coworkers.[16] As would be expected, the H1 and H11 bay or "fjord" region protons resonate furthest downfield at 8.653 and 9.161 ppm, respectively, as a direct consequence of the deshielding engendered by the aromatic ring across the "bay" or "fjord." The balance of the assignments are not particularly remarkable. Most interesting are the several long range couplings that appear in the spectrum shown in Fig. 2-9. The first is a five bond epi zig-zag coupling between H1 and H4, the latter resonating at 8.086 ppm. The second long range coupling is another epi zig-zag coupling between the H11 resonance and the H7 resonance at 7.834 ppm. Reprocessing the COSY data matrix for phenanthro[3,4-*b*]thio-phene (**2-5**) using a sinusoidal multiplication folowed by a 2.5 Hz Gaussian multiplication effectively eliminates the long range couplings cited above, leaving only the much larger vicinal couplings in the contour plot of the data matrix shown in Fig. 2-10. The 1.3 Hz four bond meta coupling between H11 and H9, which resonates at 7.656 ppm, is also eliminated, but the 1.6 Hz meta coupling between H8 and H10, 8.006 and 7.736 ppm, respectively, remains visible following Gaussian processing of the data matrix.

From the preceding example, the potential benefits of processing a given data set in several different ways are clearly highlighted. Using contemporary spectrometer systems whose computers are equipped with an array coprocessor, the investment in time to view the data following the application of more than one weighting function is minimal and certainly well warranted in view of the benefits that may be derived.

**2-5**

## Strychnine: Extending the application of COSY spectra.

Thus far, we have examined the applications of the COSY experiment to several molecules. For purposes of comparing the COSY experiment with other experiments designed to establish proton-proton connectivities discussed below, it is useful to consider the COSY spectrum of strychnine (**2-6**) and its attendant problems.

Assignment of the proton NMR spectrum of strychnine has been the subject of several papers. Carter, Luther and Long[17] reported one set of assignments in 1974, which were revised on the basis of some one dimensional nuclear Overhauser enhancement studies by Chazin, Colebrok and Edward.[18] Still more recently, strychnine has been used as a model compound for a proton double quantum INADEQUATE study by Craig and Martin,[19] which also revised the H17a/H17b geminal coupling on the basis of a 2D homonuclear J-resolved experiment. Finally, strychnine has also been used as a model by Zektzer and Martin[20] to permit direct comparison of homonuclear zero quantum coherence spectroscopy with the results obtained by Craig and Martin[19] using double quantum coherence techniques.

The COSY spectrum of the aliphatic, vinyl and aromatic regions of strychnine (**2-6**) is shown in Fig. 2-11. The data was acquired as 256 x 1K complex points and was subjected to sinusoidal multiplication prior to both Fourier transformations in addition to being zero-filled prior to the second, providing a final data matrix consisting of 512 x 512 points, which was symmetrized[10] before plotting. Resonances contained in the region of the spectrum shown run the gamut from well resolved to non-first-order geminal pairs such as H17a/b to overlapped multiplets containing responses for several resonances.

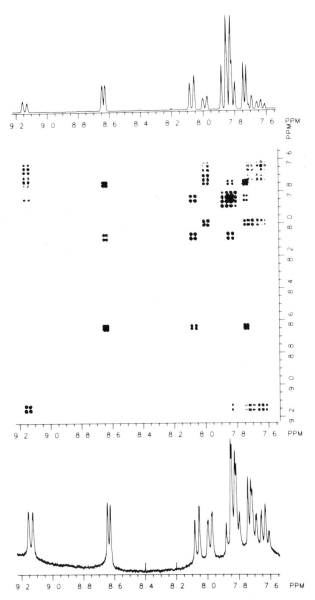

*Fig. 2-9.* Contour plot of the COSY spectrum of phenanthro[3,4-*b*]thiophene (**2-5**) recorded in deuterochlorofrom at 30°C and an observation frequency of 300.054 MHz. Long range epi zig-zag coupling between H1 and H5 and extended epi zig-zag coupling responses between H2 and H5 in the data matrix are indicated. The symmetrized data matrix contains of 256 x 256 points and was processed using sinusoidal multiplication prior to both Fourier transformations.

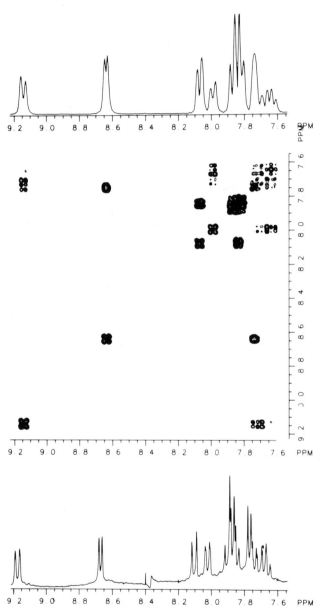

*Fig. 2-10.* Contour plot of the COSY spectrum of phenanthro[3,4-*b*]thiophene **(2-5)** identical to that shown in Fig. 2-9 except that the data was processed using a sinusoidal multiplication followed by a 2.5 Hz Gaussian multiplication prior to both Fourier transformations to eliminate the long range coupings that were highlighted in the preceding spectrum.

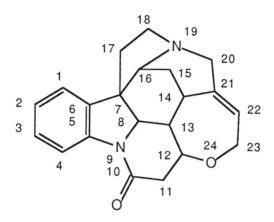

**2-6**

Beginning the analysis of the COSY spectrum of strychnine (**2-6**) again requires that we select a starting point. A convenient entry point is afforded by the H22 vinyl proton resonating at 5.88 ppm. Unfortunately, although H22 can be linked to its neighboring H23 geminal protons (4.13 and 4.05 ppm), the connectivity network terminates at H23a/b and cannot be traced any farther. In similar fashion, H20a resonating at 3.69 ppm can be linked to H20b resonating upfield at 2.71 ppm to comprise a second isolated system. Likewise, a third isolated spin system can be established consisting of the strongly coupled H17a/b pair resonating at 1.86 ppm which then can be linked to the H18a and H18b protons resonating at 3.19 and 2.86 ppm, respectively. It should be noted, however, that the latter two pairs of connectivity networks would not logically be assembled immediately after the H22-H23a/b system, since neither of them has a proton whose chemical shift provides a logical starting point from which to continue the analysis. It would, however, be logical to continue the analysis from the H12 resonance, which, as a consequence of being an oxymethine proton, resonates furthest downfield of the aliphatic protons of strychnine.

The operations just described, although they encompass a considerable number of protons as shown by **2-7**, do not allow these spin systems to be interconnected with one another to provide a larger structural fragment that could be used if the structure of an unknown were being assembled. In contrast, a large and useful structural fragment can be assembled from the remaining aliphatic protons which are represented by **2-8**.

Beginning from the furthest downfield aliphatic proton, H12, which resonates at 4.27 ppm, a strong off-diagonal response correlates it with the H11a proton resonating as a part of a complex multiplet centered at about 3.11 ppm while a weaker off-diagonal response

correlates H12 with the HIIb resonance at 2.66 ppm. The H12
resonance represents a very logical  starting point for the continuation of

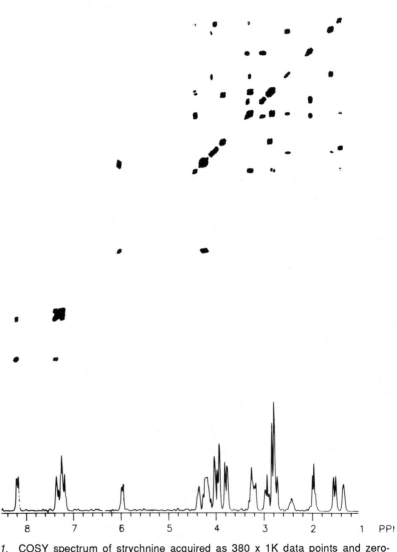

*Fig. 2-11.* COSY spectrum of strychnine acquired as 380 x 1K data points and zero-
filled during processing to afford a final data matrix consisting of 512 x
512 points, which was symmetrized prior to plotting. Sinusoidal
multiplication was used prior to both Fourier transformations.

**2-8**

**2-7**

the examination of the data shown in Fig. 2-11 since, on the basis of chemical shift, it should be next to an oxygen atom. If we were dealing with a molecule of unknown structure, we would probably have an estimate of the number of oxygen atoms it contains from a high resolution mass spectrum. In this case, there are only two oxygens, one of which is required by the carbonyl at C10, which could be deduced from infrared and/or carbon NMR data. Thus, the H22-H23a/b connectivity above would locate the second oxygen, which must then be flanked by the C12/H12 heteronuclear pair. In this fashion, although we may not directly link H12 to the H23a/b geminal methylene pair, there are strong grounds for inferring that they are in the same region of the structure. Further evidence supporting the structural proximity of H12 and H23a/b can be garnered from nuclear Overhauser enhancements, which are described below.[18]

Continuing with the elucidation of the structural fragment shown by **2-8**, the H11a/b pair, while coupled to one another, they represent a terminal point in the connectivity network (see **2-8**). Continuing in the other direction from the H12 resonance, we see in the structure, **2-8**, that it is vicinally coupled to the cis H13 proton by a 3.12 Hz coupling constant.[17] From work contained in the literature, we note that the H13 proton resonates furthest upfield at 1.25 ppm.[17,18] Unfortunately, the H12-H13 connectivity in the COSY spectrum of strychnine (**2-6**) shown in Fig. 2-11 is only very weakly visible and may, depending on the threshold used when preparing the contour plot, be visible only in slices extracted from the data matrix. This problem illustrates, however, an important point: the use of information contained in slices. In many instances, particularly in the case of noisy spectra, it may not be possible for contour plots to be prepared with a threshold level low enough to show all of the responses in the data matrix. In those cases where the structure elucidation process suggests that a response should or might be present between two protons, it is always best to confirm either the presence or absence of the response by examining the appropriate slice(s) of the symmetrized data matrix.

Couplings correlating other protons with the H13 resonance at 1.25 ppm in the COSY spectrum shown in Fig. 2-11 give rise to responses in the contour plot that are highly variable in intensity. For example, H13 is coupled to the trans H8 proton resonating at 3.85 ppm by a 10.47 Hz coupling[17] that gives rise to an intense response in the contour plot. In contrast, the coupling of H13 to the cis H14 proton resonating at 3.13 ppm is 3.10 Hz[17] and, based on the weak response between H12 and H13, would also be expected to give only a weak response, which in fact it did. A further problem associated with the H13-H14 connectivity is the fact that H14 is one of three resonances contained in a complex multiplet consisting of H14 (3.13 ppm), H11a (3.11 ppm) and H18b (3.19 ppm) (Fig. 2-12). Extricating the coupling pathway that correctly traces the further couplings of the connectivity network we are elucidating can be difficult in cases where numerous responses originate from protons with closely similar chemical shifts. Such problems are, however, conveniently overcome using the proton zero and double quantum coherence experiments, which are discussed in detail below.

Despite problems associated with spectral congestion, a further connectivity from the H14 resonance is located in the contour plot that links H14 to the H15b proton resonating at 2.34 ppm ($J_{H14H15b}$ = 4.58 Hz[17]). There is, however, no response visible in either the contour plot or the corresponding slice taken from the data matrix that correlates H14 to H15a ($J_{H14H15a}$ = 1.98 Hz[17]). The geminal H15 pair (J = 14.37 Hz) is easily correlated with one another via a strong response in the contour plot. Problems similar to those associated with the H14-H15a/b connectivities also plague the establishment of the H15a/b-H16 connectivity. Thus H15a, which is weakly coupled to H16 ($J_{H15aH16}$ = 1.82 Hz[17]), fails to give a response in the contour plot while the more strongly coupled H15b resonance ($J_{H15bH16}$ = 4.11 Hz[17]) does give a response that correlates it with the H16 resonance at 3.92 ppm, completing the establishment of the structural fragment shown by **2-8**.

Aliphatic structural fragments, which may be assembled from the COSY spectrum shown in Fig. 2-11, and results from other experiments, can be combined to assemble still larger substructures. For example, nuclear Overhauser enhancements (nOe) reported by Chazin, Colebrok and Edward[18] first confirm the relationship of H12-H23a across the 24-oxygen atom as implied above. Likewise, irradiation of the H16 resonance gave nOe's at H17, linking the main structural fragment shown by **2-8** to the H17-H18 ethylene bridge and to the H1 aromatic proton, thus orienting in the four spin aromatic system. Further irradiation at H18b gave an enhancement at H20b, linking these systems, a final irradiation at the latter completing the structure assembly through the observation of an nOe at the H22 vinyl proton.

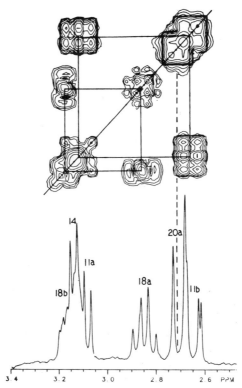

*Fig. 2-12.* Expansion of the COSY spectrum of strychine (**2-6**) in deuterochloroform along the diagonal in the region from 2.7 to 3.4 ppm showing the congestion of the H18b, H14 and H11a resonances on the diagonal.

Thus, COSY spectral data alone will not necessarily provide the means of completely assembling an unknown structure. Rather, experimental evidence derived from COSY spectra is best used in concert with other types of structural information to provide the most strongly self-consistent arguments possible for the structure being examined.

### Phenanthro[4,3-a]dibenzothiophene: A factor limiting application of the COSY experiment; spectral congestion.

As illustrated in the case of the overlapped H11a, H14 and H18b resonances of strychnine (**2-6**), which was just discussed, congestion along the diagonal may significantly complicate the interpretation and/or limit the overall utility of the COSY experiment. A final case that underscores this point even more emphatically is found with phenanthro[4,3-a]dibenzothiophene (**2-9**),[21] a small molecule that clearly illustrates the desirability for alternative methods for the establishment of proton-proton connectivities.

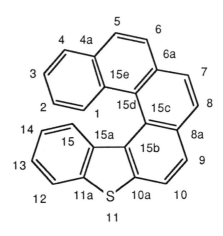

**2-9**

Structurally, phenanthro[4,3-a]dibenzothiophene (**2-9**) would be expected to be helical in nature. Although the X-ray crystal structure of **2-9** has not been determined, the structure of its 9-methyl analog has been obtained[22] and has shown that the two terminal benzene rings overlap considerably. As a function of the shielding that each ring imposes upon its counterpart at the opposite end of the helix, we would expect that substantial upfield shifts would be observed for some of the protons. The high resolution reference spectrum plotted below the COSY spectrum of **2-9** shown in Fig. 2-13 confirms this contention, five of the fourteen protons contained in the structure resonating between 7.3 and 6.6 ppm. Although it was not possible to assign these resonances from the COSY data shown in Fig. 2-13, they can be assigned from a knowledge of the carbon resonance assignments, which were derived by $^{13}C$-$^{13}C$ double quantum coherence[23] in conjunction with a proton-carbon heteronuclear chemical shift correlation spectrum.[21] In this fashion, the upfield shifted $^{1}H$ resonances were identified as H13 (7.31 ppm), H3 (7.11 ppm), H14 (6.90 ppm), H1 (6.83 ppm) and H2 (6.68 ppm). From the connectivities shown in the COSY spectrum, both of the four spin systems have connectivities to constituents that resonate in the highly congested region of the spectrum between 7.70 and 7.95 ppm, which contains responses for the nine remaining protons in the spectrum. Beyond the correlations in the well resolved upfield region of the spectrum, no further correlations may be drawn from the COSY spectrum with any degree of assurance.

Congestion inherent to the proton spectrum of **2-9** in the region from 7.95 to 7.70 ppm highlights the difficulty in applying the COSY

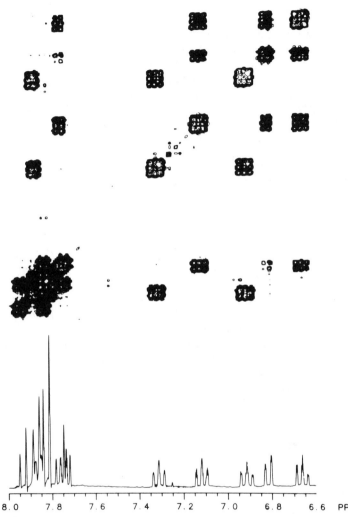

*Fig. 2-13.*    COSY spectrum of phenanthro[4,3-a]dibenzothiophene (**2-9**) in
deuterochloroform acquired at 300.068 MHz. The spectrum was
acquired as 256 x 512 complex points with 200 increments of $t_1$
actually accumulated. The data was processed with sinusoidal
multiplication prior to both Fourier transformations and was
symmetrized prior to plotting. The sweep width in the $F_2$ domain was
±325 Hz, which gives a dwell time of 2.667 msec, which was set to
the first evolution time duration. During the course of the
experiment, the evolution time duration was systematically
incremented by the dwell time to afford a final evolution time of
533.33 msec for the final (200th) block of data taken.

experiment in the case of proton spectra with a high degree of congestion. Off-diagonal responses that correlate strongly coupled resonances are located quite close to the diagonal and frequently cannot be differentiated from the diagonal responses they correlate. In situations like the one presently being considered, alternative spectral means will be necessary to make correlations in the highly congested region. Fundamentally, two alternative experiments are available for the purpose just delineated. Proton double quantum coherence experiments work well in the face of highly congested spectra, particularly when the spectra are highly overlapped but not necessarily strongly coupled. In the latter case, where the spectra are both congested and strongly coupled, homonuclear zero quantum coherence can be employed to considerable advantage. Both the proton double quantum and proton zero quantum experiments are discussed later in this chapter.

## Variants of the Basic COSY Experiment

Numerous variations of the fundamental COSY experiment are possible. These range from phase sensitive versions of the experiment, which convert the magnitude calculated result of the phase twisted line shape inherent to two-dimensional NMR spectroscopy into a pure absorption phase signal with an attendant gain in resolution, through a variety of multiple quantum filtered experiments that can be used to simplify the spectra of molecules containing numerous complex and overlapped subspectra. Advantages accrue with many of these experiments which may be beneficial in the analysis of particular types of spectral problems. The individual variations of the COSY experiment will be discussed in turn and their advantages considered with appropriate examples.

## Phase Sensitive COSY

One of the problems inherent with the COSY spectra of highly complex molecules such as those of proteins, polysaccharides or other biopolymers is resolution. When there are many spin systems contained in the NMR spectrum of a molecule with similar chemical shifts, even though they may not be coupled to one another, they will still tend to produce off-diagonal responses in the same regions. Hence, response overlap becomes a real problem that can be exacerbated by the absolute value calculation frequently made in the final stages of processing COSY spectral data. In particular, the basic two-dimensional NMR line shape can best be characterized as phase twisted, which results in the familiar four pointed star shaped peaks, which will be observed when sufficiently low contour levels are plotted. Obviously, it is desirable to minimize the undesirable characteristics of the responses by employing either sinusoidal or pseudo echo weighting during both Fourier transformations. Despite the improvement afforded by sinusoidal or pseudo-echo weighting, there will still arise circumstances where this is insufficient and it becomes preferable to resort instead to pure absorption mode line shapes.

The advantage of resorting to absorption mode line shapes and phase sensitive display of the data matrix is several fold. First, absolute value spectra can have severely distorted off-diagonal responses due to overlap of the broad individual dispersive components. Pure absorption line shapes provide effectively improved resolution and a consequently decreased tendency for the distortion of line shapes when off-diagonal responses do overlap. Further, phase sensitive COSY displays provide the means of measuring chemical shifts and coupling constants when this information would otherwise be buried in the normal spectra. This trait of the phase sensitive experiment can be especially important when dealing with biopolymers. Thus there are several excellent reasons for utilizing phase sensitive COSY experiments. This should not be taken to imply, however, that it is necessary to routinely employ phase sensitive COSY. Rather, these experiments probably should be reserved for those cases where their use is warranted. As will be seen below, the processing of the phase sensitive variants can be somewhat more complex, counterbalancing, for routine usage, the benefits they provide.

Phase sensitive COSY data may, at present, be acquired using either of two methods. The older of the two methods is generally referred to as the States-Haberkorn method,[24] while the newer procedure, developed by Marion and Wüthrich,[25] is given the acronym TPPI (Time Proportional Phase Incrementation). There are fundamental differences to these two procedures, which will be discussed in turn. The phase sensitive COSY experiments are similar, however, in that they both require a doubling of the number of $t_1$ increments relative to the conventional COSY experiment. On first consideration, this might appear to obviate the phase sensitive techniques for general usage. However, since the minimum phase cycling requirements for the phase sensitive experiments are halved relative to the conventional experiment's phase cycles shown in Table 2-1, there is no increase in total performance time for the experiment.

**States-Haberkorn phase sensitive COSY**. NMR data following Fourier transformation in a single frequency domain consists of an absorption and a dispersion component. Extension to two frequency domains and a double Fourier transformation gives a mixture of double absorption and double dispersion components in the real part and hence there are no combinations that can give rise to purely absorptive components. Thus, in order to obtain two-dimensional NMR data that has a pure absorption phase, it is necessary to perform two separate experiments with correspondingly different phase cycles. The pulse sequence used is identical to that of the basic COSY experiment shown in Fig. 2-1. The eight step phase cycles applied to the two 90° pulses are shown in Table 2-2; one set of data is collected with each of the two phase cycles specified. Care should be exercised in performing the experiments to make the initial value of the evolution time, $t_1$, as close to zero as permitted by the spectrometer. The provision just described is necessary to preclude the need for phase corrections in $F_1$ following the second Fourier transformation.

Following the completion of data acquisition, both sets of data acquired using the phase cycling scheme shown in Table 2-2 are processed in the normal fashion with respect to $t_2$. This pair of operations gives:

$$S^A(t_1,t_2) \rightarrow S^A(t_1,F_2) \qquad [2\text{-}4]$$

and

$$S^B(t_1, t_2) \rightarrow S^B(t_1,F_2) \qquad [2\text{-}5]$$

Following Fourier transformation with respect to $t_2$ in each case, the same phase correction is applied to both data sets. To obtain pure absorption line shapes following the second Fourier transformation rather than the normal double absorption/double dispersion character, it is next necessary to construct a hybrid data matrix from the two prepared according to Eqns. [2-4] and [2-5]. Specifically, the hybrid data matrix, $S^H$, is assembled from the real component of $S^A(t_1,F_2)$ and the imaginary component of $S^B(t_1,F_2)$. Fourier transformation of $S^H$ with respect to $t_1$ affords a data matrix, $S^H(F_1,F_2)$, which will have the desired pure absorption line shapes. No phasing with respect to $F_1$ should be necessary if care was taken to minimize the initial value of the $t_1$ interval.

### Time proportional phase incrementation (TPPI) COSY.
The time proportional phase incrementation or TPPI method for obtaining phase sensitive COSY spectral data was developed by Marion and Wüthrich[25] and grew out of earlier work by other research groups.[26,27] TPPI differs fundamentally from the States-Haberkorn method in that it employs a real Fourier transformation rather than the complex Fourier transform normally utilized in two-dimensional NMR spectroscopy.

### Comparative evaluation of the States-Haberkorn and TPPI methods for obtaining phase sensitive COSY data.
Regardless of the method employed,[24,25] the major advantage inherent to the phase sensitive methods is that they avoid the highly undesirable phase twist line shape normally associated with echo or antiecho selection. Unfortunately, while phase twist line shape is bad enough in and of itself, the situation is exacerbated by the calculation of the absolute value spectrum. Thus, phase sensitive spectra offer a very substantial gain in resolution accompanied by an increase in sensitivity.

Keeler and Neuhaus[28] have reporded a comprehensive analysis of the States-Haberkorn[24] and Marion and Wüthrich[25] phase sensitive methods and have concluded that they are closely related although they

Table 2-2.        Phase cycling for States-Haberkorn[24] phase sensitive COSY.

| Acquisition | Phase Cycle A | | Phase Cycle B | |
|---|---|---|---|---|
| | $\phi_1$ | $\phi_2$ | $\phi_1$ | $\phi_2$ |
| 1 | 0 | 3 | 0 | 0 |
| 2 | 1 | 2 | 1 | 3 |
| 3 | 2 | 1 | 2 | 2 |
| 4 | 3 | 0 | 3 | 1 |
| 5 | 0 | 1 | 0 | 2 |
| 6 | 1 | 0 | 1 | 1 |
| 7 | 2 | 3 | 2 | 0 |
| 8 | 3 | 2 | 3 | 3 |

may seem to be superficially quite dissimilar. Perhaps the most significant difference between the two approaches to the problem is that the States-Haberkorn method employs the generally available complex Fourier transformation while the Marion and Wüthrich approach utilizes a real Fourier transformation which is less generally available in commercial spectrometer software packages.

**A brief survey of applications of phase sensitive COSY.** The number of papers detailing applications of phase sensitive COSY experiments is burgeoning, making any sort of comprehensive survey impossible in a limited space. The range of molecules to which these experiments have been applied is diverse but is principally centered in the area of large biomacromolecules, which have included oligonucleotides and a wide variety of peptides and proteins.[29-44] Applications in the area of small molecule structure elucidation are beginning to appear as the advantages inherent to the phase sensitive method become more widely appreciated. In particular, it is to be expected that phase sensitive spectra will eventually have a significant impact in the study of congested terpene spectra.

## Long Range COSY (LRCOSY)

Occasionally, it is desirable to have access to long range coupling information. When COSY spectra are highly digitized in the second frequency domain, it is sometimes possible to obtain sufficient digital resolution to allow small couplings (< 2 Hz) to be observed. Indeed, 256 x 256 point data matrices are frequently sufficient to allow the observation of prominent long range couplings in the spectra of polynuclear aromatic systems,[45,46] since the narrower spectral widths utilized in such cases are correspondingly better digitized. However, with wider spectral widths, digital resolution in $t_1$ is normally too low to allow the observation of long range couplings, since the amounts of time necessary to obtain sufficient levels of digitization would be prohibitive. This situation clearly underscores the need for an alternative method of accessing long range coupling information, which is satisfied by a modification of the basic COSY pulse sequence which was shown in Fig. 2-1.

The intensity of off-diagonal responses corresponding to long range couplings are proportional to

$$\sin(\pi J_{AX} t_2) \exp^{(-t_2/T_2)} \qquad [2\text{-}6]$$

which will obviously be small. Further, since the components of magnetization start in an antiphase orientation, it can also be shown that the detected transfer will be small in the case of long range couplings. Assuming that $J_{AX} << (T_2)^{-1}$, the transfer process will be most efficient at time $t_2 = T_2$. This leads to optimal centering of the acquired signal at times $t_1 = T_2$ and $t_2 = T_2$, which would lead to a very large data matrix if sufficient numbers of $t_1$ increments and high enough digitization in $t_2$ are employed to meet these optima. Practically, these conditions cannot be conveniently satisfied. These requirements can, however, be met by inserting fixed delays, $\Delta$, after the incremented evolution period, $t_1$, and prior to the acquisition time, $t_2$.[47] The modified pulse sequence for detection of long range coupling information (LRCOSY) is shown in Fig. 2-14. Location of the center of the pseudo echo in accord with the

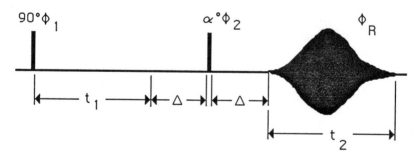

Fig. 2-14 Pulse sequence for observation of long range couplings in a COSY experiment. The fixed delays, $\Delta$, allow the optimal positioning of the echo at times corresponding to $t_1 = T_2$ and $t_2 = T_2$ when the conditions $t_1 = T_2 - \Delta$ and $t_2 = T_2 - \Delta$ are satisfied. Usefully, the LRCOSY pulse sequence limits the number of increments and $t_1$ and digitization in $t_2$, which would otherwise be required to observe long range couplings.

constraints delineated above is achieved by chosing the duration of $\Delta$ such that the conditions $t_1 = T_2 - \Delta$ and $t_2 = T_2 - \Delta$ are satisfied. The flip angle of the second pulse, $\alpha$, can be advantageously set to 60° to improve sensitivity.

**A brief survey of applications of LRCOSY**.  The LRCOSY experiment would conceptually seem to be quite useful.  Various applications of the LRCOSY experiment appear in the literature. Bax and Freeman[47] have applied the long range COSY experiment to a tricyclododecane derivative with considerable success. Wynants and Van Binst and coworkers[48-50] have used the LRCOSY experiment to observe long range couplings in a number of cyclic peptides.  Sydnes and Skjetne[51] have employed LRCOSY in a study of nitro benzo[e]pyrenes. Sakai, Higa and Kashman have utilized the LRCOSY experiment in the elucidation of the structure of the novel antitumor macrolide misakinolide-A.[52]

**Alternatives to LRCOSY**. In our own experience, we have found the utilization of the proton multiple quantum coherence NMR experiments described below and two-dimensional homonuclear Hartmann-Hahn (HOHAHA) experiments (see Chapter 4) to be potentially superior alternatives when it is necessary to track small coupling constants.  Several studies[11,19,53-55] have confirmed the abilities of the proton double quantum coherence experiment in observing very small coupling pathways.  A description of the proton double quantum NMR experiment and examples of its applications are presented below.

### Broadband Proton Decoupled
### COSY (HDCOSY)

In some instances, it may be beneficial to either suppress or reduce scalar couplings in one frequency domain.  Broadband decoupling in the $F_1$ frequency domain of a COSY experiment is possible by a simple modification of the basic COSY pulse sequence shown in Fig. 2-1.  By inserting a movable (sliding) 180° pulse in the evolution period, $t_1$, it is possible either to completely suppress couplings in $F_1$[47] or to downscale them.[56] The HDCOSY pulse sequence is shown in Fig. 2-15.  The overall duration of the evolution period is fixed.  The delay preceding the 180° pulse is incremented in the usual fashion, while that following it is systematically decremented, effectively relocating the 180° pulse through the evolution period to provide the decoupling in $F_1$.  Phase cycling for the experiment is given in Table 2-3.

**A brief survey of applications of the HDCOSY experiment**. The HDCOSY pulse sequence or variants have been used with some degree of success, and there are a number of examples contained in the literature. Hosur and coworkers have reported several applications relating to chemical shift scaling[57,58] and J-scaling.[59] Ernst and coworkers[60] have discussed $F_1$ decoupling in theoretical terms for both COSY and SECSY experiments and have applied them, in conjunction with phase sensitive data accumulation, to the study of proteins. Brown[61] in similar fashion has applied $F_1$ decoupling to advantage with a polyvinyl alcohol model compound. Reveco and coworkers have applied the HDCOSY experiment to the study of the complex spectra of a cyclometalated complex of ruthenium.[62] Preliminary indications would seem to suggest that HDCOSY will be generally useful experiment, though it remains to be seen whether applications will continue to appear.

## Super COSY

One problem with two-dimensional NMR experiments is, of course, low digital resolution, especially in the case of large biomolecules. If sufficient digital resolution is available, the signal-to-noise (S/N) ratio of cross or off-diagonal peaks relative to diagonal peaks will be the same. Unfortunately, only in rare instances will it be

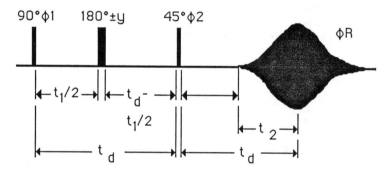

Fig. 2-15.    Pulse sequence to provide broadband homonuclear decoupling in the $F_1$ frequency domain. The 180° pulse is systematically repositioned through the evolution period.

possible to attain adequate digital resolution and hence, off-diagonal responses will frequently be weak. Diagonal responses in COSY spectra have in-phase character which tends to augment signal intensities, while off-diagonal responses have antiphase character, which cancels intensity. Bax and coworkers[63] have addressed this problem from the standpoint of data processing. Kumar, Hosur and Chandrasekhar[64] have alternatively devised a pulse sequence that has been christened super COSY which removes antiphase character from the off-diagonal responses, transferring it into the diagonal peaks.

Table 2-3.   Phase Cycling for the $F_1$ broadband decoupled COSY (HDCOSY) experiment. The minimum phase cycle requires 32 steps.

| $\phi_1$ | 0000 | 0000 | 1111 | 1111 | 2222 | 2222 | 3333 | 3333 |
|---|---|---|---|---|---|---|---|---|
| $\phi_2$ | 1111 | 3333 | 2222 | 0000 | 3333 | 1111 | 0000 | 2222 |
| $\phi_3$ | 0123 | 0123 | 1230 | 1230 | 2301 | 2301 | 3012 | 3012 |
| $\phi_R$ | 0202 | 0202 | 1313 | 1313 | 2020 | 2020 | 3131 | 3131 |

The super COSY pulse sequence devised by Kumar and coworkers[64] is shown in Fig. 2-16. The concept of transferring antiphase components into phase by adding delay periods with 180° pulses midway through them was first applied to homonuclear correlation by Kumar and coworkers but was borrowed from heteronuclear chemical shift correlation.[65,66] For an AX spin system, Kumar and coworkers have shown the optimum duration of the fixed delays, $\Delta$, to be set to $\Delta = 1(4/J)$, while Burum and Ernst[65] have shown $\Delta$ to be optimally set to $\Delta = 1/(2J)$ for $AX_2$ or $1/(3J)$ for $AX_3$ spin systems. Preliminary work using uracil as an AX model system has shown the pulse sequence to be capable of producing spectra in which off-diagonal peak intensities are greater than those of the diagonal. Application of super COSY to a decapeptide gave a significantly greater number of responses with much better intensity than could be obtained with the conventional COSY experiment.

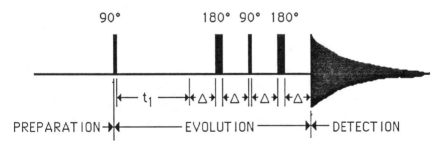

Fig. 2-16.   Super COSY pulse sequence developed by Kumar and coworkers.[64] Optimization of the fixed delay, $\Delta$, is discussed in the text for specific cases but a compromise setting of $\Delta = 0.3/J$ should be used when there are a variety of different types of spin system present in the spectrum.

**A brief survey of applications of the super COSY experiment**. Several applications of the super COSY experiment have appeared. Confirming the contention in the original report that super COSY experiments could be used for long range correlations by choosing appropriate durations for the fixed delays ($\Delta$), Mayor and Hosur[67] were able to observe five bond couplings in a simple dipeptide model system when $\Delta = 0.15$ sec. In a similar fashion, Gundhi, Chari and Hosur have also applied the super COSY experiment to the observation of four bond scalar couplings in dinucleotides.[68] Based upon the success of applications that have already appeared, it seems probable that usage of the super COSY pulse sequence will continue to expand.

### Edited COSY Spectra Incorporating BIRD Pulses

In a series of papers, Rutar, Wong and coworkers[69,70] have investigated the subtleties of the bilinear rotation decoupling or BIRD pulse cluster initially described by Garbow, Weitekamp and Pines.[71] These efforts have led to a further modification of the basic COSY pulse sequence (Fig. 2-1) as shown in Fig. 2-17. Using the modified pulse sequence shown in Fig. 2-17, Rutar and Wong[72] have demonstrated that it is possible to "edit" the COSY spectrum, identifying terminal spins of linear spin systems. A complementary version of the experiment, which differs only in the phase of the constituents of the BIRD pulse cluster provides signals for inner spins of a linear system flanked by vicinal neighbors. Phase cycling schedules for the two variants of the experiment are shown in Table 2-4.

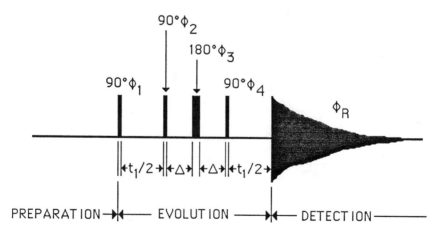

*Fig. 2-17.* MIS(T)/(I) pulse sequence devised by Rutar and Wong.[72] The MIS(T) phase cycling selects for the terminal spins of linear spin systems, while the MIS(I) phase cycling selects for the inner spins of linear systemse flanked by vicinal neighbors on both sides. The intensity of off-diagonal or cross peaks is proportional to $\frac{1}{2}\sin(\pi J2\Delta)$, which is maximal at $1/(4J)$.

Results obtained for a four spin linear system using the MIS(T) pulse sequence shown in Fig. 2-17 are presented in Fig. 2-18. Spin systems of the type shown schematically in Fig. 2-18 are commonly encountered in the spectra of polynuclear aromatics, oligosaccharides and similar molecules. The MIS(T) spectrum of phenanthro-[3,4-*b*]thiophene (**2-5**) are shown in Fig. 2-19. The long term utility of the MIS(T/I) pulse sequences remains to be determined and will probably depend upon the success of future applications. It is clear, however, that the range of applications for this experiment will be limited to a few special cases such as polynuclear aromatics and oligosaccharides, which contain the linear spin systems upon which it can be beneficially executed.

Table 2-4. Minimum phase cycling for the MIS(T) and MIS(I) pulse sequence of Rutar and Wong.[72] The MIS(T) sequence selects for the terminal spins of linear spin systems while the MIS(I) pulse sequence selects for inner spins of linear systems that are coupled to a vicinally neighboring spin on both sides. The phase cycling shown is sufficient to suppress unwanted resonances. Artifacts introduced by quadrature imperfections may be eliminated by repeating each step four times with a 90° incrementation of the receiver phase, thus affording a sixteen step phase cycle.

| MIS(I) Acquisition | $\phi_1$ | $\phi_2$ | $\phi_3$ | $\phi_4$ | $\phi_R$ |
|---|---|---|---|---|---|
| 1 | 0 | 0 | 1 | 2 | 0 |
| 2 | 0 | 1 | 2 | 3 | 0 |
| 3 | 0 | 2 | 3 | 0 | 0 |
| 4 | 0 | 3 | 0 | 1 | 0 |

| MIS(T) Acquisition | $\phi_1$ | $\phi_2$ | $\phi_3$ | $\phi_4$ | $\phi_R$ |
|---|---|---|---|---|---|
| 1 | 0 | 0 | 1 | 0 | 0 |
| 2 | 0 | 1 | 2 | 1 | 0 |
| 3 | 0 | 2 | 3 | 2 | 0 |
| 4 | 0 | 3 | 0 | 3 | 0 |

## Spin-Echo Correlated Spectroscopy (SECSY)

Autocorrelated proton or COSY experiments that employ a square data matrix may in some instances provide space in the matrix for correlations that will seldom, if ever, be observed. One example of such a correlation would be between an aromatic proton and an aliphatic methyl group. With the rational intent of increasing the utilization of the data matrix, Nagayama, Wüthrich and Ernst[73] described a modification of the COSY experiment in 1979 in which the second 90° pulse was moved to the center of the evolution period as shown in Fig. 2-20. A more detailed comparison of the SECSY and COSY experiments is found in another paper by Nagayama and coworkers,[74] but it suffices to note that the experiment functions to produce a spin echo with a consequent rotation of the diagonal in the data matrix. A comparison of the correlation of resonances in the COSY relative to the SECSY experiment is presented in Fig. 2-21. At present, the SECSY experiment is seldom used. Phases are cycled in a fashion analogous to that used in the COSY experiment and shown in Table 2-1.

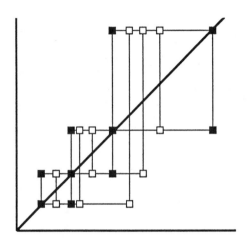

*Fig. 2-18.* Contour plot schematic illustrating the results obtained using COSY and the MIS(T) pulse sequence of Rutar and Wong.[72] The solid points on the diagonal indicate the locations of the four spins of a linear spin system. Solid off-diagonal points represent the off-diagonal or cross peak locations which would be obtained using the conventional COSY pulse sequence (Fig. 2-1). Nonzero couplings are assumed only between vicinal neighbors in this schematic; long range couplings are ignored. Open, off-diagonal points represent responses which that be obtained using the MIS(T) pulse sequence shown in Fig. 2-17.

## Multiple Quantum Filtered COSY Experiments

### Multiple Quantum Filtration

The concept of multiple quantum filtration (MQF) must be credited to Ernst and coworkers[75] who described several variants of the technique in a 1982 communication. Quite simply, MQF provides us with the means of selectively removing certain components of a COSY spectrum. For example, methyl singlets generally provide minimal molecular connectivity information, with perhaps the exception of vinyl methyls of E-trisubstituted double bonds, which frequently show correlations to their respective vinyl protons. Indeed, methyl resonances, because of their intensity, frequently obscure much weaker proton multiplets, and the availability of a means of filtering these responses from the COSY spectrum would thus be quite desirable. It is precisely this ability that the communication by Piantini, Sørensen and Ernst[75] addressed and provided.

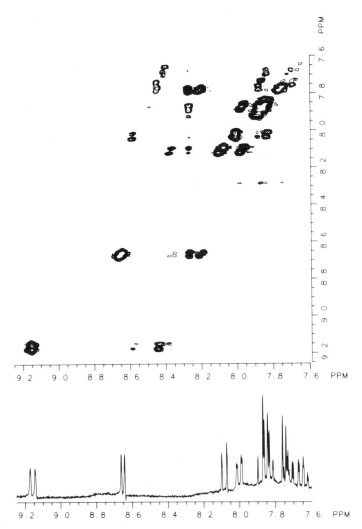

*Fig. 2-19* .MIS(T) spectrum of phenanthro[3,4-*b*]thiophene **(2-5)** in deuterochloroform recorded at 300.042 MHz.

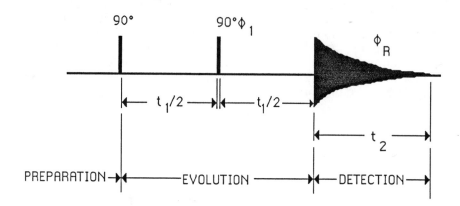

*Fig. 2-20.* Spin-echo correlated spectroscopy (SECSY) pulse sequence. By placing the second 90° pulse midway through the evolution period, the size of the data matrix may be somewhat reduced. Overall sensitivity relative to more conventional COSY pulse sequences is lower.

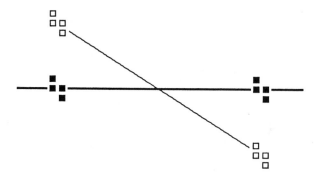

*Fig. 2-21.* Schematic representation of the correlation of an AX spin system obtained using the SECSY pulse sequence shown in Fig. 2-20. The location of the second 90° pulse midway through the evolution period, $t_1$, produces an echo. As a consequence of this modificaiton of the pulse sequence, the diagonal axis is rotated as shown. Responses that are scalar (J) coupled to one another are correlated about an axis that has an angle of 135° relative to the $F_1 = 0$ Hz axis along which the normal spectrum now resides. Responses that are the equivalent of the diagonal responses in the COSY experiment are shown as solid squares, the responses equivalent to off-diagonal responses are shown as open squares. See Fig. 2-5 for comparison.

Multiple quantum filtering relates to the complexity of the spin systems contained in a given spectrum in the following way: singlets will not have available to them double or higher quantum transitions and hence a double quantum filter (DQF) will select for spin systems that minimally have a double quantum transition, eg, two spin systems. Double quantum filtration is proably the most widely applicable since it employs routine 90° phase shifts, which may be generated with an contemporary NMR spectrometer. Triple quantum filtered (TQF) spectra will eliminate singlets and all AB or AX components, since these spectral components do not have a three quantum transition. Unfortunately, triple and higher quantum filtering requires the implementation of non-90° phase shift capabilities, which have only recently become available on commercial spectrometers. The orders of multiple quantum coherence that will be observed for various receiver phases are shown in Table 2-5.[76]

Table 2-5. Possibilities for the selective detection of multiple quantum transitions as a function of the phase of coadded experiments. For overscored phases, the resulting signals must be subtracted.

| Phase ($\phi$) of coadded experiments | | | | | | Observed order of MQT | | | | | | | | | |
|---|---|---|---|---|---|---|---|---|---|---|---|---|---|---|---|
| | | | | | | 0 | 1 | 2 | 3 | 4 | 5 | 6 | 7 | 8 | 9 |
| 0 | | | | | | 0 | 1 | 2 | 3 | 4 | 5 | 6 | 7 | 8 | 9 |
| 0 | 180 | | | | | 0 | | 2 | | 4 | | 6 | | 8 | |
| 0 | $\overline{180}$ | | | | | | 1 | | 3 | | 5 | | 7 | | 9 |
| 0 | 90 | 180 | 270 | | | 0 | | | | 4 | | | | 8 | |
| 0 | $\overline{90}$ | 180 | $\overline{270}$ | | | | | 2 | | | | 6 | | | |
| 0 | $\overline{60}$ | 120 | $\overline{180}$ | 240 | $\overline{300}$ | | | | 3 | | | | | | 9 |

Thus, three and higher quantum filtering of COSY experiments will probably be somewhat slower to gain in popularity than the DQF-COSY variant.

## Double quantum filtered COSY

***The pulse sequence and parameter selection.*** Double quantum filtering of a COSY spectrum is quite simple to implement in terms of modification of the basic pulse sequence. The modification to provide double quantum filtering of a COSY spectrum is shown by the pulse sequence in Fig. 2-22. The normal COSY pulse sequence (see Fig. 2-1) is followed by a minimal duration fixed delay and then a

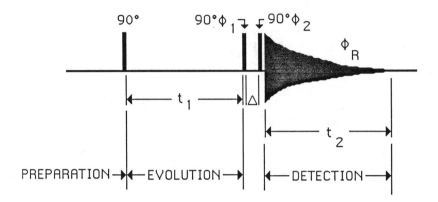

*Fig. 2-22.* Double quantum filtered COSY pulse sequence. Double quantum filtering is provided by the third 90° pulse following the normal COSY pulse sequence (see Fig. 2-1). Phases are cycled according to Table 2-5 to provide echo or antiecho selection as required.

third 90° pulse whose phase is cycled independently of that of the second 90° pulse according to the scheme shown in Table 2-5. Selection of either the echo or the antiecho may be made, depending upon the phase used for the final 90° pulse in the sequence. The fixed delay is inserted to allow only a sufficient period of time for the spectrometer to execute the phase shift. Generally, a fixed duration delay in the range of about 10 µsec will be sufficient, although this may vary somewhat from one instrument to the next.

***Phase cycling.*** Phase cycling in the DQF-COSY experiment is no more complicated than the phase cycling encountered in the "normal" COSY experiment. Recourse is available to a minimal four step phase cycle, various sixteen step phase cycles and phase cycles for the acquisition of phase sensitive double quantum filtered data. The basic four and sixteen step phase cycles for the DQF-COSY experiment are shown in Table 2-6.[77]

Conversion of double quantum coherence into observable single quantum coherence is accomplished by the additional pulse, while the "filtering" of single and higher order multiple quantum coherences is provided by the phase cycling of the final pulse and the receiver.

Acquisition of phase sensitive double quantum filtered data entails the same increase in complexity as was encountered above in making the transition from absolute value to phase sensitive COSY spectra. Once again, two separate phase cycles are employed; these are shown in Table 2-7. The results of the two separate experiments are combined in a manner analogous to that described above for the phase sensitive COSY experiment.[77,78]

Table  2-6.     Phase cycling schemes for double quantum filtering of a COSY
spectrum.  A minimal four step phase cycle may be used although
better quality spectra may be obtained by using a sixteen step phase
cycle.

| | Four Step - Echo Selection | | | |
|---|---|---|---|---|
| $\phi_1$ | 0000 | $\phi_2$ | 0123 | $\phi_R$ | 0321 |

| | Sixteen Step - Echo Selection | | | |
|---|---|---|---|---|
| $\phi_1$ | 0000 | 1111 | 2222 | 3333 |
| $\phi_2$ | 3210 | 2103 | 1032 | 0321 |
| $\phi_R$ | 0123 | 0123 | 0123 | 0123 |

| | Sixteen Step - Antiecho Selection | | | |
|---|---|---|---|---|
| $\phi_1$ | 0000 | 1111 | 2222 | 3333 |
| $\phi_2$ | 3210 | 0321 | 1032 | 2103 |
| $\phi_R$ | 0123 | 0123 | 0123 | 0123 |

Table 2-7. Phase cycling for phase sensitive double quantum filtered COSY.[77,78]

| Phase Cycle A | | | | | | | |
|---|---|---|---|---|---|---|---|
| $\phi_1$ 0000 | 1111 | 2222 | 3333 | 0000 | 1111 | 2222 | 3333 |
| $\phi_2$ 0000 | 1111 | 2222 | 3333 | 2222 | 3333 | 0000 | 1111 |
| $\phi_R$ 3210 | 1023 | 3210 | 1023 | 1023 | 3210 | 1023 | 3210 |

| Phase Cycle B | | | | | | | |
|---|---|---|---|---|---|---|---|
| $\phi_1$ 0000 | 1111 | 2222 | 3333 | 0000 | 1111 | 2222 | 3333 |
| $\phi_2$ 1111 | 2222 | 3333 | 0000 | 3333 | 0000 | 1111 | 2222 |
| $\phi_R$ 3210 | 1023 | 3210 | 1023 | 1023 | 3210 | 1023 | 3210 |

***Parameter selection.***  The selection of parameters for the
double quantum filtered COSY experiment is fundamentally the same
as that entailed in acquiring conventional COSY data.  The dwell time
and interpulse delays are chosen in a manner identical to that for the
normal COSY experiment.  The sole parameter that does not enter into
consideration in the conventional experiment is the fixed duration delay,
$\Delta$, which separates the final two 90° pulses in the DQF-COSY pulse
sequence shown in Fig. 2-22.  In principle, the $\Delta$ interval may be
eliminated, assuming that the spectrometer can instantaneously perform
the requisite phase shift required in going from the second to the third
90° pulse.  In actual practice, however, phase shifters take a finite period

of time to operate, which entails a short fixed delay ranging from 10 to 25 μsec.

**_Data Processing_**. Considerations implicit in the processing of a conventional COSY spectrum apply to DQF-COSY spectra. Consequently, no additional discussion of the processing will be given. Readers are instead referred to the discussion on data processing in the section dealing with the basic COSY experiment above.

**Application of double quantum filtered COSY spectra to the alkaloid brucine**. To illustrate the differences in the data obtained from the double quantum filtered COSY experiment relative to the "normal" COSY experiment, the spectrum of the alkaloid brucine (**2-10**) will be considered.

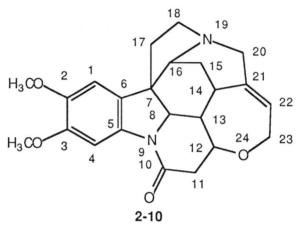

**2-10**

Structurally, brucine (**2-10**) differs only very slightly from strychnine (**2-6**), which has already been employed as a model compound. The sole difference is the replacement of two of the aromatic protons at the 2- and 3-positions by methoxyl groups. The aliphatic region of the brucine spectrum contains resonances for 25 protons within a region 4.2 ppm wide (see high resolution reference spectrum plotted beneath the contour plot in Fig. 2-23). Quite clearly, the most intense signals in the spectrum are due to the two methoxyl groups, which may be expected to introduce a substantial dynamic range problem. The intense methoxyl resonances are problematic in that they may tend to obscure responses in close proximity to their location along the diagonal. Further, the intensity of the methoxyl resonances will make it more difficult to plot spectra showing the weaker responses of individual protons. Compare the normal COSY spectrum shown in Fig. 2-23 to the double quantum filtered COSY specttrum shown in Fig. 2-24. In the latter, the intensity of the methoxyl resonances is substantially reduced with a consequent improvement in the dynamic range of the data matrix. The dynamic range difference is illustrated quite nicely by the projections of the COSY and double

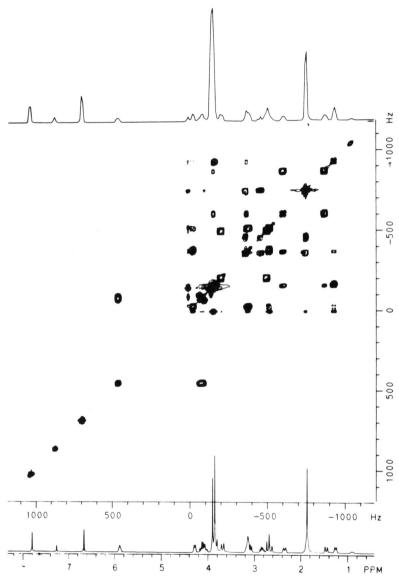

*Fig. 2-23.* COSY spectrum of 10 mg of the alkaloid brucine (**2-10**) recorded in deuterochloroform at an observation frequency of 300.068 MHz. The data was acquired as 256 x 512 points to yield a matrix consisting of 256 x 256 points after processing. The contour plot shown was symmetrized prior to plotting. Total acquisition time was 15 hours (overkill). The high resolution spectrum is plotted beneath the contour plot, while the projection is plotted above.

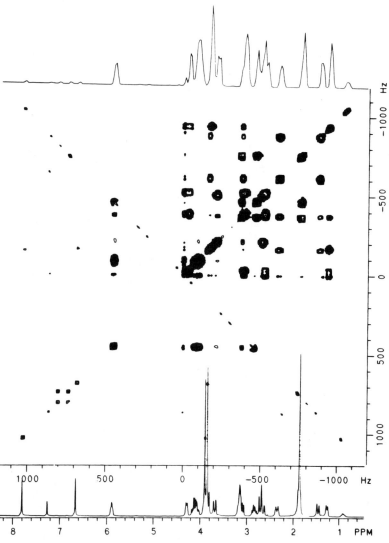

*Fig. 2-24.*    Double quantum filtered COSY spectrum of a 10 mg sample of brucine
(**2-10**) in deuterochloroform recorded at 300.068 MHz. The data was
acquired as 256 x 512 points to yield a matrix consisting of 256 x 256
points after processing.  The contour plot shown was symmetrized prior
to plotting.   Total acquisition time was 15 hours (overkill). The high
resolution spectrum is plotted beneath the contour plot, while the
projection is plotted above.

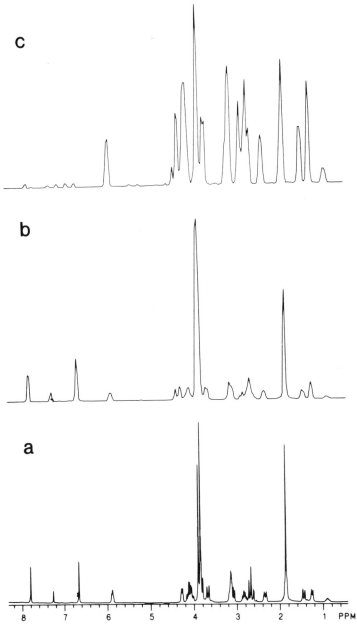

*Fig. 2-25.* Comparison spectra of brucine (**2-10**). (a) Normal high resolution proton spectrum. (b) 0° projection of the normal COSY spectrum shown in Fig. 2-23. (c) 0° projection of the double quantum filtered COSY spectrum shown in Fig.2-24. Note the very large difference in the intensity of the methoxyl resonances in the reference spectrum (a) and the normal COSY projection (b) relative to the double quantum filtered projection (c).

quantum filtered COSY (DQF-COSY or 2QCOSY) spectra shown above the brucine reference spectrum in Fig.2-25. In the normal COSY spectrum, the methoxyl resonances dominate the spectrum. In the DQF-COSY spectrum, in contrast, the methoxyl resonances are substantially reduced in intensity. A further advantage of the DQF-COSY spectrum, which is seen only on comparison of the contour plots, is the reduction in the overall intensity of the diagonal responses. The latter point is quite important when it is necessary to observe off-diagonal responses in close proximity to the diagonal, eg, responses correlating an AB spin system. In general, the advantages just cited make the DQF-COSY experiment quite useful as an initial experiment for assembling vicinal proton-proton ($^3J_{HH}$) connectivity networks.[79]

In contrast to the advantages inherent to the DQF-COSY experiment, which are substantial, the drawbacks associated with this experiment are really quite minimal. First, when extremely deep contour levels are plotted, an antidiagonal may become visible. (See Fig. 2-24.) Second, there is some loss in the sensitivity of the DQF-COSY experiment relative to the conventional COSY experiment,[78] although this would appear to be amply compensated by the improvement in the dynamic range of the experiment provided by the suppression of the singlets and the undesired diagonal elements.

### Application of double quantum filtered COSY to the cyclic depsipeptide didemnim-B.

The ability of double quantum filtration to enhance access to the information contained in a COSY spectrum has led to increased usage of the experiment in studies of complex molecular structure, namely peptides. Several properties of the double quantum filtered COSY experiment mentioned above make it a logical choice when dealing with compounds that contain several groups of methyl protons. For example, consider the cyclic depsipeptide didemnim-B (**2-11**), which contains fifteen methyl groups. The four N-methyl resonances, which appear as intense singlets in the proton reference spectrum plotted below the contour plot in Fig.2-26 in the region from 2.5 to 3.6 ppm, are detrimental to the interpretation of the spectrum due to their large diagonal responses. In contrast, it will be noted from the projection shown above the contour plot, that the signals for the N-methyl resonances are largely suppressed by the DQF-COSY experiment, thereby decreasing potential dynamic range problems that would otherwise be associated with the experiment.

### Acquisition of spectra in water.

The ability of the DQF-COSY experiment to filter out resonances due to methyl singlets is not the only beneficial effect of the multiple quantum filter. Specifically, the multiple quantum filter also has the capacity to filter out signals of solvents, the most noteworthy being that of water. When used in conjunction with presaturation, DQF-COSY can easily be employed in the collection of COSY spectra with samples dissolved in a 90% $H_2O$:10% $D_2O$ solvent mixture, which is also quite useful in studying peptide spectra.[75]

R = -CH(CH₃)CH₂CH₃

**2-11**

**Phase sensitive double quantum filtered COSY**.  When the data are accumulated in the phase sensitive mode, the full power of DQF-COSY can be realized.  Comparable sensitivity for the phase sensitive versions of DQF-COSY and conventional COSY are obtained due to the reduction in the size of the diagonal elements. The phase sensitive COSY experiment suffers from the presence of a diagonal with dispersion line shape, whereas the DQF-COSY experiment produces a diagonal with absorption line  shape. Synergistic benefits resulting from the combination of phase sensitive data acquisition and processing, in conjunction with double quantum filtration, afford an extremely powerful experiment.  Perhaps the greatest strength of the experiment resides in the very high resolution obtained in the off-diagonal responses.   In principal, the resolution afforded by the off-diagonal or cross peaks in the DQF-COSY experiment can be used to measure accurate chemical shifts and coupling constants when adequate digitization is available. This benefit is significant in the analysis of spectral details of complex molecules whose one-dimensional spectra are intractable to analysis.  It is probably safe to assume that we will see the DQF-COSY experiment used in the elucidation of the structures of progressively more complex biomacromolecules.[80]

## Higher Order Multiple Quantum Filtered COSY

Filtration using multiple quantum filters of orders higher than 2 is possible but thus far has been limited to a relatively small number of laboratories.  While the pulse sequences for triple and higher quantum filtration are no more complex than for double quantum filtration, they require the use of $\pi/n$ phase shifts, which are only now becoming

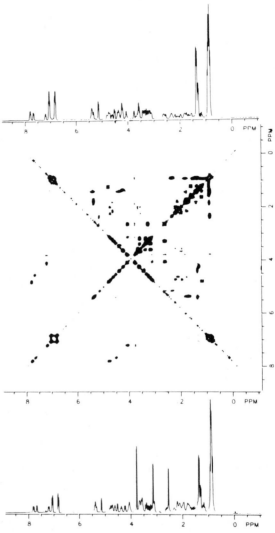

*Fig. 2-26.*     Double quantum filtered COSY spectrum of the cyclic depsipeptide didemnim-B **(2-11)** recorded in deuterochloroform at 300.068 MHz. The normal high resolution proton spectrum is plotted below the contour plot while the 0° projection is plotted above the contour plot. Didemnim-B contains fifteen methyl groups in its structure, which give rise to the complex, very intense methyl responses at about 1 ppm in the normal proton spectrum as well as the N- and O-methyl singlets observed downfield. In contrast, the projection of the DQF-COSY spectrum shows that the intensity of the methyl responses compares much more favorably with the resonances due to single protons in the spectrum, leading to a substantial improvement in the dynamic range and a consequent ease of contour plot preparation.

routinely available on commercial NMR spectrometers. A comprehensive description of the appearance of multiple quantum filtered spectra ranging from double through seventh order has been reported by Wüthrich and coworkers,[81] to which the interested reader is referred for further detail. Examples of high order multiple quantum filtration are also found in the work of Dabrowski and coworkers,[86] Rance and Wright[80] and Homans et al. [82]

## Applications of Multiple Quantum Filtered COSY

Since the first report of the double quantum filtered variant of the COSY experiment by Ernst and coworkers,[75] the technique has been utilized in the assignment of the proton NMR spectra of sesqui-terpenes,[83] peptides, oligomers of DNA and oligosaccharides[84-86] and polynuclear aromatics.[87] Doubtless additional applications in the elucidation of novel natural product structures will continue to appear as more and more workers develop an appreciation for the advantages offered by multiple quantum filtration.

## OTHER QUANTUM METHODS FOR ESTABLISHING PROTON-PROTON CONNECTIVITIES

Jeener's visionary experiment, homonuclear correlated spectroscopy or COSY, has had a massive impact on the way in which we conduct spectral assignment and structure elucidation studies. As illustrated in the preceding sections of this chapter, it is a versatile experiment but not entirely without shortcomings and limitations. There are, however, many instances when the COSY experiment cannot provide us with complete connectivity information. Fortunately, however, we have recourse to a number of other two-dimensional NMR experiments, some of which circumvent the difficulties encountered with the COSY experiment. The following sections of this chapter will focus on "alternate" quantum experiments for the establishment of proton-proton connectivities .

### Proton Double Quantum Coherence

Originally, the pulse sequence we now employ to provide proton double quantum coherence was intended for the observation of $^{13}C$-$^{13}C$ couplings at natural abundance in the presence of isolated $^{13}C$ nuclide signals with 200 times greater intensity.[88,89] Preceding the appearance of pulse sequences for the observation of natural abundance carbon-carbon connectivities by several years, Wokaun and Ernst described the means for selective detection of multiple quantum transitions via two-dimensional NMR techniques.[77] In a more recent and comprehensive study, Braunschweiler, Bodenhausen and Ernst discussed multiple quantum coherence including double and higher order coherences.[53] In a nearly simultaneous paper, Mareci and Freeman demonstrated the utilization of the double quantum INADEQUATE pulse sequence in establishing molecular connectivities via proton double quantum coherence.[90] These seminal papers provided the spawning ground for the now growing number of papers that utilize proton double quantum coherence two-dimensional NMR spectroscopy.

The earliest applications of the proton double quantum coherence experiment were in NMR studies of proteins, which continue to offer a fertile area of research.[54,55,91] Proton double quantum spectroscopy is well suited to proteins, since nominally there are few responses located along the diagonal. Thus it is possible to establish connectivities in the congested upfield aliphatic and aromatic regions of protein spectra.[92,93] Applications of proton double quantum coherence in the area of natural products structure elucidation and spectral assignment have appeared more recently, the first in mid-1985.[11,94] Here again, traits of the proton double quantum coherence experiment lend themselves well to natural products. First, absence of diagonal responses is highly beneficial with terpenoidal molecules.[11,15,95,96] Second, the proton double quantum experiment is well suited to elucidating coupling pathways where J ~ 0 Hz.[11,19,53-55] Finally, it has recently been demonstrated that the double quantum pulse sequence can be optimized for small scalar couplings (eg, J = 1.75 Hz) but will provide good response intensity for even large geminal couplings, since their coupling constants are harmonic multiples of the optimization.[19]

Thorough understanding of the workings of the proton double quantum experiment is not trivially attained. Classical spin vector diagrams are incapable of completely accounting for the creation of double quantum coherence. Recourse to density matrix formalism is necessary. Practically, however, before even a cursory consideration of the density matrix description of the double quantum coherence, it is useful to consider selective creation of double quantum coherence. For purposes of the following discussion on selective creation of double quantum coherence, we will refer to the simple AX enegy level diagram shown in Fig. 2-27, from which the zero quantum transition between energy levels 2 and 3 is intentionally omitted.

Double quantum coherence between levels 1 and 4 in the energy level diagram (ELD) shown in Fig. 2-27 can be selectively created by first applying a selective 90° pulse to the A spin 1→2 transition. This will produce a signal in the receiver with the corresponding frequency of the transition in the usual fashion. Next, after allowing the A spin 1→2 coherence to evolve for a finite period of time, we may apply a selective 180° pulse to the X spin 2→4 transition, which interchanges the states of energy levels 2 and 4. At this point, since we had already established the 1→2 coherence, the exchange of energy levels 2 and 4 leads to the establishment of coherence between energy levels 1 and 4, which is, of course, double quantum coherence. It is important to note that as a function of this operation, the signal that was previously observable in the receiver will no longer be observed despite the fact that the 1→4

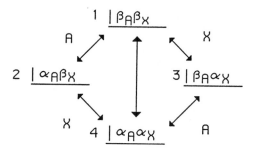

*Fig. 2-27.*    Energy level diagram (ELD) for a simple AX spin system. Single quantum transitions connect adjacent energy levels in the usual sense. The double quantum transition connects energy levels 1 and 4. The zero quantum transition, which normally would connect energy levels 2 and 3 is omitted for clarity.

coherence does indeed exist. If we allow the 1→4 coherence to evolve for some period of time, $t_1$, which is defined in the usual sense of an evolution time, and then apply a second 180° pulse to the 2→4 transition, we will reconvert the evolved double quantum or 1→4 coherence back into observable single quantum 1→2 coherence. It is important to note, however, that the observed single quantum coherence recreated in the fashion just described has a phase that is a function of the frequency of the double quantum coherence and the evolution time. By systematic variation of the duration of $t_1$ in successive experiments, it is thus possible to determine the double quantum, 1→4, frequency.

   Selective creation of double quantum coherence, although a useful instructional tool, is not generally applicable. Preferable means of exciting double or for that matter multiple quantum coherence require that the sequence of pulses be offset independent. Two alternative schemes for exciting multiple quantum coherences in an offset independent manner are available; both require a 180° pulse centered in the creation interval, $\tau$. Using the pulse train shown in Eqn. [2-7] in which the two 90° pulses have

$$90°(x) - \tau - 180°(y) - \tau - 90°(x) \qquad [2\text{-}7]$$

the same phase, only even orders of coherence will be created. The pulse train shown in Eqn. [2-8], in which the two 90° pulses are 90° out of phase, creates only odd order coherence:

$$90°(x) - \tau - 180°(x) - \tau - 90°(y) \qquad [2\text{-}8]$$

The discussion that follows utilizes the train of pulses defined in Eqn. [2-7].

Two-dimensional proton double quantum coherence or, for that matter, $^{13}C$-$^{13}C$ double quantum INADEQUATE experiments, use the pulse sequence shown in Fig. 2-28 which is discussed later in this chapter. The pulse sequence shown in Fig. 2-28, which is derived from Eqn. [2-7], creates double quantum coherence in an offset independent fashion and is hence generally useful. Mateescu and Valeriu[97] have lucidly described this experiment using density matrix calculations in a stepwise manner. The preliminary density matrix for an AX spin system is given by

$$
\begin{array}{ccccc}
 & I{+}{+} & I{-}{+} & I{+}{-} & I{-}{-} \\
I{+}{+} & P_1 & 1Q_A & 1Q_X & DQ_{AX} \\
I{-}{+} & & P_2 & ZQ_{AX} & 1Q_X \\
I{+}{-} & & & P_3 & 1Q_A \\
I{-}{-} & & & & P_4
\end{array}
$$

[2-9]

$$IAX$$

where $P_1$-$P_4$ define the populations of the four energy levels (see also Fig. 2-27) and the off-diagonal elements (coherences) connect pairs of different spin states. Single quantum transitions are denoted 1Q, zero quantum as ZQ and double quantum as DQ. Hence, the double quantum transition, $1 \rightarrow 4$, corresponds to the inversion of both the A and X spins while the zero quantum transition corresponds to the flip of both spins with conservation as $I{-}{+} \rightarrow I{+}{-}$. It should also be noted that the zero quantum transition does not generally represent a zero energy process.

Although it is entirely beyond the scope and intent of this chapter to deal with the density matrix description of double quantum coherence in any detail, some useful points derive from a minimal treatment. Upon completion of the train of pulses shown in Eqn. [2-7], the density matrix of an AX spin system is defined by Eqn. [2-10] where $c = -\cos 2\pi J\tau$ and $s = -i \sin^2 \pi J\tau$. It is apparent from the matrix above that double quantum coherence exists to the exclusion of single and zero quantum coherences. Double quantum coherence will evolve during the $t_1$ period and no signal will be detectable in the receiver. After completion of the evolution period, $t_1$, the final "read" pulse will convert the matrix from the form shown above to one in which only single quantum coherence exists. As described in the case of selective creation of double quantum coherence, the single quantum matrix elements

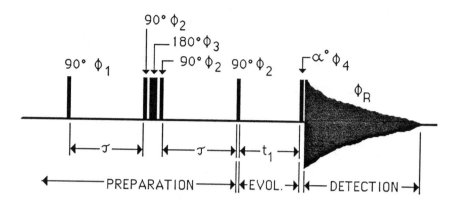

*Fig. 2-28 .*    Offset independent proton double quantum pulse sequence incorporating a composite 180° pulse midway through the creation period (2τ). The duration of the interval τ is normally set to values ranging from $1/4(J_{HH})$ to $1/8(J_{HH})$. The evolution period, $t_1$, is normally set as a function of the dwell time/4.

$$\begin{matrix} 1-c & 0 & 0 & s \\ 0 & 1 & 0 & 0 \\ 0 & 0 & 1 & 0 \\ s^* & 0 & 0 & 1+c \end{matrix} \qquad [2\text{-}10]$$

now carry information about the double quantum evolution period. By performing a series of such experiments in which the duration of the evolution period is systematically incremented, the double quantum frequency information is encoded into the modulation of the detectable single quantum coherences. Extraction of the double quantum frequency information is thus possible following completion of the second Fourier transformation.

## Location of Responses in Proton
## Double Quantum 2D-NMR Spectra

Before considering the pulse sequence and performance of the proton double quantum 2D-NMR experiment, it is appropriate to focus our attention on the nature and location of responses that may be observed. The comprehensive treatment of the double quantum NMR experiment by Braunschweiler, Bodenhausen and Ernst[53] details three basic types of response. Responses which we will generally be most interested in will be those due to direct connectivity. In an AX spin system where $J_{AX} \neq 0$, we will observe a direct connectivity response.

**Type I responses**. Braunschweiler, Bodenhausen and Ernst[53] have defined responses that correlate directly coupled spins as Type I responses. Defining the chemical shifts of spins A and X relative to the transmitter or carrier frequency as $\Omega_A$ and $\Omega_X$, respectively, responses due to the scalar coupling of these spins will be observed in the double quantum frequency domain ($F_1$ or $\omega_1$) at:

$$\omega_1 = \pm(\Omega_A + \Omega_X) \qquad\qquad [2\text{-}11]$$

It should be noted from Eqn. [2-11] that responses in the double quantum frequency domain, $F_1$ or $\omega_1$, will be symmetically disposed about the axis $\omega_1 = 0$ Hz. Responses due to direct correlation will appear in the $F_2$ or $\omega_2$ frequency domain at the chemical shifts of the coupled pair at $\Omega_A$ and $\Omega_X$. Finally, Type I responses will be symmetrically disposed about a "skew" diagonal which is defined by $F_1 = \pm 2F_2$. Schematically, the Type I responses of an AX spin system are illustrated in Fig. 2-29. In the case where the final read pulse ($\alpha$) in the

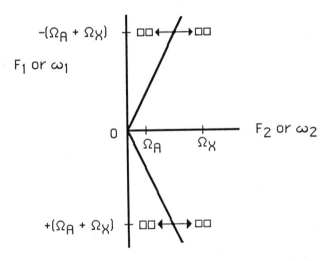

Fig. 2-29.    Schematic presentation of the proton double quantum spectrum of an AX spin system acquired using a 90° read pulse and a phase cycling scheme limited to 90° phase shifts as shown in Table 2-8 below. Responses occur in the double quantum frequency domain, $F_1$ or $\omega_1$, at the algebraic sum of the offsets of the coupled resonances from the transmitter, $\pm(\Omega_A + \Omega_X)$. In the case where the read pulse is < 90°, the intensity of the response located at $\omega_1 = -(\Omega_A + \Omega_X)$ will be > $+(\Omega_A + \Omega_X)$. When the read pulse is > 90° the converse will be true.

pulse sequence shown in Fig. 2-28 is 90°, the responses contained in the two quadrants of the double quantum spectrum will exhibit equal intensity. If the flip angle of the read pulse is < 90°, the responses in the top half of the contour plot ($F_1$ < 0 Hz) will exhibit greater intensity than their counterparts ($F_1$ > 0 Hz). When the flip angle of the read pulse is > 90° the converse will be true.

In the case of more complex spin systems, for example an AMX system, additional responses will also be observed at the $F_1$ frequency of the Type I responses. Specifically, at the double quantum frequency that correlates the A and M spins $[F_1 = \pm(\Omega_A + \Omega_M)]$; an additional response will be observed at $\Omega_X$. The response at $F_2 = \Omega_X$ and $F_1 = \pm(\Omega_A + \Omega_M)$ is due to transfer of magnetization to the "passive" X spin of the system, while the A and M spin responses at $[F_2 = \Omega_A, F_1 = \pm(\Omega_A + \Omega_M)]$ and $[F_2 = \Omega_M, F_1 = \pm(\Omega_A + \Omega_M)]$, respectively, are referred to as the "active" spins by Mareci and Freeman.[90] Schematic presentation of the proton double quantum spectrum of an AMX spin system illustrating passive transfer is shown in Fig. 2-30. We will adhere to the descriptor suggested by Mareci and Freeman in referring to responses due to transfer as passive

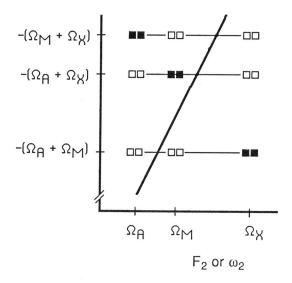

*Fig. 2-30.* Proton double quantum spectrum of an AMX spin system. Direct connectivities are denoted by open squares; responses due to passive transfer are denoted by solid squares. Only the $F_1$ < 0 Hz region of the spectrum is shown. In cases where the flip angle of the read pulse, $\alpha$, is < 90° the intensity of the $F_1 \leq 0$ Hz region will exhibit greater response intensity. When the flip angle of the read pulse is > 90° the converse will be true.

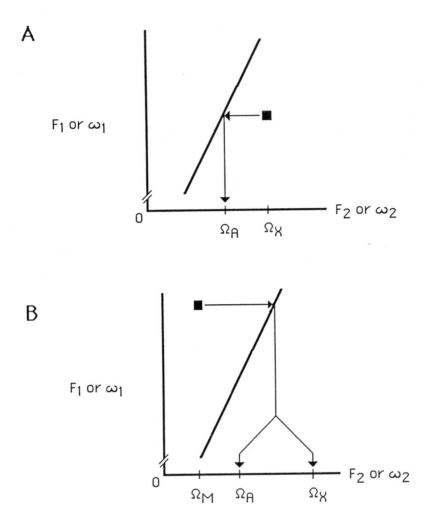

*Fig. 2-31*     Schematic presentation of responses in a proton double quantum spectrum due to (A) magnetic equivalence (Type II) and (B) remote connectivity (Type III). In both cases only the $F_1 < 0$ Hz region of the spectrum is shown. Unlike Type I responses for direct connectivity, the Type II and Type III responses do not exhibit a flip angle dependence and hence have equal intensity in both quadrants of the data matrix. Type II responses are distinguished by an isolated response whose $F_1$ frequency intersects the skew diagonal at the $F_2$ frequency of a spectral response. Type III responses are distinguished by an isolated response whose $F_1$ frequency intersects the skew diagonal at an $F_2$ frequency midway between two chemical shifts.

spins. Responses due to passive transfer of magnetization exhibit dependence on the flip angle of the read pulse. Utilization of a 90° flip angle for the final read pulse will result in an equal transfer of intensity to the passive response and the single quantum transitions having an energy level in common with the double quantum transitions. Optimal settings of the read pulse flip angle are 45° or 135°, which results in a considerable simplification of the spectrum. The intensity of the passive transfer responses is substantially attenuated but, in cases of sufficiently high S/N ratios, these responses may still be observed and utilized to identify other members of a spin system when sufficient contour levels are plotted.

**Type II responses**. Magnetically equivalent nuclides, as in the case of an $A_2X$ spin system, can give rise to Type II responses. Unlike Type I responses, the Type II response will appear as a lone peak at a double quantum frequency $(\omega_1)$ that intersects the skew diagonal at a chemical shift $\omega_2 = \Omega_A$, indicating the existence of the two magnetically equivalent A nuclei. In the presence of strong coupling or chemical equivalence, signals may also appear that reside on the skew diagonal. Finally, Type II responses are symmetric about $\omega_1 = 0$ Hz regardless of the flip angle of the read pulse employed in the collection of the data. A schematic example of a Type II response is shown in Fig. 2-31.

**Type III responses**. Responses that arise due to a remote connectivity have been labeled Type III responses by Braunschweiler, Bodenhausen and Ernst.[53] Type III responses will appear in a proton double quantum spectrum as a lone response at a double quantum frequency, $\omega_1$, which intersects the skew diagonal at an $F_2$ or $\omega_2$ frequency that does not correspond to any of the chemical shifts of the resonances in the spectrum. Responses categorized as Type III arise when the remote nuclei that are not coupled to one another have a common spin coupling partner. Information garnered from Type III responses is equivalent to that provided by relayed coherence transfer experiments.[98] A schematic example of a Type III response is also shown in Fig. 2-31.

## The Proton Double Quantum Pulse Sequence and Phase Cycling

Although double quantum transitions are populated using the basic COSY pulse sequence shown in Fig. 2-1, this is not the most efficient means of creating double quantum coherence. While numerous schemes for creating double quantum coherence exist,[99] the most commonly utilized pulse sequence is shown in Fig. 2-28. The pulse sequence shown in Fig. 2-28 differs from that originally reported by Mareci and Freeman[90] only in the replacement of the 180° pulse by a composite $180°$ pulse[100-102] to compensate for inaccuracies in the measurement of the $180°$ pulse.

From the early work of Wokaun and Ernst,[77] selection of double quantum coherence to the exclusion of all but six quantum coherences can be attained by appropriate phase cycling as shown in Table 2-5. The specific phase cycling prescription used in conjunction with the proton double quantum pulse sequence shown in Fig. 2-28 is given in Table 2-8. The phase cycling routine specified effectively suppresses single quantum coherence, although residual single quantum signals may occasionally be observed along the axis $F_1 = F_2$ when very deep contour levels are plotted. Additionally, the phase cycle shown in Table 2-8 does not, in and of itself, provide quadrature detection in the double quantum ($F_1$ or $\omega_1$) frequency domain. Quadrature detection in the second frequency domain of a proton double quantum spectrum requires either the use of non-$90°$ phase shifting capabilities, which have become available only on the newest generation of commercial spectrometers, or through the usage of a $45°$ or $135°$ read pulse as suggested in the work of Mareci and Freeman.[103]

Table 2-8.      Phase cycling for proton double quantum coherence two-dimensional NMR spectroscopy using the pulse sequence shown in Fig. 2-28. The phase cycling shown rejects all orders of multiple quantum coherence below sixth (see Table 2-5) except for the desired double quantum coherence.[76,90] The phase cycling shown does not provide quadrature detection in the second ($F_1$ or $\omega_1$) frequency domain. The phase cycle assumes that the transmitter will be set upfield or downfield of the region of the spectrum containing responses so that all responses will have the same algebraic sign relative to the transmitter.

| $\phi_1$ | 00002222 | 11113333 | 22220000 | 33331111 |
|---|---|---|---|---|
| $\phi_2$ | 00000000 | 11111111 | 22222222 | 33333333 |
| $\phi_3$ | 11111111 | 22222222 | 33333333 | 00000000 |
| $\phi_4$ | 01230123 | 12301230 | 23012301 | 30123012 |
| $\phi_R$ | 03212103 | 10323210 | 21030321 | 32101032 |

To illustrate the benefits that accrue from using quadrature detection in both frequency domains, we will first consider the experiment when performed without quadrature detection in $F_1$ and the problems that this constraint imposes on data storage and processing. Using the proton double quantum pulse sequence and the phase cycling scheme shown in Fig. 2-28 and Table 2-8, respectively, requires, if data in $F_1$ is not to be taken in quadrature, that the transmitter be located downfield of the lowest field resonance. Locating the transmitter downfield of the spectral responses ensures that all responses will have the same algebraic sign, thus obviating the requirement of distinguishing positive and negative frequencies, which is necessary with quadrature detection. Next, we will consider quadrature detection in $F_1$ and the savings in disc storage and the beneficial effects on digital resolution this provides. Performance of the proton double quantum experiment and parameter selection are discussed in the following section.

## Parameter Selection and Experiment Execution

Practically, the proton double quantum experiment requires selection of only a few parameters. We will consider each of the necessary parameters in turn.

**Evolution time incrementation.** Let us consider first the requirements associated with single phase detection in $F_1$. Assuming that data is taken in quadrature in $F_2$ even though there will be no responses downfield of the transmitter, the sweep width can be established in the usual fashion. The spectral width in the double quantum frequency domain will, of necessity, be twice that in the first frequency domain. Further, since double quantum responses will be symmetically disposed about the axis, $F_1 = 0$ Hz in the absence of quadrature detection in $F_1$, it is necessary to actually digitize a frequency range four times that of the spectral window containing responses in $F_2$. Thus, the evolution time in the proton double quantum experiment will be a function of half the dwell time in this case just described.

Quadrature detection in the $F_1$ frequency domain improves the situation somewhat. Rather than locating the transmitter downfield of the lowest field resonance, the transmitter may be located at any convenient position where no proton resonates. Effectively, this simple change halves the frequency range in $F_1$ that must be considered.

As a consequence of the need to digitize a greater frequency range in the second frequency domain in a proton double quantum experiment, the number of increments of the evolution time also requires additional consideration. Generally, it will be advisable to utilize 512 increments of the evolution time. Because of the wide spectral width in $F_1$ it may also be beneficial to zero fill the data matrix at least once prior to the second Fourier transformation. Even under these conditions, digitization in the second frequency domain of the proton double quantum spectrum may be somewhat coarse.

**The $\tau$ interval used in the creation of double quantum coherence**. Generally, the duration of the $\tau$ interval is governed by the spin coupling constant and is normally set as a function of $1/4(J_{HH})$ or $1/8(J_{HH})$. At this point, it remains to determine the value of J to be employed. For the acquisition of survey proton double quantum spectra, 7 Hz is a good compromise. Usefully, a 7 Hz optimization will allow the observation of smaller vicinal couplings simultaneously with responses due to larger geminal couplings. Selection of a smaller coupling constant upon which to optimize will, of course, allow potential access to long range coupling information as well. In addition to providing access to long range couplings, optimization at 1.75 Hz in the case of strychnine (**2-6**), for example, also provided good response intensity for a wide range of couplings, which included the large geminal couplings in the range of 15 Hz.[19] On first thought, the preceding statement may seem incorrect. However, because of the duration of the $\tau$ interval when optimized for 1.75 Hz (142.85 msec), larger couplings will oscillate in and out of phase several times as a function of their higher frequencies. Consequently, couplings that are integer harmonic multiples of 1.75 Hz will be in phase at the appropriate points during the creation of double quantum coherence in the portion of the pulse seqeunce shown in Fig. 2-28 that corresponds to Eqn. [2-7]. Hence, components of magnetization due to couplings of ~ 3.5, 7 and 14 Hz, among others, will be converted into double quantum coherence along with those components associated with the smaller couplings. Optimization using a single $\tau$ value corresponding to a small coupling such as 1.75 Hz is by no means as elegant as the uniform excitation approach recently reported by Ernst and coworkers,[104] one example of which employed eleven uniformly spaced $\tau$ values ranging from 40 to 240 msec. In a practical sense, however, the benefits that accrue by optimizing at a small coupling constant are significant and much simpler to implement than resorting to a uniform excitation scheme. A practical comparison of 1.75 and 7 Hz optimization with the pulse sequence shown in Fig.2-28 is presented in Fig. 2-32, which shows the double quantum or $F_1$ frequency domain projections of the proton double quantum spectrum of strychnine (**2-6**) acquired using these optimizations.

**Read pulse flip angle**. The flip angle of the read pulse is an important consideration. The utilization of a 90° read pulse is by no means advantageous. A 90° read pulse will provide equivalent response intensities in both quadrants of the $F_1$ frequency domain while simultaneously allowing transfer of magnetiztion to passive spins which is undesirable. Furthermore, the use of a 90° read pulse precludes quadrature detection in $F_1$ in the absence of non-90° phase shifts.[103] Hence, it is preferable to use either a 45 or a 135° read pulse.[19,90]

**Digitization in the $F_2$ frequency domain**. Selection of a level of digitization for the normal chemical shift or $F_2$ frequency domain will largely depend upon the amount of disc storage space available on

the spectrometer system being used. In the first example, in which data
will intentionally not be taken in quadrature in $F_1$, this represents a much

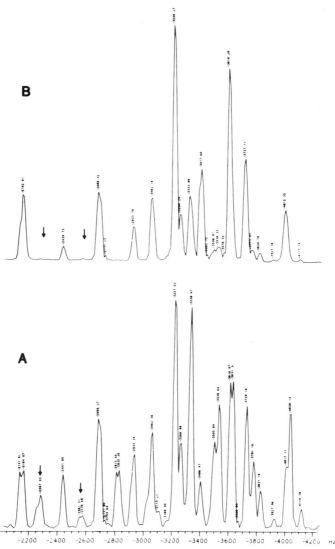

*Fig. 2-32 .*     Projections through F1 of the proton double quantum spectra strychnine
(2-6) recorded with the duration of $\tau = 1/4(J_{HH})$ and optimized at 1.75
(A) and 14 Hz (B) (142.85 and 17.8 msec, respectively).    The arrows
located over both spectra denote the locations of H14-H22 and H20b-
H22 long range couplings.   Note that while these long range couplings
have good intensity in the 1.75 Hz optimized spectrum (A), they are
essentially lost in the 14 Hz optimized spectrum (B) despite excellent
the excellent signal-to-noise ratio.[19]

larger problem. If data is being taken in quadrature with the transmitter located downfield of the lowest field response, it is necessary to pay somewhat greater attention to the level of digitization employed. In such circumstances, it is almost mandatory to employ 2K points, which after complex Fourier transformation will afford 512 points in the region of the double quantum spectrum containing responses. This is the minimum number of points adequate for this experiment. In conjunction with the numbers of increments of the evolution time and zero filling prior to the second Fourier transformation, these operations provide levels of digitization that are comparable to those of a 512 x 512 point COSY spectrum when the greater dispersion of responses in $F_1$ is taken into account. In contrast, when data is taken in quadrature in $F_1$, we have more choice in how we utilize digitization. Taking 1K complex points probably affords the minimum acceptable level of digitization in $F_2$.

## Proton Double Quantum Data Processing

Following the acquisition of the proton double quantum spectral data set, we must next consider processing requirements. Here the impact of our decision on how data is accumulated in $F_1$ becomes most apparent. The discussion of the data processing that follows assumes that the phase cycle specified in Table 2-8 was employed and that the data was taken in quadrature in $F_2$ with the transmitter located downfield of the lowest field resonance. This treatment illustrates quite graphically the benefits that accrue from having the ability to take quadrature data in both frequency domains. The processing regimen necessary when quadrature data acquisition in $F_1$ is employed is described below.

As with the COSY data processing, which was discussed above, it is preferable to use sinusoidal multiplication or its equivalent prior to both Fourier transformations to curtail tailing of the responses. Sinusoidal multiplication followed by the first Fourier transformation of the data matrix will afford a series of modulated spectra in which all of the data is upfield of the carrier. Normally, it would be advantageous in terms of disc storage space to transpose only the half of the data matrix containing responses if this approach were being used. However, to illustrate the content of data in the matrix and to provide a graphic example of the effect of the read pulse flip angle on the intensity of responses in the matrix, the entire matrix will be considered in this first example.

The first Fourier transformation followed by transposition provides matrices that are identical to those described by Eqns. [2-1] and [2-2]. Sinusoidal multiplication of the interferograms followed by zero filling and then the second Fourier transformation obeys Eqn. [2-3]. At this point, the $S(F_2,F_1)$ data matrix generated by the second Fourier transformation will have the $F_1$ = 0 Hz axis running vertically through the matrix. To facilitate plotting of the data, we find it convenient to subject the matrix to a second transposition, which is defined in Eqn. [2-12].

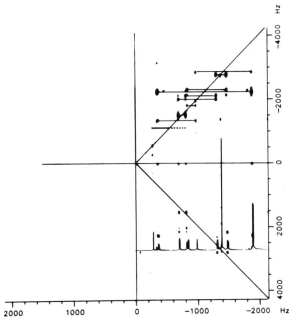

Fig. 2-33. Proton double quantum spectrum of the sesquiterpene plumericin (2-1) in deuterochloroform. The spectrum was recorded using the pulse sequence shown in Fig. 2-28 and the phase cycling contain in Table 2-8. The duration of was set as a function of 7 Hz (35.7 msec) and the read pulse flip angle was set to 45°. The entire data matrix was processed to afford the spectrum shown. The normal spectrum is plotted across the right half of the matrix to show the correspondence between responses in the spectrum and the data matrix. To conserve space, the matrix is plotted square, giving the appearance of skew diagonals with a slope of ±1 rather than ±2. Note the increased intensity in the region $F_1$ < 0 Hz relative to the $F_1$ > 0 Hz region, which is a function of the 45° flip angle of the read pulse. Using a read pulse with a flip angle > 90° would afford a spectrum with greater response intensity in the F1 > 0 Hz quadrant.

$$S(F_2, F_1) \quad \rightarrow \quad S(F_1, F_2) \qquad \qquad [2\text{-}12]$$

The proton double quantum $F_1 F_2$ matrix of plumericin (2-1) recorded and processed in the manner just described is shown in Fig. 2-33. From the contour plot shown, it should be noted first that the quadrant $F_2 < 0$, $F_1 < 0$ Hz has responses with considerably greater intensity than the $F_1 > 0$ Hz quadrant, illustrating the flip angle dependence of the responses in the matrix. Second, it will be noted that there is no information contained downfield of the transmitter ($F_2 > 0$). Quite obviously, this approach squanders three-fourths of the digital resolution capacity of the experiment, which is hardly desirable.

Data processing when quadrature detection in $F_1$ is employed requires considerably less mental gymnastics. The data set may be subjected to sinusoidal multiplication and the first Fourier transformation as above. Since the transmitter may be centrally located in the spectral window when quadrature $F_1$ detection is employed, the entire data matrix may be transposed and subjected to sinusoidal multiplication and the second Fourier transformation, affording an $S(F_2, F_1)$ matrix as above. Finally, for convenience in plotting it is desirable to perform the final transposition shown in Eqn. [2-12]. At this point, the data contained in the matrix would be equivalent to the upper right quadrant of data shown in Fig. 2-33, assuming that a 45° "read" pulse was employed. No digitization has been wasted in the other three non-informative quadrants as in the "simpler" approach using single phase detection in $F_1$!

## Presentation of Proton Double Quantum Spectra

Proton double quantum coherence, as described above, yields Type I responses symmetrically disposed about a "skew diagonal" with a slope $\pm 2$.[53] Generally, plots of double quantum spectra will be produced which preserve the relationship of $F_1 = 2F_2$ (eg, see Fig. 2-30). There have, however, been several papers in the literature that have suggested means of reorienting the proton double quantum spectrum to give the more familiar appearance of a COSY spectrum. The first paper to suggest manipulation of double quantum coherence spectra into a COSY-type format was a communication by Blümich in 1984.[105] More recently, Zuiderweg[106] has made a similar suggestion. While there may be advantages inherent to these altered presentations, we find it convenient to employ the more standard presentation, which preserves the relationship of $F_1 = 2F_2$ and will be used throughout the discussion that follows.

## Quadrature Data Acquisition in the Second Frequency Domain

Having seen that single phase detection in the second frequency domain quadruples the data storage requirements with a commensurate increase in processing time, we now need to consider data acquisition with quadrature detection in both frequency domains. Referring to Fig. 2-33, rather than positioning the transmitter downfield of the H6 proton we can instead relocate it to the center of the spectrum. In the second frequency domain quadrature detection allows us the luxury of discriminating positive from negative frequencies and hence, although the effective total frequency range remains the same, we have the benefit of twice the digital resolution that we had in Fig. 2-33. Indeed, all of the data points taken are utilized, and we will be restricted to the frequency ranges of the upper or lower right quadrant depending upon the choice of the flip angle of the read pulse (see Fig. 2-28).

Having considered the benefits, the next appropriate question is, how do we manage to collect data in quadrature in both frequency domains? Here there are several choices. Given a new instrument with 45° phase shift capabilities, it is possible to modify the phase cycling shown in Table 2-8 so that after each acquisition a second is coadded with a 45° increment of the receiver phase. By using modified phase shifting, we will obtain quadrature detection in the second frequency domain. A second method which has been proposed by Bax, entails the use of a 45° composite z-pulse.[107] Here, the first experiment is acquired normally and added to memory. The second experiment is then performed with a 45° composite z-pulse following the creation of double quantum coherence, where the composite pulse takes the form shown in Eqn. [2-13], accompanied by a 90° clockwise rotation of the receiver reference phase.

$$90°(-x) - 45°(-y) - 90°(x) \qquad [2\text{-}13]$$

The net result of this pair of acquisitions is identical to that achieved using a 45° phase shift on the second acquisition described immediately above. Finally, Mareci and Freeman[103] have shown that the utilization of a 135° read pulse provides the means of discriminating between the coherence transfer echo and antiecho. Carrying this approach a step further, Lallemand[108] and coworkers have devised a "double selection" experiment in which the excitation sequence shown in Eqn. [2-7] is replaced by one in which a 135° pulse is substituted for the final 90° pulse, giving the sequence shown in Eqn. [2-14], followed

$$90°(x) - \tau - 180°(y) - \tau - 135°(x) \qquad [2\text{-}14]$$

by a 135° read pulse. Here again we achieve quadrature detection in both frequency domains.

Of the three techniques of recording double quantum spectra in quadrature in both dimensions, the first requires 45° phase shift capabilities and thus will not be amenable to all instruments. The method suggested by Bax[107] has not been demonstrated experimentally, leaving only the methods of Mareci and Freeman[103] and Lallemand and coworkers,[108] which do work, as an alternative choice. While the choice between Mareci and Freeman's single selection experiment and the double selection experiment of Lallemand and coworkers is perhaps a subtle one, we feel that the latter affords potentially superior performance in that there will be a substantialy diminished chance of obtaining quadrature images when the data is plotted.

## Interpretation of Proton Double Quantum Spectra

### Application of the proton double quantum experiment to plumericin (2-1).

Having discussed the processing and some aspects of presenting proton double quantum spectra, it is appropriate to continue into a discussion of the interpretation of these very useful spectra. For purposes of this discussion, we shall employ the proton double quantum spectrum of plumericin (2-1) shown in Fig. 2-34.[11] Recall that the COSY spectrum of 2-1 (Fig. 2-4) assembled several structural fragments, shown in Fig. 2-7, but was incapable of linking these fragments together. A normal COSY spectrum generally will not have the levels of digitization necessary to provide small coupling information such as the couplings $^4J_{H4aH6}$ and $^4J_{H4aH3a}$, which are necessary to link fragments **A** and **B** to **C** to afford **D**. As an alternative, it is possible to perform a second COSY experiment optimized for long range couplings. There is, however, a potentially preferable option available to us in the 7 Hz (35.7 msec) optimized proton double quantum experiment illustrated in Fig. 2-34.

One attribute that makes the proton double quantum experiment particularly attractive is its ability to uncover small couplings where J is near 0.[19,53-55] Hence, we would expect that there might be a somewhat better chance that the proton double quantum experiment would be able to provide us with the connectivity information necessary to link the structural fragments of plumericin that may be assembled from COSY data into the more comprehensive structural fragment illustrated by **D** in Fig. 2-7.

It is instructive to invest the time necessary to reconstruct structural fragments **A-C** shown in Fig. 2-7 using the proton double quantum spectrum shown in Fig. 2-34. Type I proton double quantum responses appear in the $F_1$ frequency domain at the algebraic sum of the offsets of the coupled resonances relative to the transmitter frequency. We may begin, once again, with the H6 resonance, which resonates furthest downfield at 7.423 ppm. If, as in the COSY spectrum, H6 were to exhibit no couplings to other protons, there would be no responses observed at the $F_2$ frequency of H6 in the $F_1$ domain.

However, note that a Type I response is observed at $F_1$ = -1133 Hz, which correlates H6 with the H4a proton resonating at 5.525 ppm. Hence, one of the linkages critical to the assembly of structural fragment

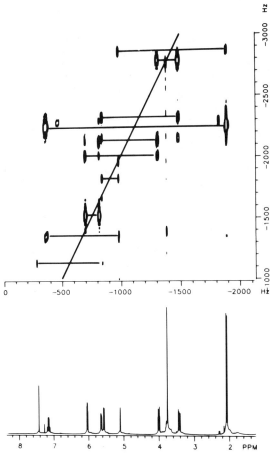

*Fig. 2-34.*    Proton double quantum spectrum of plumericin (**2-1**) (35 mg) recorded in deuterochloroform at 300.068 MHz.[11] The spectrum was recorded as 512 x 2K data points using the phase cycling shown in Table 2-8 with the transmitter located downfield of the H6 resonance, which was observed at 7.423 ppm. The experiment was optimized with $\tau$ = $1/4(^3J_{HH})$ = 7 Hz (35.71 msec). Total acquisition time was 14 hours (overkill). The data was processed using sinusoidal multiplication prior to both Fourier transformations with zero filling prior to the second to afford the final 512 x 512 point data matrix which is presented as a 10 level contour plot. The normal high resolution proton spectrum (32K) of plumericin (**2-1**) is plotted below the contour plot. Chemical shifts given in ppm are downfield of TMS. The axis directly beneath the contour plot gives chemical shifts upfield of the transmitter in hertz.

**D** shown in Fig. 2-7 is readily provided by the proton double quantum spectrum.

Using H4a as the next point of departure in the assignment of the proton double quantum spectrum of **2-1**, we have the protons contained in structural fragment **B** to work with. The five spin linear system comprised of H4a, H9b, H7a, H8 and H9 has protons that resonate at 5.525, 3.403, 3.948 5.618 and 6.013 ppm, respectively, and constitutes fragment **B**. First, consider the H4a-H9b connectivity. Using the $F_2$ frequencies of H4a and H9b, the calculated $F_1$ frequency for the Type I response correlating these resonances is -2332 Hz. The response in $F_1$ observed at -2331 Hz confirms this coupling. Continuing, H9b is coupled to H7a, their $F_2$ frequencies giving a calculated $F_1$ transition frequency of -2800 Hz. The response correlating these resonances was observed at $F_1$ = -2798 Hz. Correlating H7a to the H8 vinyl proton resonance requires a Type I response at $F_1$ = -2136 Hz; the observed response is found at $F_1$ = -2135 Hz. Finally, the vinyl protons, H8 and H9, have a calculated double quantum Type I response frequency of $F_1$ = -1526 Hz while the observed double quantum frequency was -1525 Hz. Thus, through the series of operations just described, structural fragment B is readily assembled. In addition to the connectivities necessary to link the vicinal neighbors, there are longer range couplings that have discernible double quantum transitions. The complete listing of double quantum transitions observed for plumericin (**2-1**) is collected in Table 2-9.

Table 2-9.  Proton chemical shifts and double quantum coherence response frequencies of plumericin (**2-1**) determined from the spectrum shown in Fig. 2-34.

| | | Double quantum responses | | |
|---|---|---|---|---|
| Proton | $\delta_H$ | Connectivity | Calc'd (Hz) | Obs'd (Hz) |
| 3a | 5.093 | 3a/4a | -1793.8 | -1798 |
| | | 3a/10 | -1316.7 | -1318 |
| | | 3a/11 | -2839.8 | -2841 |
| 4a | 5.525 | 4a/6 | -1130.9 | -1133 |
| | | 4a/9b | -2331.8 | -2331 |
| 6 | 7.423 | | | |
| 7a | 3.948 | 7a/8 | -2135.5 | -2135 |
| | | 7a/9 | -2017.1 | -2018 |
| | | 7a/9b | -2799.6 | -2798 |
| 8 | 5.618 | 8/9 | -1525.5 | -1525 |
| | | 8/9b | -2297.7 | -2303 |
| 9 | 6.013 | | | |
| 9b | 3.403 | | | |
| 10 | 7.116 | 10/11 | -2260.5 | -2259 |
| 11 | 2.034 | | | |
| 13 | 3.753 | Type II | -2752.2 | -2755 |

To complete the assembly of structural fragment **D** it is necessary to construct fragment **A** and then to link **A** to **D**. The ethylidine fragment comprising **A** consists of the doubled quartet resonating at 7.116 ppm and the methyl resonance at 2.034 ppm, both of which are coupled in turn to the H3a proton resonating at 5.093 ppm. The Type I response linking H10 and H3a has a calculated $F_1$ frequency of -1317 Hz and was observed at $F_1$ = -1318 Hz; the response linking H10 to the H11 methyl doublet has a calculated $F_1$ frequency of -2261 Hz and was observed at $F_1$ = -2259 Hz, completing the assembly of fragment **A**. Linking **A** to **D** is accomplished via the H3a-H4a connectivity, which has a calculated $F_1$ frequency of -1794 Hz. The H3a-H4a Type I response was observed at $F_1$ = -1798 Hz thus completing the assembly of structural fragment **D** (Fig. 2-7).

### Application of proton double quantum coherence to strychnine.
Strychnine provides an interesting testing ground for the proton double quantum experiment in that it contains resonances for seventeen nonequivalent proton resonances in the region from 4.3 to 1.2 ppm.[19] Furthermore, the NMR spectrum of strychnine has a broad range of vicinal and geminal coupling constants, the former as small as 2.0 Hz, while the latter are as large as 17.4 Hz.

The 7 Hz (35.7 msec) optimized proton double quantum spectrum of the H22 vinyl proton and the aliphatic region of strychnine (**2-6**) is shown in Fig. 2-35. Responses due to the larger vicinal couplings and some of the geminal couplings were easily discerned. At this optimization, however, some of the vicinal coupling connectivities could not be observed. Specifically, the H12-H13 (3.1 Hz) and H11b-H12 (3.3 Hz) coupling responses were relatively weak. Responses for H13-H14 (3.1 Hz), H14-H15a (4.0 Hz), H14-H15b (1.0 Hz), H15a-H16 (4.9 Hz) and H15-H16 (2.0 Hz) were not visible in Fig. 2-35.

It is useful to compare the data contained in Fig. 2-35 with the connectivities assembled for strychnine (**2-6**) from the COSY spectrum discussed above. First, the H22-H23a and H22-H23b responses are readily observed at $F_1$ = -2142 and 2164 Hz, respectively. The H20a-H20b geminal coupling was observed at $F_1$ = -3227 Hz. Finally, the H17a-H17b and H18a-H18b geminal couplings and the vicinal couplings within this ethylene bridge are shown in the expansion of the spectrum plotted in Fig. 2-36. These connectivities provide structural information identical to that portrayed by the structural fragment shown by **2-7**. It should be reemphasized that these three spin systems are not linked to one another by the 7 Hz optimized proton double quantum spectrum, nor were they linked by the conventional COSY spectrum.

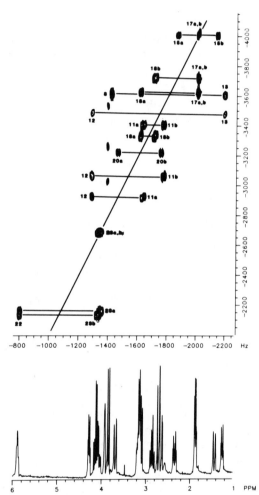

*Fig. 2-35.* Proton double quantum spectrum of strychnine (**2-6**) (10 mg) in deuterochloroform recorded at 300.068 MHz.[19] The experiment was optimized with $\tau = 1/4(^3J_{HH}) = 7$ Hz (35.71 msec). The data was accumulated as 512 x 2K points and was processed using a sinusoidal multiplication prior to both Fourier transformations and zero filling prior to the second to give the final 512 X 512 point matrix, which is shown in the contour plot. Only the aliphatic and vinyl regions of the spectrum are shown. Chemical shifts plotted below the reference spectrum are given in ppm downfield of TMS. Shifts along the axis directly below the contour plot specified in hertz give resonance locations relative to the transmitter which was located downfield of the aromatic region. Acquisition time was 6 hours.

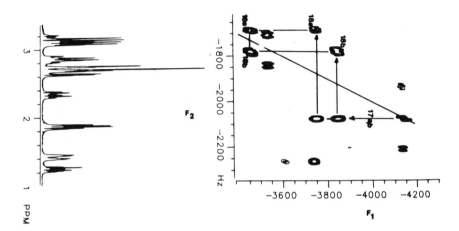

*Fig. 2-36* . Expansion of the 7 hertz optimized proton double quantum spectrum of strychnine (**2-6**)[19] (Fig. 2-35) to show the connectivities of the H17-H18 protons of the ethylene bridge.

Returning to Fig. 2-35, we may next consider the structural fragment shown by **2-8**. Type I responses in the 7 Hz optimized spectrum shown in Fig. 2-35 establish the following connectivities: H11a-H11b, H11a-H12, H11b-H12, H12-H13, H13-H8 and H15a-H15b. There unfortunately is a break in the assembly of the structural fragment in that the H13-H14 and H14-H15a/b connectivities were not observed in Fig. 2-35, nor either of the H15a/b-H16 connectivities. Hence, the assembly of structure **2-8** cannot be completed from the 7 Hz optimized proton double quantum spectrum shown in Fig. 2-35.

Clearly, the absence of the connectivities necessary to complete **2-8** points to a variability of the proton double quantum experiment to reveal small couplings despite the success with plumericin (**2-1**)[11] and in other cases.[53-55] To solve this problem, we have recourse to two alternative strategies. First, we may employ a series of τ values to achieve uniform excitation in the manner described by Rance et al.[104] Implementation of the uniform excitation scheme of Rance et al.,[104] would require, at a minimum, rewriting the pulse sequence and possible modifications to the operating software of the spectrometer, changes that may make this solution to the problem unattractive for many users. Realizing that optimization for a small coupling constant will bring larger couplings in and out of phase several times during the evolution period, τ , Craig and Martin[19] have demonstrated that the judicious choice of a small coupling optimization can provide "pseudo-uniform optimization."

While optimizing for a small τ value as a method of obtaining a broader optimal response is by no means as elegant as the method of Rance et al.[104] nor as efficient, since there still may be responses missed due to phase considerations, this second alternative has the advantages of being trivially implemented and requiring no modifications to either the pulse program or the operating software.

The proton double quantum spectrum of strychnine (**2-6**) optimized for 1.75 Hz (142.85 msec) is shown in Fig. 2-37. It will immediately be noted that there are considerably more responses observed with the 1.75 Hz optimization than were obtained using the 7 Hz optimization shown in Fig. 2-35. Perhaps the most immediately notable region of difference is associated with the responses correlating H22. While the 7 Hz optimized spectrum provided only the H22/H23a and H22/H23b correlations, three additional connectivities can be established from the 1.75 Hz optimized spectrum. First, correlations are observed linking H22 to both the H20a and H20b resonances, these appearing at $F_1$ = -2287 and -2568 Hz, respectively. Perhaps more important still was the H22/H14 connectivity, which links two structural fragments that could not be linked from the COSY experimental data discussed above to afford **2-12**. The H22/H14 connectivity was observed at $F_1$ = -2442 Hz in Fig. 2-37.

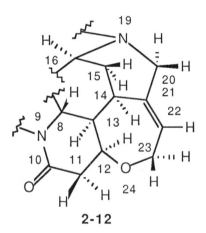

**2-12**

Regarding the connectivities missing when the assembly of structural fragment **2-8** was attempted from the 7 Hz optimized proton double quantum spectrum shown in Fig. 2-35, a substantially different picture is observed when the 1.75 Hz optimized spectrum in Fig. 2-37 is considered. Type I responses linking H13-H14, H14-H15a, H14-H15b, H15a-H16 and H15b-H16, which were missing from the 7 Hz spectrum, are all readily observed in the 1.75 Hz optimized spectrum. When considered in conjunction with the long range couplings of the H22 resonance discussed above, it is possible to assemble the large structural fragment represented by **2-12**.

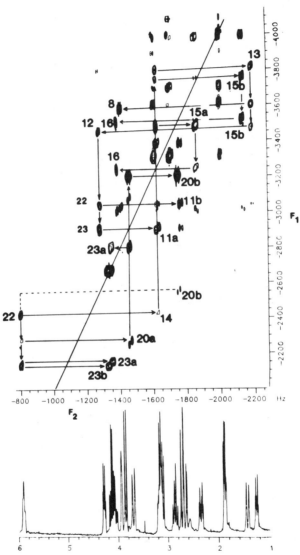

*Fig. 2-37.*          Proton double quantum spectrum of strychnine $(2-6)$[19] recorded in deuterochloroform at 300.068 MHz. Conditions were identical to those describewd in Fig. 2-35 except that the experiment was optimized with $\tau = 1/4(^3J_{HH}) = 1.75$ Hz (142.85 msec). It will be noted that there are a substantially greater number of responses contained in this spectrum than in that shown in Fig. 2-35. Compare also the projections through the double quantum frequency domain of this spectrum and that in Fig. 2-35, which are compared in Fig. 2-32.

## Strategy in Recording Proton
## Double Quantum Spectra

Overall, perhaps the most useful strategy to consider would be the acquisition of two proton double quantum spectra, one nominally optimized for a 7 Hz vicinal coupling, the other optimized for a smaller, long range coupling, for example the 1.75 Hz used successfully with strychnine.[19] The former used singly would not necessarily provide all of the vicinal couplings of a natural product, while the latter would likely provide a mixture of vicinal and longer range couplings that could not easily be sorted from the single experiment.    However, two optimizations, when used in concert, provide a powerfully synergistic tool for structure elucidation.    As an alternative, it would be possible to perform a single experiment using the uniform excitation pulse sequence of Rance and coworkers[104] described in the following section, which should be capable of providing all coupling pathways.    The problem of interpreting the potentially very complex  resulting spectrum would remain.

## Alternative Proton Double
## Quantum Pulse Sequences

In addition to the proton double quantum pulse sequence shown in Fig. 2-28, several other pulse sequences are available for the creation of proton double quantum coherence. An assortment  of older sequences is described by Bax[99] but generally these have not been used to any appreciable extent. Probably the most versatile alternative pulse sequence is described in the work of Rance et al. [104] Specifically, the uniform excitation concept of Rance and coworkers[104] utilizes a modified proton double quantum pulse sequence that is symmetric about the evolution period, $t_1$, as shown in Fig. 2-38.  The experiment may be variously implemented using either a purging pulse (lower trace) or a z -filter (upper trace) to improve the suppression of antiphase magnetization.    The $45°_y$ purging pulse is applied uniformly in all experiments of the sequence.  The final pulse of the z-filter is cycled in 90° steps with the receiver phase.  The duration of the interval $\tau_z$ may be varied within the z-filter in an otherwise identical sequence of experiments to suppress zero quantum coherence present during $\tau_z$. Complete details of the phase cycling of the symmetric excitation pulse sequence shown in Fig. 2-38 are contained in the work of Rance and coworkers.[104]

In addition to the sequence shown in Fig. 2-38, Rance and coworkers[104] have described a sequence that eliminates the 180° ($\pi$) pulses.  Normally such a modification would be expected to reintroduce chemical shift dependence.  However, the effects of frequency offset have been shown to be removed from the modified sequence when an additional phase cycle is used.  Two accumulations are made with the phase of the final 90° pulse (equivalent to upper trace in Fig. 2-38 minus the 180° refocusing pulses) incremented by 90°. The results of the two experiments are then subtracted.

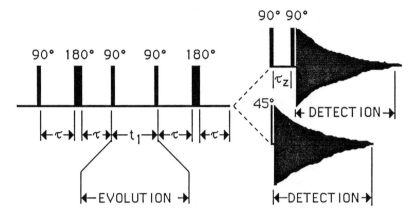

*Fig. 2-38 .*     Uniform excitation proton double quantum coherence pulse sequence described by Rance and coworkers.[104] The $\tau$ interval may be uniformly varied to allow the excitation of a broad range of coupling constants. In their original report, Rance and coworkers[104] varied the duration of the $\tau$ interval from 40-240 msec in 11 uniform steps, which could be expected to excite double quantum transitions ranging from in excess of 10 Hz to about 1 Hz. Two alternatives are also offered prior to data acquisition. The lower trace shows the use of a 45° purge pulse, which is repeated with each experiment performed. The upper trace shows a z-filter that may be used to improve the suppression of antiphase components of magnetization. In addition, the duration of the $\tau_z$ interval may also be varied in a series of otherwise identical experiments to suppress zero quantum coherence present during $\tau_z$.

## A Brief Survey of Proton Double Quantum Applications

The most numerous applications of proton double quantum are thus far in the area of protein structure determination. The earliest application of which we are aware was that of Wagner and Zuiderweg to BPTI.[55] Contributions from various other research groups have also addressed problems in the assignment of the proton NMR specta of proteins and/or peptides by double quantum coherence.[54,55,91-93,109-115] Applications of the proton double quantum experiment to other classes of compounds are somewhat less numerous. Several successful applications in the elucidation and/or spectral assignment of complex antibiotic structures have appeared; these include polymyxin B, which is complicated by the presence of six diaminobutyric acid residues[110,111];

ristocetin[94] and most recently teicoplanin $A_2$.[116]  A number of applications to terpenoidal compounds have also been reported including plumericin,[11] jeunicin,[15] planaxool[95] and cycloseychellene.[96] Thus far, only the model study of strychnine (**2-6**) by Craig and Martin[19] has applied the proton double quantum experiment to an alkaloid. Very recently, a paper describing the identification of constituent sugar residues in oligosaccharide using double quantum coherence has also appeared.[44]  Doubtless additional applications in the classes of compounds mentioned as well as applications in other classes of compounds will appear.

## Proton Zero Quantum Coherence

In addition to demonstrating the means of exciting selected multiple quantum coherences, early work by Wokaun and Ernst[77] also suggested the means of exciting zero quantum coherences (ZQC). As shown in Table 2-5, using simple 90° phase shifts, it is possible to excite either two and six quantum coherence or zero, four and eight quantum coherence depending on whether successive experiments are co-added or subtracted.  Early development of the capabilities of zero quantum coherence was slowed by the observation that it is difficult to achieve uniform excitation of ZQC throughout a complex spin system.[53] Indeed, ZQC was essentially relegated to the class of a quantum mechanical nuisance with little mention other than the early study of ZQC by Pouzard, Sukumar and Hall that appeared in 1981.[117]

The major drawbacks to the creation of ZQC are dependence on both spin-spin coupling constant size and chemical shift.[3,117] Despite the inherent difficulties associated with its creation, ZQC has several useful attributes. First, zero quantum coherence is insensitive to magnetic field inhomogeneities, which permits the acquisition of high resolution spectra in inhomogeneous magnetic fields.[77]  In normal work with a highly homogeneous magnetic field, exceedingly well resolved spectra may be obtained. Second, using Turner's analogy,[118] in which single quantum coherence may be thought of as a rotating dipole and double quantum coherence as a rotating quadrupole, zero quantum coherence may be perceived as a rotating monopole, insensitive to the phase of pulses applied to a spin system. The latter point is important in that it permits the use of a minimal four step phase cycle in conjunction with a homogeneity spoiling pulse to suppress higher order coherences.[76,119] In light of the attributes just described, it is somewhat surprising that zero quantum coherence NMR has developed as slowly as it has.  Nevertheless, ZQC offers potential advantages over both COSY and proton double quantum coherence in some cases, which are illustrated below.

Zero quantum coherence is represented by the transition connecting levels 2 and 3 in the energy level diagram shown in Fig. 2-39.  Compare this with the double quantum transition that connects energy levels 1 and 4 as shown in Fig. 2-27. Another interesting contrast between zero and double quantum coherence is provided when the location of responses in the $F_1$ frequency domain is considered. The $F_1$ frequency for a double quantum response of an AX spin system is derived from the algebraic sum of the offsets of the resonances relative to the transmitter (see Eqn. [2-11]). In contrast, the $F_1$ frequency for a zero quantum transition in an AX spin system is an algebraic difference defined by:

$$\omega_1 = \pm(\Omega_A - \Omega_X) \qquad\qquad [2\text{-}13]$$

Hence, proton zero quantum response locations may be considered in a manner analogous to proton double quantum responses. The specific nature of responses contained in a zero quantum coherence spectrum is described in further detail below.

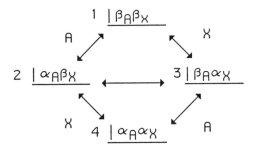

Fig. 2-39.   Energy level diagram (ELD) for a simple AX spin system. Single quantum transitions connect adjacent energy levels in the usual sense. The zero quantum transition connects energy levels 2 and 3. The double quantum transition, which normally would connect energy levels 1 and 4 is omitted for clarity. It should be noted that the zero quantum transition does not necessarily represent a zero enegy process.

## The Proton Zero Quantum Pulse
## Sequence and Phase Cycling

Excitation of zero quantum coherence, as mentioned above, is somewhat problematic in that zero quantum transitions are dependent on both spin-spin couplings and chemical shifts.[3,117] The standard pulse train used to excite double quantum coherence shown in Eqn. [2-7] will not excite zero quantum transitions. The simpler sequence shown in Eqn. [2-15] does excite the zero quantum transitions, albeit with the aforementioned dependence on spin-spin couplings and chemical shift.[3,117] Coupling and chemical shift difficulties are overcome, however, by the sequence suggested by Müller,[119] which is shown in Eqn. [2-16].

$$90°(x) - \tau - 90°(x) \qquad\qquad [2\text{-}15]$$

$$90°(x) - \tau - 180°(y) - \tau - 45°(y) \qquad\qquad [2\text{-}16]$$

Müller's[119] sequence allows a more uniform excitation of zero quantum coherence since the chemical shift terms are suppressed by the 180°(y) pulse in the sequence; the 45°(y) pulse generates the desired zero quantum coherence from the spin terms of the density operator. Following its creation using Eqn. [2-16], zero quantum coherence is allowed to evolve before being converted back into observable single quantum coherence. The pulse train shown in Eqn. [2-16] serves as the fundamental basis for the zero quantum pulse sequence shown in Fig. 2-40.

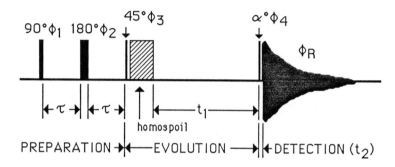

*Fig. 2-40.* Zero quantum pulse sequence proposed by Müller.[119] Zero quantum coherence is created with the completion of the first 45° pulse, the 180° pulse serving to remove sensitivity to chemical shift differences. The duration of the τ interval is optimized as a function of $1/4(J_{HH})$, typically 35.7 msec for an average 7 Hz vicinal coupling. The duration of the evolution time, $t_1$, is generally set as a function of dwell time/0.75. The use of a homospoil pulse is optional. With the minimal four step phase cycling, the homospoil pulse can be employed to destroy all higher order coherences that occur in < 100 msec.[76] When more extensive phase cycling schemes are employed, the homospoil pulse needn't be utilized.

Following the creation of zero quantum coherence, another pair of options is available to the user of this experiment. Advantage may be taken of the relative insensitivity of the zero quantum coherence to magnetic field inhomogeneities, and a homogeneity spoiling pulse may be applied to destroy all single and higher quantum coherences.[77] Exercise of this option allows the utilization of a minimal four step phase cycle, which is presented below. Alternately, the user may opt not to employ the homogeneity spoiling pulse, in which case a more complex 16 or 64 step phase cycle, also shown below, should be employed.

Finally, options are also available in performing the reconversion of zero to observable single quantum coherence. First, a 90° pulse may be employed. As with double quantum coherence experiments discussed above, a 90° reconversion or "read" pulse permits passive transfer.[11,53,90] In contrast, the preferred use of a 45° reconversion pulse ($\alpha = 45°$, Fig. 2-40), reduces passive response intensity and preferentially transfers the zero quantum coherence into one quadrant of the resulting two-dimensional spectrum, making possible quadrature phase detection of the zero quantum coherence.[119] The 45° reconversion pulse has the further beneficial effect of halving the number of responses in the data matrix, since the intensity ratio of a peak to its mirror image in $F_1$ is proportional to $(\tan \alpha/2)^2$ as shown schematically in Fig. 2-41.[119]

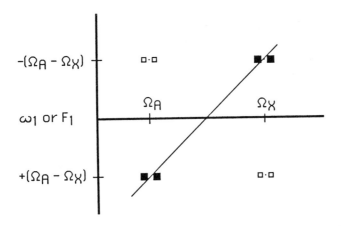

*Fig. 2-41.* Schematic representation of zero quantum responses for an AX spin system acquired using the pulse sequence shown in Fig. 2-40 with the read pulse, $\alpha$, set to 45°. The large, solid squares represent the pair of correlated responses that will normally be observed. The smaller, open squares represent the $F_1$ mirror image response suppressed as a function of the read pulse, the observed intensity a function of $(\tan \alpha/2)^2$. Changing the read pulse to 135° will reverse the sense of the responses shown. Responses appear in $F_1$ at the algebraic difference of the offsets of the correlated resonances relative to the transmitter $(\pm(\Omega_A - \Omega_X))$. The slope of the axis correlating the responses shown is $F_1 = +2F_2$. In systems showing multiple correlations, there will be a series of parallel correlation lines with equal slopes. There is no single correlation axis like the diagonal of the COSY experiment or the skew diagonal of the proton double quantum experiment.

Phase cycling options available for use with the zero quantum coherence pulse sequence are shown in Table 2-10. The minimal four step phase cycle shown is intended to be employed with a homogeneity spoiling pulse of sufficient duration to suppress single and double quantum artifacts. In instances where the capability of applying a homospoil pulse is not available, the more comprehensive 16 and 64 step phase cycles shown may be employed. Generally, we have found the former to give satisfactory performance, although the latter may be used to some advantage when it is necessary to take larger numbers of acquisitions.

*Table 2-10.*    Phase cycling prescriptions for proton zero quantum coherence using the pulse sequence shown in Fig. 2-40. The phase cycling shown selects for zero, four and eight quantum orders of coherence. The homogeneity spoiling pulse normally used with the four step phase cycle suppresses all but zero qquantum coherence.

| Four step phase cycle | | | | | | | | | |
|---|---|---|---|---|---|---|---|---|---|
| $\phi_1$ | 0000 | $\phi_2$ | 1313 | $\phi_3$ | 1133 | $\phi_4$ | 1111 | $\phi_R$ | 0022 |

| Sixteen step phase cycle | | | | |
|---|---|---|---|---|
| $\phi_1$ | 0000 | 0000 | 0000 | 0000 |
| $\phi_2$ | 1111 | 3333 | 1111 | 3333 |
| $\phi_3$ | 1111 | 1111 | 3333 | 3333 |
| $\phi_4$ | 1230 | 1230 | 1230 | 1230 |
| $\phi_R$ | 0123 | 0123 | 2301 | 2301 |

| Sixty-four step phase cycle | | | | |
|---|---|---|---|---|
| $\phi_1$ | 00000000 | 00000000 | 11111111 | 11111111 |
| | 22222222 | 22222222 | 33333333 | 33333333 |
| $\phi_2$ | 11113333 | 11113333 | 22220000 | 22220000 |
| | 33331111 | 33331111 | 00002222 | 00002222 |
| $\phi_3$ | 11111111 | 33333333 | 22222222 | 00000000 |
| | 33333333 | 11111111 | 00000000 | 22222222 |
| $\phi_4$ | 12301230 | 12301230 | 23012301 | 23012301 |
| | 30123012 | 30123012 | 01230123 | 01230123 |
| $\phi_R$ | 01230123 | 23012301 | 12301230 | 30123012 |
| | 23012301 | 01230123 | 30123012 | 12301230 |

## Location of Responses in Proton
## Zero Quantum 2D-NMR Spectra

Unlike proton double quantum two-dimensional NMR experiments in which more than one type of double quantum coherence may be observed, there is but one type of proton zero quantum response. In addition, however, particularly if insufficient homogeneity spoiling is applied, it is possible for proton single and double quantum coherence to appear as an artifact.

**Proton zero quantum responses**. Proton zero quantum responses appear at the $F_2$ frequencies of the coupled resonances displaced from the zero axis in $F_1$ according to Eqn. [2-15]. In the case where the reconversion pulse is set to 90°, an AX spin system will give rise to four responses in the zero quantum data matrix located as shown in Fig. 2-41, half of which are redundant. It should be noted, however, that in the case where the reconversion pulse is 90° that all four of the responses will have equal intensity. When the reconversion pulse employed is reset to 45°, the intensity of the pair of responses shown as open squares in Fig. 2-41 is reduced and proportional to $(\tan\ \alpha/2)^2$. In most instances the pair of responses denoted by the open squares may not even be visible in the contour plot.

**Single and double quantum artifacts**. Generally speaking the phase cycling employed in the zero quantum experiment should eliminate most if not all artifact responses. However, when minimal four step phase cycling in conjunction with a homogeneity spoiling pulse is employed, there is some chance that single or double quantum artifact responses may be observed, particularly if the power applied during the spoiling process is insufficient.

Single and double quantum artifacts are easily located in zero quantum two-dimensional NMR spectra, since the former are associated with a diagonal and the latter are a skew diagonal. Conveniently, the diagonal and skew diagonal originate from the location of the transmitter in both $F_1$ and $F_2$. Using the zero quantum spectrum phenanthro-[3,4:3',4']phenanthro[2,1-*b*]thiophene (**2-13**),[120] we may quickly locate the single and double quantum artifacts contained in the spectrum shown in Fig. 2-42, which was recorded using a four step phase cycle and a 25 msec homogeneity spoiling pulse during which the spoiling pulse power was intentionally zeroed to allow artifacts to appear.

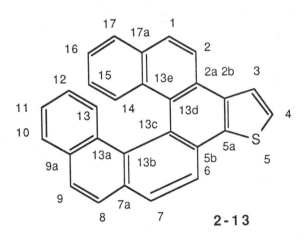

**2-13**

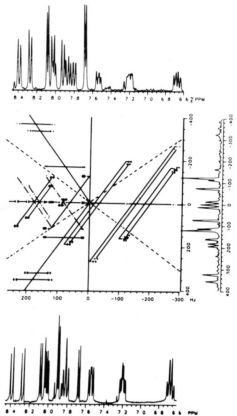

*Fig. 2-42.* Proton zero quantum spectrum of phenanthro[3,4:3',4']phenanthro-[2,1-*b*]thiophene (**2-13**) recorded in deuterochloroform at an observation frequency of 300.068 MHz at 20°C.[120]  The spectrum was recorded using the pulse sequence shown in Fig. 2-40 and the four step phase cycle shown in Table 2-10. The duration of the homogeneity spoiling interval was 25 msec. The homogeneity spoiling pulse power, however, was intentionally zeroed to allow artifacts to appear in the spectrum. The high resolution proton spectrum (4K points), labeled in ppm downfield from TMS, is plotted below the contour plot. The projection through the zero quantum ($F_1$) frequency domain is plotted vertically while the projection of the normal spectrum through $F_2$ is plotted horizontally across the top of the contour plot. Axes bounding the contour plot show offsets relative to the transmitter in hertz. The data were collected as 512 x 1K points and were processed with sinusoidal multiplication prior to both Fourier transformations. The dashed line ($F_1 = \pm F_2$) passing through the origin ($F_1 = F_2 = 0$ Hz) denote the single quantum diagonal correlation axes. There are no single quantum artifacts in the spectrum. The solid lines, $F_1 = \pm 2F_2$, passing through the origin denote the double quantum skew diagonal axes. Double quantum responses correlating the three AX spin systems are observed at $F_1 = \pm 165$, 322 and 356 Hz.

The transmitter is located at $F_1$ and $F_2 = 0$ Hz and is denoted by the horizontal and vertical lines, respectively, shown in the contour plot. The dashed lines, $F_1 = \pm F_2$, passing through the origin ($F_1 = F_2 = 0$ Hz) show the diagonal along which single quantum artifacts will appear or about which they will be disposed. Other than a scant few responses that may be only noise, there are no significant single quantum artifcats contained in the spectrum. The double quantum skew diagonal axes also pass through the origin, $F_1 = \pm 2F_2$. Symmetrically disposed about the skew diagonal are responses for the three AX spin systems contained in the spectrum of **2-13**, which are observed at $F_1 = \pm 165$, 322 and 356 Hz.

## Parameter Selection and Experiment Execution

As with the proton double quantum experiment discussed above, there are only a few parameters that must be chosen to perform the proton zero quantum NMR experiment. Individual parameters will be considered in turn.

### Evolution time incrementation.
Proton zero quantum two-dimensional NMR spectra, unlike their proton double quantum counterparts, can be performed with the transmitter located in the middle of the spectrum in $F_2$. Utilization of a 45° reconversion ($\alpha$) pulse preferentially transfers zero quantum coherence into one quadrant of the two-dimensional NMR spectrum, thus providing the means for quadrature phase detection of the zero quantum spectrum.

To establish the evolution time, $t_1$, as a function of the dwell time, we must next consider probable couplings in the spectrum to be studied. In the case of aralkyl molecules, there is little chance that couplings will occur between aromatic resonances and their aliphatic counterparts. Hence, all couplings will be between like resonances, and the necessary range of frequencies in $F_1$ that must be digitized will be small. In this instance, it is useful to set the initial evolution time equal to the dwell time and to increment the evolution time by this interval. In contrast, when a purely aromatic compound, for example **2-13**, is being studied, there is a substantial probability that protons at the extreme upfield and downfield ends of the spectral range may be coupled to one another. The same obviously holds true for molecules such as terpenes. In these cases, it is necessary to digitize a somewhat broader range of frequencies in $F_1$. For survey spectra of these types of molecules using the four step phase cycle shown in Table 2-10, we have found it useful to employ an initial setting of the evolution time of 0.75 x dwell time.

### The $\tau$ interval used in the creation of zero quantum coherence.
The duration of the $\tau$ interval is governed by the spin-spin

couplings of interest and is determined as a function of $1/4(J_{HH})$. For general survey zero quantum coherence spectra, we have found a 7 Hz optimization to be an excellent choice for both phenanthro-[3,4:3',4']phenanthro[2,1-*b*]thiophene (**2-13**)[120] and strychnine (**2-6**).[121] Attempts at exploiting the harmonic properties of the evolution time by using small couplings, eg, 1.75 Hz, which were quite successful with the proton double quantum study of strychnine (**2-6**) recently reported[19], were also successful with the zero quantum experiment confirming Müller's original suggetions of the possibility of small coupling optimization in his original work.[119]

**Read pulse flip angle**. The flip angle of the reconversion or "read" pulse is important. Here again, it is inadvisable to employ a 90° read pulse. For reasons that were already discussed, it is advisable to employ a 45° read pulse for the proton zero quantum experiment. Insofar as we are aware, there is nothing to be gained through the utilization of a 135° read pulse.

**Digitization in the $F_1$ and $F_2$ frequency domains**. The range of frequencies in the $F_1$ domain of zero quantum coherence spectra that may contain responses is comparable to that of the proton double quantum coherence experiment. Consequently, we have generally found it most useful to digitize the $F_1$ frequency domain by performing 512 increments of the evolution period. The persistence of the signal as it evolves as zero quantum coherence is generally quite good. Unlike the COSY experiment in which the signal will have degraded substantially after only several hundred increments of the evolution time, the signal in the zero quantum experiment may still be quite strong in the final blocks of data taken. Thus, it is generally advisable to allow the spectrometer to complete the acquisition of as many blocks of data as possible before terminating the experiment.

Like a COSY experiment, digitization of the $F_2$ frequency domain with 1K complex points, which will yield a spectrum digitized with 512 real points after Fourier transformation, is generally satisfactory.

## Proton Zero Quantum
## Data Processing

Processing a proton zero quantum data matrix is not at all unlike the processing of any of the experiments described above. Once again, it is generally preferable to employ sinusoidal multiplication or its equivalent prior to both Fourier transformations to take advantage of the sharp lines offered by zero quantum coherence. Following the second Fourier transformation, 0° projection of the data matrix will afford the information contained in the zero quantum frequency domain while 90° projection will recover the normal proton spectrum. Personally, we have found it more convenient to subject the data matrix to one final transposition prior to plotting. Hence, the content of the 0° and 90° projections will be reversed relative to what was just described.

## Interpretation of Proton
## Zero Quantum Spectra

**Application of the proton zero quantum experiment to phenanthro[3,4:3',4']phenanthro[2,1-_b_]thiophene (2-13)**. The proton zero quantum spectrum of **2-13** recorded with the power of the homogeneity spoiling pulse intentionally zeroed was shown above in Fig. 2-42 to illustrate the location of single and double quantum artifact responses.[120] The zero quantum spectrum recorded under normal conditions using the 64 step phase cycle specified in Table 2-10 is shown in Fig. 2-43. It should be noted from even the most cursory inspection of Fig. 2-43 that the artifacts contained in the spectrum shown in Fig. 2-42 have been completely suppressed. The parallel soild lines $(F_1 = +2F_2)$ denote normal zero quantum coherences, in most cases, between vicinally coupled neighbor protons. The only long range connectivity information observed in the spectrum is afforded by the very weak responses that correlate H10 and H17 (7.81 and 7.84 ppm, respectively) to their meta coupling partners, H12 and H15 (6.68 and 6.70 ppm, respectively), which give rise to the zero quantum response centered at $F_1 = \pm342$ Hz. Finally, the dashed parallel lines $(F_1 = -2F_2)$ locate the largely suppressed mirror image responses of the three AX spin systems of **2-13**.

Recalling that zero quantum responses will be located in $F_1$ at the algebraic difference of the coupled resonances relative to the transmitter, let us consider the specific example of one of the AX spin systems of **2-13** which is shown in the expansion in Fig. 2-44. This particular AX spin system corresponds to the H5 and H6 protons, which resonate at 8.28 and 8.04 ppm, respectively. Relative to the transmitter, the resonances are located at $v = +200$ and $+126$ Hz, respectively. The algebraic difference between the coupled resonances is 74 Hz. Consequently, we would expect to see zero quantum responses located at $F_1 = \pm74$ Hz, which is exactly the case.

Perhaps one of the most useful features of the proton zero quantum cohrence 2D-NMR experiment is its intrinsically high resolution. The high degree of resolution afforded by the experiment can be illustrated by examining the correlation of the AB components of the two ABMX spin systems of **2-13**. The expansion of the AB components in question is shown in Fig. 2-45. The H11 and H16 protons resonating at approximately 7.2 ppm differ in chemical shift by only 0.01 ppm. Their H12 and H15 counterparts resonating at 6.69 ppm differ by only 0.02 ppm. Despite the significant degree of overlap between these pair of protons, it is still possible to distinguish quite clearly the connectivities between the individually coupled pairs as shown in Fig. 2-45. Independent confirmation of the connectivities in shown in Fig. 2-45 was provided by a long range heteronuclear chemical shift correlation spectrum (see Chapter 3 for a discussion of this technique).[122] These same connectivities (H11-H12 and H15-H16) can be inferred from the symmetrized COSY spectrum of **2-13** which is shown in Fig. 2-46 for comparison.[120] However, the differences

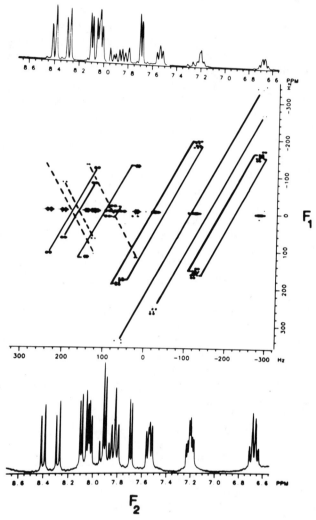

*Fig. 2-43.* Proton zero quantum spectrum of **2-13** acquired using the 64 step phase cycle specified in Table 2-10. The high resolution proton spectrum (4K) is plotted beneath the contour plot, the 0° projection is plotted above. The data was acquired as 512 x 1K points and was processed using sinusoidal multiplication prior to both Fourier transformations. Connectivities are denoted by solid lines (suppressed redundant connectivities are indicated by dashed lines). Chemical shifts below the reference spectrum and the projected spectrum are specified in ppm downfield from TMS. The axis plotted immediately below the contour plot shows the location of responses relative to the transmitter (0 Hz). The axis plotted vertically flanking the contour plot gives zero quantum frequency in hertz.

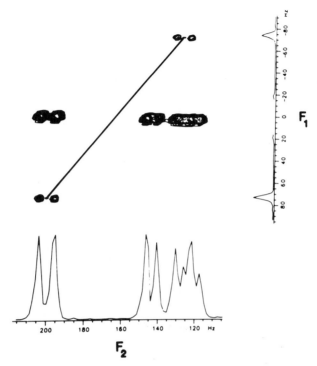

*Fig. 2-44.* Contour plot shown the correlation of an AX spin system using zero quantum coherence. The data was acquired using a 45° read pulse to suppress responses along the axis $2F_2 = -F_1$. Relative to the transmitter, the resonances are located at $v = +200$ and $+126$ Hz, respectively. The algebraic difference between the coupled resonances is 74 Hz. Consequently, we would expect to see zero quantum responses located at $F_1 = \pm 74$ Hz.

between the H11 and H16 are so small (0.01 ppm) as to make the discrimination of the coupling between H11 and H10 and H16 and H15 virtually impossible from the COSY spectrum. In contrast, these connectivities are easily established from the zero quantum spectrum shown in Fig. 2-43.

Finally, although there is undesired signal along the $F_1 = 0$ Hz axis in the spectrum shown in Fig. 2-43, it is still possible to establish correlations between the components of AB spin systems using the proton zero quantum experiment. For example, the H7 and H8 resonances of **2-13** constitute an AB spin system, resonating at 7.93 and 7.90 ppm, respectively, with offsets relative to the transmitter of $v = +99$ and $+90$ Hz, respectively. Despite this relatively small separation, the zero quantum responses correlating these resonances which are located at $F_1 = \pm 9$ Hz are easily distinguished. In contrast, in the COSY

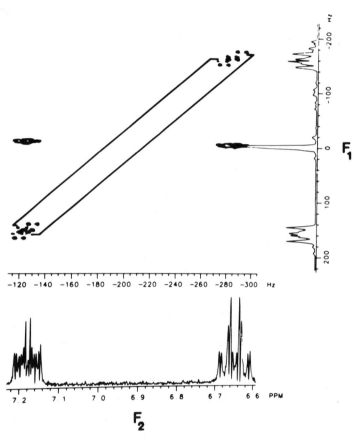

Fig. 2-45. Expansion of the proton zero quantum spectrum shown in Fig. 2-43 to show the detail in the region of the ABMX spin systems. Connectivities linking the H11/H12 and H15/H16 pairs are shown. The H11 and H16 resonances differ in chemical shift by 0.011 ppm; the H12 and H15 resonances by 0.02 ppm. Despite these relatively minor differences in chemical shift, the respective connectivities linking these two noncoupled pairs are much more easily established from the zero quantum spectrum than from the COSY spectrum shown in Fig. 2-46.

spectrum of **2-13** which is shown in Fig. 2-46, the off diagonal responses which correlate H7 and H8 are indistinguishable from the diagonal responses.

**Application of the proton zero quantum experiment to strychnine (2-6).** The proton zero quantum spectrum of the aliphatic and vinyl region of strychnine (**2-6**) is shown in Fig. 2-47.[121] The spectrum was recorded using a 7 Hz optimization. The experiment was performed using the four step phase cycle shown in Table 2-10 in conjunction with a 25 msec homospoil pulse. Under these conditions, there were only very minimal residual proton single and double quantum responses visible in the spectrum, the latter arising from the H11a-H12 and H11b-H12 coupling pathways.

Information contained in the proton zero quantum spectrum shown in Fig. 2-47 was largely comparable to the 7 Hz optimized proton double quantum spectrum (for comparison, see Fig. 2-35). However, as a result of the improved resolution afforded by the zero quantum coherence, the H22-H23a and H22-H23b responses were individually resolved in the zero quantum spectrum but not in the proton double quantum spectrum. At the 7 Hz optimization used to record the spectrum a relatively wide range of spin couplings gave responses: from H11b-H12 (3.3 Hz) to H11a-H11b (17.4 Hz). Notably absent from the spectrum, however, were the H12-H13, H13-H14, H14-H15b and H15b-H16 vicinal responses. The absence of these responses leaves a gap in the structural fragment represented by **2-8** that can be assembled from COSY or proton double quantum coherence spectra. Optimizing for smaller couplings obviates the difficulty of the missing vicinal couplings but still does not provide the long range coupling information that could be garnered from the 1.75 Hz optimized proton double quantum spectrum. It is useful to note, however, that the zero quantum spectrum shown in Fig. 2-47 was recorded in only 45 minutes while the proton double quantum spectrum optimized for 1.75 Hz was acquired overnight.

## Presentation of Proton
## Zero Quantum Spectra

Proton zero quantum spectra are generally comparable to COSY spectra and may be plotted in similar fashion as shown in Fig.s 2-43 and 2-47. To render proton zero quantum spectra similar in appearance to COSY spectra, Blümich[105] has suggested a method for further manipulation of the data. An experimental method that produces a "COSY-type" result has been proposed by Santoro, Bermejo and Rico.[123]

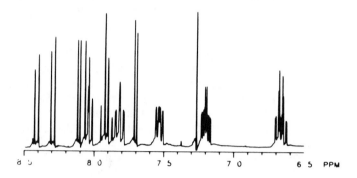

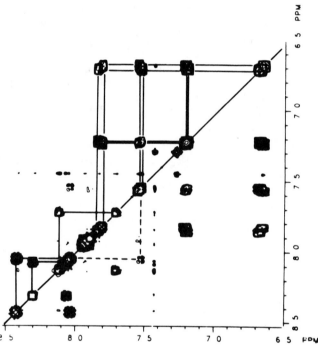

*Fig. 2-46.* COSY spectrum of phenanthro[3,4:3',4']phenanthro[1,2-_b_]thiophene (**2-13**) in deuterochloroform recorded at 300.068 MHz. The spectrum was acquired using the standard COSY pulse sequence (Fig. 2-1) as 256 x 512 data points and was zero-filled an subjected to sinusoidal multiplication prior to both Fourier transforms to afford the 512 x 512 point data matrix shown in the contour plot. The data was also symmetrized[10] prior to plotting. The high resolution proton spectrum (8K) is plotted above the contour plot. Individual spin system connectivities are shown above the diagonal. The only long range coupling information that may be gleaned from the spectrum is the 5 bond epi zig-zag coupling between H1 and H14 denoted by the dashed line below the diagonal.

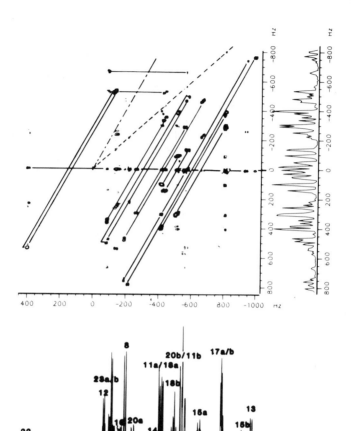

*Fig. 2-47.*     Proton zero quantum spectrum of strychnine (**2-6**) in deuterochloroform recorded at 300.068 MHz. The experiment was performed
using the pulse sequence shown in Fig. 2-40 with the read pulse, $\alpha$ =
45°, and the minimal four step phase cycle specified in Table 2-10.
The experiment was optimized using $\tau = 1/4(^3J_{HH}) = 35.7$ msec (7 Hz).
The data was recorded as 512 x 1K points and was processed using
sinusoidal multiplication prior to both Fourier transformations. The
acquisition time was 45 minutes. The high resolution proton spectrum
(32K) is plotted below the contour plot. Chemical shifts specified in
ppm are downfield of TMS. The axis beneath the contour plot (Hz)
shows the location of resonances relative to the transmitter. The axis
plotted vertically gives zero quantum frequncies. The 90° projection
showing the zero quantum frequency domain is plotted vertically
flanking the contour plot. The dashed line (- - -) shows the single
quantum correlation axis; the broken line ( -- · -- · -- ) represents the
skew diagonal along which double quantum responses are
symmetrically disposed at $F_1$ ~ 500 and 625 Hz.

## Alternative Zero Quantum
## Pulse Sequences

The first alternative to Müller's zero quantum pulse sequence[119] was a variant proposed by Santoro, Bermejo and Rico.[123] Their pulse sequence, which is shown in Fig. 2-48, uses the nucleus of Müller's sequence[119] followed by evolution in the usual fashion (see Fig. 2-40). However, after the first half of the evolution period, the read pulse (135° in this case) is applied to the evolving magnetization, converting it back into observable single quantum coherence. Rather than detecting the magnetization at this point, the pulse sequence described by Santoro and co-workers[123] instead allows the magnetization to undergo the second half of the evolution period as single quantum coherence, which is finally detected. Santoro and co-workers have christened their pulse sequence SUCZESS (SUCcessive ZEro quantum Single quantum coherences for Spin correlation). Although interesting, other than the initial report, the SUCZESS pulse sequence has not been utilized. Phase cycling for the SUCZESS pulse sequence is shown in Table 2-11.

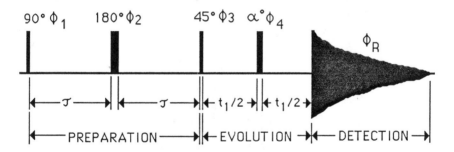

Fig. 2-48. Pulse sequence for successive zero quantum single quantum coherences for spin correlation (SUCZESS) described by Santoro, Bermejo and Rico.[123] The pulse sequence as originally reported, used a pulse, $\alpha = 135°$. In principle, however, it should also be possible to utilize a 45° read pulse as in the original work of Müller.[119]

Table 2-11. Phase cycling for the SUCZESS pulse sequence shown in Fig. 2-48.

| $\phi_1$ | 0000 | 0000 | 2222 | 2222 | 0000 | 0000 | 2222 | 2222 |
|---|---|---|---|---|---|---|---|---|
| $\phi_2$ | 1111 | 1111 | 1111 | 1111 | 3333 | 3333 | 3333 | 3333 |
| $\phi_3$ | 1313 | 1313 | 1313 | 1313 | 1313 | 1313 | 1313 | 1313 |
| $\phi_4$ | 1331 | 2002 | 3113 | 0220 | 1331 | 2002 | 3113 | 0220 |
| $\phi_R$ | 0022 | 1133 | 0022 | 1133 | 0022 | 1133 | 0022 | 1133 |

A second group of modifications is based on the SECSY[73,74] variant of the standard COSY experiment and, in this sense, is also related to the SUCZESS pulse sequence just described. In early 1986 Hall and Norwood[124] first communicated an alternative zero quantum pulse sequence that has some similarities to the symmetrical proton double quantum pulse sequence of Rance and coworkers[104] shown in Fig. 2-38. Specifically, as shown in Fig. 2-49, Hall and Norwood[124] create zero quantum coherence using a the sequence shown in Eqn. [2-17]

$$90°(x) - \tau - 180°(y) - \tau - 90°(x) \qquad [2\text{-}17]$$

after which zero quantum coherence is allowed to evolve in the usual sense. It should be noted here that their creation sequence differs from that employed by Müller (see Fig. 2-40).[119] After the reconversion of zero quantum coherence to observable single quantum coherence, the Hall and Norwood[124] sequence shown in Fig. 2-49 employs a refocusing scheme, the second half of an echo ultimately acquired and stored.

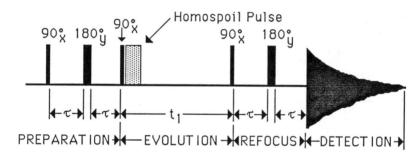

Fig. 2-49.   Zero quantum pulse sequence with a refocusing interval proposed by Hall and Norwood.[124]   Zero quantum coherence is created during the preparation interval, allowed to evolve during $t_1$, refocused and the second half of the echo recorded. The homogeneity spoiling pulse may be considered optional in high resolution applications in a homogeneous field.

A second modification of the sequence shown in Fig. 2-49 was reported by Hall and Norwood later in 1986.[125] By using a 180° pulse systematically repositioned through the evolution period, broadband decoupling was added to the zero quantum experiment. The broadband decoupled zero quantum pulse sequence just described is shown in Fig. 2-50.

The length of the evolution period, $t_d$, is held constant in the pulse sequence shown in Fig. 2-50. Since the extent of scalar coupling evolution is unaffected by the positioning of the 180° pulse located in the evolution period, scalar couplings will be invariant with respect to the evolution time. However, the position of the 180° pulse does affect

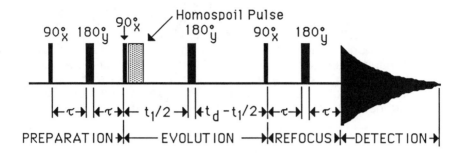

*Fig. 2-50.* Zero quantum pulse sequence proposed by Hall and Norwood[125] incorporating broadband decoupling through the variable refocusing provided by the 180° pulse systematically repositioned through the evolution period. The homogeneity spoiling pulse may be considered optional for high resolution applications in a homogeneous field.

chemical shift evolution. Thus, the final data set will be modulated with respect to chemical shift as a result of the repositioning of the 180° pulse in the evolution period. In contrast, since the evolution period, $t_d$, has a constant duration, the data set will not be modulated with respect to scalar couplings. Double Fourier transformation of the data set will afford a decoupled zero quantum spectrum.

It should be noted that the pulse shown in Fig. 2-50 following the evolution period reconverts evolving zero quantum coherence into single quantum coherence. In the case where the reconversion pulse is 90° as shown in Fig. 2-50, only that ZQC which is not antiphase with respect to a passive spin of the same kind as either of those active in the coherence will be converted back into observable single quantum coherence. In contrast, if a 45° reconversion pulse is employed, all zero quantum coherences that were initially excited will be observed to some degree. These features allow the elimination of some features of zero quantum spectra providing, in a fashion, the means of editing spectra.

Following the broadband decoupled experiment described by Hall and Norwood[125] a third variant was reported which was given the acronym ZECSY (ZEro quantum Correlated SpectroscopY).[126] The ZECSY pulse sequence is shown in Fig. 2-51 and the phase cycling is given in Table 2-12.

Hall and Norwood[126] in their description of the ZECSY experiment utilized the pulses $\beta = 90°$ and $\alpha = 45°$. The ZECSY (Fig. 2-52) spectrum of phenanthro[3,4:3',4']phenanthro[2,1-*b*]thiophene (**2-13**), was taken using parameters, digitization, concentration and processing identical to that used in preparing Fig. 2-43.

The zero quantum spectrum of **2-13** shown in Fig. 2-43 recorded using the Müller (Fig. 2-40) pulse sequence[119] readily establishes all of the vicinal connectivities. Responses are observed at $\pm(\Omega_A - \Omega_x)$ in the zero quantum ($F_1$) frequency domain. In the ZECSY spectrum, responses should be observed, according to Hall and Norwood[126] at

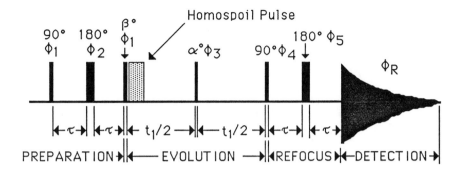

*Fig. 2-51.*   ZECSY pulse sequence originally proposed by Hall and Norwood.[126] In
their original work, Hall and Norwood utilized $\beta$ = 90° and $\alpha$ = 45°.
More recently, Hall and Norwood[127] have reported a variant of the
ZECSY pulse sequence that employs $\beta = \alpha = 45°$. Both experiments utilize
the phase cycling regimen shown in Table 2-12.


Table 2-13.   Phase cycling scheme for the ZECSY experiments described by Hall and
Norwood.[124,127] The phase cCycle shown can either be used in its
entirety or limited to the first sixteen steps.

| $\phi_1$ | 00000000 | 00000000 | 11111111 | 11111111 |
|---|---|---|---|---|
| | 22222222 | 22222222 | 33333333 | 33333333 |
| $\phi_2$ | 11111111 | 11111111 | 22222222 | 22222222 |
| | 33333333 | 33333333 | 00000000 | 00000000 |
| $\phi_3$ | 00001111 | 22223333 | 11112222 | 33330000 |
| | 22223333 | 00001111 | 33330000 | 11112222 |
| $\phi_4 = \phi_R$ | 01230123 | 01230123 | 12301230 | 12301230 |
| | 23012301 | 23012301 | 30123012 | 30123012 |
| $\phi_5$ | 12301230 | 12301230 | 23012301 | 23012301 |
| | 30123012 | 30123012 | 01230123 | 01230123 |

frequencies of $(\Omega_A + \Omega_x)/2$ and $(\Omega_A - \Omega_x)/2$ for an AX spin system. In the
spectrum shown in Fig. 2-52, responses for the AX spin systems are
observed at the latter frequency but not at the former. Rather, responses
are observed at $\pm(\Omega_A - \Omega_x)$ and $\pm(\Omega_A - \Omega_x)/2$. In the more complex spin
systems shown in Fig. 2-52, the ZECSY spectrum also gives responses
for passive transfer analogous to those observed in proton double
quantum spectra as discussed above. This feature can be quite useful
in tracking complex connectivity networks. The spectrum acquired using
Müller's pulse sequence and shown in Fig. 2-43 does not offer this
potential advantage. Unfortunately, as will be noted from Fig. 2-52, the
responses corresponding to the two four spin systems of **2-13** are so
numerous in the spectrum as to render the connectivity information very
difficult to extract. Finally, it will also be noted that there are numerous

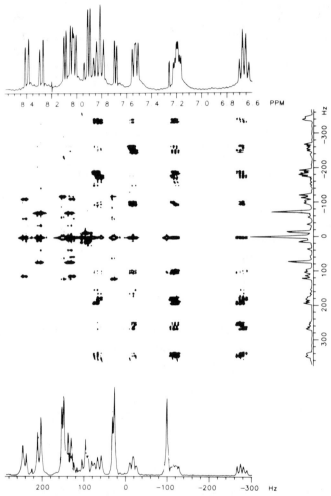

*Fig. 2-52.*    ZECSY spectrum of phenanthro[3,4:3',4']phenanthro[2,1-*b*]thiophene
(**2-13**) recorded using the pulse sequence shown in Fig. 2-51 with β =
90° and α = 45°.   The spectrum was recorded with the τ interval
optimized for an assumed 7 Hz vicinal coupling (35.7 msec).   The data
was acquired as 512 x 1K points and was processed using sinusoidal
multiplication prior to both Fourier transformations.    The high
resolution spectrum is plotted above the contour plot and the
projection through the $F_2$ frequency domain below.   Chemical shifts in
ppm are downfield of TMS.   Chemical shifts in hertz are relative to the
transmitter.    The projection through the $F_1$ frequency domain is
plotted vertically flanking the contour plot.

artifacts along the axis $F_1 = 0$ which tend to obscure the connectivities of the AB spin system contained in the spectrum of **2-13**. This problem can be reduced, however, by acquiring the data without spinning the sample.

Overall, the choice of which zero quantum pulse sequence to employ will probably be dependent upon the goals of the investigation. In those cases where there is a high degree of congestion, Müller's experiment[119] provides a spectrum containing far fewer responses with a correspondingly lower probability of overlap. Long range connectivities do not, however, give rise to "passive transfer" responses, and it is thus necessary to establish long range connectivities in the same manner as their vicinal counterparts. The ZECSY[126] and modified ZECSY[127] offer the advantage of "passive transfer" responses but do so at the expense of a much more congested spectrum with correspondingly greater probability of response overlap.

## A Brief Survey of Proton
## Zero Quantum Applications

Proton zero quantum coherence has been much slower to gain in popularity than its double quantum analog. Applications contained in the literature are very few in number. Aside from the phenanthro-[3,4:3',4']phenanthro[2,1-*b*]thiophene[120] and strychnine[121] examples discussed above, Ruessink, De Kanter and MacLean have used zero quantum coherence to determine $^2H$-$^{14}N$ coupling constants[128], Müller and Pardi[129] have described a proton zero quantum coherence relay experiment that was applied to a peptide and Hall and Norwood[124-126] have examined amino acid spectra in inhomogeneous magnetic fields using zero quantum. In a very early study, Pouzard, Sukumar and Hall reported the spectral analysis of some simple spin systems by zero quantum coherence.[117] The first applications of proton zero quantum coherence in natural products structure elucidation have also begun to appear. Cardellina and coworkers[130] have reported the utilization of the proton zero quantum experiment of Müller[119] in the elucidation of the structures of a trio of novel brominated sesquiterpenes from a marine alga (see also Problem 6-2). The Müller sequence has also been utilized to disentangle the proton NMR spectrum of benzo[*b*]tri-phenyleno[1,2-*d*]thiophene[131] and to establish proton-proton connectivities in a novel dithiepin analog.[132]

## REFERENCES

1.      M. L. Martin, J.-L. Delpuech and G. J. Martin, *Practical NMR Spectroscopy*, Heyden, London, 1980, pp. 200-205.

2.      J. Jeener, Ampere International Summer School, Basko Polje, Jugoslavia, 1971.

3.      W. P. Aue, E. Bartholdi and R. R. Ernst, *J. Chem. Phys.*, **64**, 2226 (1976).

4.      K. Nagayama, *Adv. Biophys.*, **14**, 139 (1981).

5.      K. Wüthrich, *NMR of Proteins and Nucleic Acids*, Wiley, New York, 1986.

6.      A. Bax, R. Freeman and G.A. Morris, *J. Magn. Reson.*, **42**, 164 (1981).

7.   R. R. rnrst, G. Bodenhausen and A. Wokaun, *Principles of Nuclear Magnetic Resonance in One and Two Dimensions*, Clarendon Press, Oxford, 1987.

8.   O. W. Sørensen, G. Eich, M. H. Levitt, G. Bodenhausen and R. R. Ernst, *Prog. Nucl. Magn. Reson. Spectrosc.*, **16**, 163 (1983).

9.   D. I. Hoult and R. E. Richards, *Proc. Roy. Soc., London*, **A344**, 311 (l975).

10.  R. Baumann, G. Wider, R. R. Ernst and K. Wüthrich, *J. Magn. Reson.*, **44**, 402 (1981).

11.  G. E. Martin, R. Sanduja and M. Alam, *J. Org. Chem.*, **50**, 2383 (1985).

12.  G. Wider, S. Macura, A. Kumar, R. R. Ernst and K. Wüthrich, *J. Magn. Reson.*, **56**, 207 (1984).

13.  A. Bax, R. Freeman and G. A. Morris, *J. Magn. Reson.*, **43**, 333 (1981).

14.  D. van der Helm, E. L. Enwall, A. J. Weinheimer, T. K. B. Karns and L. S. Ciereszko, *Acta Crystallography*, **B32**, 1558 (1976).

15.  R. Sanduja, G. S. Linz, M. Alam, A. J. Weinheimer, G. E. Martin and E. L. Ezell, *J. Heterocyclic Chem.*, **23**, 529 (1986).

16.  M. J. Musmar, M. R. Willcott, G. E. Martin, R. T. Gampe, Jr., M. Iwao, M. L. Lee, R. E. Hurd, L. F. Johnson and R. N. Castle, *J. Heterocyclic Chem.*, **20**, 1661 (1983).

17.  J. C. Carter, G. W. Luther and T. C. Long, *J. Magn. Reson.*, **15**, 122 (1974).

18.  W. J. Chazin, L. D. Colebrok and J. T. Edward, *Can. J. Chem.*, **61**, 1749 (1983).

19.  D. A. Craig and G. E. Martin, *J. Nat. Prod.*, **49**, 456 (1986).

20.  A. S. Zektzer and G. E. Martin, *J. Nat. Prod.*, **50**, 455 (1987).

21.  M. J. Musmar, G. E. Martin, M. L. Tedjamulia, H. Kudo, R. N. Castle and M. L. Lee, *J. Heterocyclic Chem.*, **21**, 929 (1984).

22.  M. J. Quast, G. E. Martin, V. M. Lynch, S. H. Simonsen, J. G. Stuart, M. L. Tedjamulia, R. N. Castle and M. L. Lee, *J. Heterocyclic Chem.*, **23**, 1115 (1986).

23.  M. J. Musmar, G. E. Martin, R. T. Gampe, Jr., M. L. Lee, R. E. Hurd, M. L. Tedjamulia, H. Kudo and R. N. Castle, *J. Heterocyclic Chem.*, **22**, 219 (1985).

24.  D. J. States, R. A. Haberkorn and D. J. Ruben, *J. Magn. Reson.*, **48**, 286 (1982).

25.  D. Marion and K. Wüthrich, *Biochem. Biophys Res. Commun.*, **113**, 967 (1983).

26.  A. G. Redfield and S. Kunz, *J. Magn. Reson.*, **19**, 250 (1975).

27.  G. Bodenhausen, R. Freeman, G. A. Morris, R. Niedermeyer and D. L. Turner, *J. Magn. Reson.*, **25**, 559 (1977).

28.  J. Keeler and D. Neuhaus, *J. Magn. Reson.*, **63**, 454 (1985).

29.  M. P. Williamson, D. Marion and K. Wüthrich, *J. Mol. Biol.*, **173**, 341 (1984).

30.  R. M. Scheek, R. Boelens, N. Russo, J. H. van Boom and R. Kaptein, *Biochemistry*, **23**, 1371 (1984).

31.  D. Neuhaus, G. Wagner, M. Vasak, J. H. R. Kagi and K. Wüthrich, *Eur. J. Biochem.*, **143**, 659 (1984).

32.  J. Feigon, W. Leupin, W. A. Denny and D. R. Kearns, *Biochemistry*, **22**, 5943 (1983).

33.  J. Feigon, A. H.-J. Wang, G. A. van der Marel, J. H. Van Boom and A. Rich, *Nucl. Acids Res.*, **12**, 1243 (1984).

34.  M. A. Weiss, D. J. Patel, R. T. Sauer and M. Karplus, *Proc. Natl. Acad. Sci., USA*, **81**, 130 (1984).

35.  M. A. Weiss, D. J. Patel, R. T. Sauer and M. Karplus, *J. Am. Chem. Soc.*, **106**, 4269 (1984).

36.  M. S. Borido, G. Zon and T. L. James, *Biochem. Biophys. Res. Commun.*, **119**, 663 (1984).

37.  C. L. Perrin and R. K. Gripe, *J. Am. Chem. Soc.*, **106**, 4036 (1984).

38.  C. T. W. Moonen, R. M. Scheek, R. Boelens and F. Muller, *Eur. J. Biochem.*, **141**, 323 (1984).

39.  P. L. Weber, D. E. Wemmer and B. R. Reid, *Biochemistry*, **24**, 4553 (1985).

40.  A. J. Wand and S. W. Englander, *Biochemistry*, **24**, 5290 (1985).

41.  S. W. Fesik, W. H. Holleman and T. J. Perun, *Biochem. Biophys. Res. Commun.*, **131**, 517 (1985).

42.  M. R. Kumar, R. V. Hosur, K. B. Roy, H. T. Miles and G. Govil, *Biochemistry*, **24**, 7703 (1985).

43.  K. V. R. Chary, S. Srivastava, R. V. Hosur, K. B. Roy and G. Govil, *Eur. J. Biochem.*, **158**, 323 (1986).

44.  J. Dabrowski, A. Ejchart, M. Kordowicz and P. Hanfland, *Magn. Reson. Chem.*, **25**, 338 (1987).

45.  G. E. Martin, S. L. Smith, W. J. Layton, M. R. Willcott, M. Iwao, M. L. Lee and R. N. Castle, *J. Heterocyclic Chem.*, **20**, 1367 (1983).

46.  M. J. Musmar, R. T. Gampe, Jr., G. E. Martin, W. J. Layton, S. L. Smith, R. D. Thompson, M. Iwao, M. L. Lee and R. N. Castle, *J. Heterocyclic Chem.*, **21**, 225 (1984).

47.  A. Bax and R. Freeman, *J. Magn. Reson.*, **44**, 542 (1981).

48.  C. Wynants and G. Van Binst, *Biopolymers*, **23**, 1799 (1984).

49.  C. Wynants, G. Van Binst and H. R. Loosli, *Int. J. Peptide Protein Res.*, **25**, 608 (1985); *ibid.*, **25**, 615 (1985).

50.  C. Wynants, G. Van Binst, A. Michel and J. Zanen, *Int. J. Peptide Protein Res.*, **26**, 561 (1985).

51.  L. K. Sydnes and T. Skjetne, *Magn. Reson. Chem.*, **24**, 317 (1986).

52.  R. Sakai, T. Higa and Y. Kashman, *Chem. Lett.*, 1499 (1986).

53.  L. Braunschweiler, G. Bodenhausen and R. R. Ernst, *Mol. Phys.*, **48**, 535 (1983).

54.  S. W. Homans, R. A. Dwek, D. L. Fernandes and T. W. Rademacher, *Biochim. Biophys. Acta*, **798**, 78 (1984).

55.  G. Wagner and E. Zuiderweg, *Biochem. Biophys. Res. Commun.*, **113**, 854 (1983).

56.  L. R. Brown, *J. Magn. Reson.*, **57**, 513 (1984).

57.  R. V. Hosur, M. R. Kumar and A. Sheth, *J. Magn. Reson.*, **65**, 375 (1985).

58.  P. Gundhi, K. V. R. Chary and R. V. Hosur, *FEBS Lett.*, **191**, 92 (1985).

59.  R. V. Hosur, K. V. R. Chary and K. M. Ravi, *Chem. Phys. Lett.*, **116**, 105 (1985).

60.  M. Rance, G. Wagner, O. W. Sørensen, K. Wüthrich and R. R. Ernst, *J. Magn. Reson.*, **59**, 250 (1984).

61.  L. R. Brown, *J. Magn. Reson.*, **57**, 513 (1984).

62.  P. Reveco, J. H. Medley, A. R. Garber, N. S. Bhacca and J. Selbin, *Inorg. Chem.*, **24**, 1096 (1985).

63.  A. Bax, R. A. Byrd and A. Aszalos, *J. Am. Chem. Soc.*, **106**, 7632 (1984).

64.  A. Kumar, R. V. Hosur and K. Chandrasekhar, *J. Magn. Reson.*, **60**, 143 (1984).

65.  D. P. Burum and R. R. Ernst, *J. Magn. Reson.*, **39**, 163 (1980).

66.  O. W. Sørensen, G. W. Eich, M. H. Levitt, G. Bodenhausen and R. R. Ernst, *Prog. Nucl. Magn. Reson. Spectrosc.*, **16**, 163 (1983).

67.  S. Mayor and R. V. Hosur, *Magn. Reson. Chem.*, **23**, 470 (1985).

68.  P. Gundhi, K. V. R. Chary and R. V. Hosur, *FEBS Lett.*, **191**, 92 (1985).

69.  V. Rutar, W. Guo and T. C. Wong, *J. Magn. Reson.*, **69**, 100 (1986).

70.  V. Rutar and T. C. Wong, *J. Magn. Reson.*, **71**, 75 (1987)

71.  J. R. Garbow, D. P. Weitekamp and A. Pines, *Chem. Phys. Lett.*, **93**, 504 (1982).

72.     V. Rutar and T. C. Wong, *J. Magn. Reson.*,**74**, 275 (1987).
73.     K. Nagayama, K. Wüthrich and R. R. Ernst, *Biochem. Biophys. Res. Commun.*, **90**, 305 (1979).
74.     K. Nagayama, A. Kumar, K. Wüthrich and R. R. Ernst, *J. Magn. Reson.*, **40**, 321 (1980).
75.     U. Piantini, O. W. Sørensen and R. R. Ernst, *J. Am. Chem. Soc.*, **104**, 6800 (1982).
76.     M. Rance, O.W. Sørensen, G. Bodenhausen, G. Wagner, R.R. Ernst and K. Wüthrich, *Biochem. Biophys. Res. Commun.*, **117**, 479 (1983).
77.     A. Wokaun and R. R. Ernst, *Chem. Phys. Lett.*, **52**, 407 (1977).
78.     G. A. Morris, *Magn. Reson. Chem.*, **24**, 371 (1986).
79.     M. Edwards and A. Bax, *J. Am. Chem. Soc.*, **108**, 918 (1986).
80.     M. Rance and P. E. Wright, *J. Magn. Reson.*, **66**, 372 (1986).
81.     N. Müller, R.R. Ernst and K. Wüthrich, *J. Am. Chem. Soc.*, **108**, 6482 (1986).
82.     S. W. Homans, R. A. Dwek, J. Boyd, M. Mahmoudian, W. G. Richards and T. W. Rademacher, *Biochemistry*, **25**, 642 (1986).
83.     M. Pais, C. Fontaine, L. Dominique, S. LaBarre and E. Guittet, *Tetrahedron Lett.*, 1409 (1987).
84.     M. C. Fournie-Zaluski, J. Belleney, B. Lux, C. Durieux, D. Gerard, B. Maigret and B. P. Roques, *Biochemistry*, **25**, 3778 (1986).
85.     W. Leupin, W. J. Chazin, S. Hyberts, W. Denny and K. Wüthrich, *Biochemistry*, **25**, 5902 (1986).
86.     J. Dabrowski, A. Ejchart, M. Kordowwicz and P. Hafland, *Magn. Reson. Chem.*, **25**, 338 (1987).
87.     D. S. Williamson, P. Cremonesi, E. Cavalieri, D. L. Nagel, R. S. Markin and S. M. Cohen, *J. Org. Chem.*, **51**, 5210 (1986).
88.     A. Bax, R. Freeman, T. A. Frenkiel and M. H. Levitt, *J. Magn. Reson.*, **43**, 478 (1981).
89.     A. Bax, R. Freeman and T. A. Frenkiel, *J. Am. Chem. Soc.*, **103**, 2102 (1981).
90.     T. H. Mareci and R. Freeman, *J. Magn. Reson.*, **51**, 531 (1983).
91.     J. Boyd, C. M. Dobson and C. Redfield, *J. Magn. Reson.*, **55**, 170 (1983).
92.     J. Boyd, C. M. Dobson and C. Redfield, *J. Magn. Reson.*, **62**, 543 (1985).
93.     C. Dalvit, P. E. Wright and M. Rance, *J. Magn. Reson.*, **71**, 539 (1987).
94.     S. W. Fesik, T. J. Perun and A. M. Thomas, *Magn. Reson. Chem.*, **23**, 645 (1985).
95.     R. Sanduja, G. S. Linz, A.J. Weinheimer, M. Alam and G.E. Martin, *Magn. Reson. Chem.*, **26**, in press (1988).
96.     G. E. Martin and S. C. Welch, *Spectrosc. Lett.*, **21**, 77 (1988).
97.     G. D. Mateescu and A. Valeriu, "Teaching the New NMR: A Computer Aided Introduction to the Density Matrix Treatment of Double Quantum Spectrometry," in *Magnetic Resonance: Introduction, Advanced Topics and Applications to Fossil Energy*, L. Petrakis and J. P. Fraissard, Eds., D. Reidel, Boston, 1984, pp. 501-524.
98.     G. Eich, G. Bodenhausen and R. R. Ernst, *J. Magn. Reson.*, **104**, 3731 (1982).
99.     A. Bax, *Two-Dimensional Nuclear Magnetic Resonance in Liquids*, D. Reidel, Boston, 1982, pp. 132-134.
100.    M. H. Levitt and R. Freeman, *J. Magn. Reson.*, **33**, 473 (1979).
101.    M. H. Levitt, *J. Magn. Reson.*, **48**, 234 (1982).
102.    M.H. Levitt, *J. Magn. Reson.*, **50**, 95 (1982).
103.    T. H. Mareci and R. Freeman, *J. Magn. Reson.*, **48**, 158 (1982).
104.    M. Rance, O. W. Sørensen, W. Leupin, H. Kogler, K. Wüthrich and R. R. Ernst, *J. Magn. Reson.*, **61**, 67 (1985).
105.    B. Blümich, *J. Magn. Reson.*, **60**, 122 (1984).
106.    E. R. P. Zuiderweg, *J. Magn. Reson.*, **66**, 153 (1986).

107. A. Bax, *Two-Dimensional Nuclear Magnetic Resonance in Liquids*, D. Reidel, Boston, 1982, pp. 170-171.
108. D. Piveteau, M. A. Delsuc, E. Guittet and J. Y. Lallemand, *Magn. Reson. Chem.*, **23**, 127 (1985).
109. C. C. Hanstock, J. W. Lown, *J. Magn. Reson.*, **58**, 167 (1984).
110. S. Macura, N. G. Kumar and L. R. Brown, *Biochem. Biophys. Res. Commun.*, **117**, 486 (1983).
111. S. Macura, N. G. Kumar and L. R. Brown, *J. Magn. Reson.*, **60**, 99 (1984).
112. R. Roesch and K. H. Gross, *Z. Naturforsch.*, **39c**, 738 (1984).
113. G. Otting and K. Wüthrich, *J. Magn. Reson.*, **66**, 359 (1986).
114. M. Rance and P. E. Wright, *J. Magn. Reson.*, **66**, 372 (1986).
115. C. Dalvit, M. Rance and P. E. Wright, *J. Magn. Reson.*, **69**, 356 (1986).
116. S. L. Heald, L. Mueller and P. W. Jeffs, *J. Magn. Reson.*, **72**, 120 (1987).
117. G. Pouzard, S. Sukumar and L. D. Hall, *J. Am. Chem. Soc.*, **103**, 4209 (1981).
118. D. L. Turner, *J. Magn. Reson.*, **65**, 169 (1985).
119. L. Müller, *J. Magn. Reson.*, **59**, 326 (1984).
120. A. S. Zektzer, G. E. Martin and R. N. Castle, *J. Heterocyclic Chem.*, **24**, 879 (1987).
121. A. S. Zektzer and G. E. Martin, *J. Nat. Prod.*, **50**, 456 (1987).
122. A. S. Zektzer, J. G. Stuart, G. E. Martin and R. N. Castle, *J. Heterocyclic Chem.*, **23**, 1587 (1986).
123. J. Santoro, F. J. Bermejo and M. Rico, *J. Magn. Reson.*, **64**, 151 (1985).
124. L. D. Hall and T. J. Norwood, *J. Chem. Soc., Chem. Commun.*, 44, (1986).
125. L. D. Hall and T. J. Norwood, *J. Magn. Reson.*, **69**, 391 (1986).
126. L. D. Hall and J. T. Norwood, *J. Magn. Reson.*, **69**, 585 (1986).
127. L. D. Hall and J. J. Norwood, *J. Magn. Reson.*, **74**, 171 (1987).
128. B. H. Ruessink, F. J. J. deKanter and C. MacLean, *J. Magn. Reson.*, **62**, 226 (1984).
129. L. Müller and A. Pardi, *J. Am. Chem. Soc.*, **103**, 4209 (1985).
130. D. E. Barnekow, J. H. Cardellina, II, A. S. Zektzer and G. E. Martin, *J. Am. Chem. Soc.*, submitted (1988).
131. A. S. Zektzer, L. D. Sims, R. N. Castle and G. E. Martin, *Magn. Reson. Chem.*, **26**, 287 (1988).
132. M. J. Musmar, S. R. Khan, A. S. Zektzer, G. E. Martin, and K. Smith, *J. Heterocyclic Chem.*, submitted (1988).

# CHAPTER 3
## HETERONUCLEAR CHEMICAL SHIFT CORRELATION

### CLASSICAL METHODS FOR ESTABLISHING HETERONUCLEAR CORRELATION

Just as the relationship of protons relative to one another in a molecular structure helps to define the structure, so does the location of a given proton relative to a particular carbon. At the simplest level, it is useful to establish the identity of protons that are directly bound to a particular carbon. At the next more complex level, it is useful to identify the presence or absence of other protons in the local environment of the carbon in question. Finally, at perhaps the most complex level, it is very desirable to identify specific protons and the individual associations with a given carbon. The series of operations just described, which increase substantially in complexity from one to the next, allows the molecular connectivity and hence the structure of the molecule to be assembled.

Conventional techniques permit the identification of specific protons that are directly attached to a given carbon by single frequency on- or off-resonance decoupling (SFORD) with $^{13}$C observation. By "walking" the decoupler through the proton spectral window, proton coupled carbon spin multiplets will progressively collapse as the decoupler proximity to the directly attached proton increases. Successive experiments beyond the one in which the $^{13}$C resonance is observed as a singlet will lead to the resumption of heteronuclear spin coupling, with the multiplet finally reattaining its full coupling when the decoupler is sufficiently far removed from the directly attached proton.[1] The residual coupling observed during the series of experiments just described is governed by the expression:

$$^rJ = {}^1J_{CH}\Delta F/(\gamma H_2/2\pi) \qquad [3\text{-}1]$$

where $^rJ$ is the observed residual coupling, $^1J_{CH}$ is the normal direct or one bond heteronuclear spin coupling constant, $\Delta F$ is the offset of the decoupler in hertz from the chemical shift of the directly attached proton and $(\gamma H2/2\pi)$ is the decoupler power measured in hertz (see Chapter 1). The approximation has validity where $(\gamma H2/2\pi) \gg \Delta F$ and $^1J_{CH}$.

Deviations from first order character may be observed when the protons attached to a given carbon are strongly anisochronous.[2] Unfortunately, off-resonance patterns in complex molecules are not always clearly resolved into individual multiplets, which severely limits the utility of this approach.

Protons in the local environment of a given carbon can be probed by an examination of the proton coupled carbon spectrum. Nuclear Overhauser enhancement, which makes the observation easier, can be built up and maintained by decoupling during the interpulse delay. The heteronuclear spin couplings can then be observed by simply gating off the decoupler immediately prior to acquisition. Since protons located three bonds away from a given carbon give rise to relatively large long range couplings ($^3J_{CH}$ ~ 7 Hz), the presence or absence of a proton in this position relative to a carbon may be inferred from the splitting observed for that carbon. As an example, the proton coupled $^{13}$C-NMR spectrum of pyrrolo[3,2,1-$kl$]phenothiazine (**3-1**) is shown in Fig. 3-1.[3] The C4 resonance, which has no proton located three bonds away, since the C2a position is occupied by the fusion of the pyrrole ring to the phenothiazine nucleus, appears as a doublet comprised of the two most intense resonances in the spectrum. Hence, C4 may be assigned by the absence of a three bond coupling. Unfortunately, in the case of a molecule of unknown structure, the utility of proton coupled carbon spectra may be substantially diminished. Furthermore, even regions of the spectrum shown in Fig. 3-1 are showing overlap, which may become intractable with molecules of only slightly greater complexity. The problem of resonance overlap may be circumvented by selective excitation using a DANTE sequence,[4-6] but this approach becomes cumbersome when more than a few carbons must be observed.

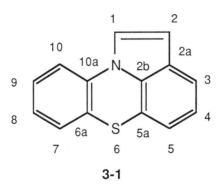

**3-1**

As an alternative to observing long range heteronuclear spin couplings, protons in the vicinity of quaternary carbons may be evaluated from spin-lattice ($T_1$) relaxation time measurements.[3,7] These studies may, however, be complicated by the presence of high abundance nuclides such as $^{14}$N[8] or $^{75}$As.[9] Spin-lattice relaxation measurements, unfortunately, are also time consuming and require extensive sample preparation, these factors serving to limit the utility of this approach as well. Furthermore, relaxation times give information only about the numbers of protons in the local environment and the distance they are removed; they do not provide any information about the specific identity of the protons involved.

Finally, before discussing the experiments that set the stage for the two-dimensional heteronuclear chemical shift correlation experiment, it is possible to identify specific protons relative to a particular carbon by selectively irradiating a proton while observing the carbon spectrum.[10] Here again, the utility of this method will be limited by the complexity of the molecule being studied. Molecules of sufficient complexity to be interesting will almost invariably be intractable to study by this method.

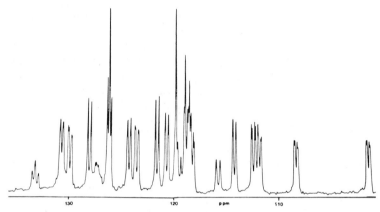

*Fig. 3-1* . Proton coupled $^{13}$C-NMR spectrum of pyrrolo[3,2,1-*kl*]phenothiazine recorded in deuterochloroform at 25.2 MHz.[3]

## SPIN POPULATION INVERSTION/ TRANSFER (SPI/SPT) EXPERIMENTS

Experiments designed to selectively transfer magnetization have been variously termed spin population inversion (SPI)[11] and spin population transfer (SPT).[12] Spin population inversion experiments were, in a sense, the forerunners of the modern heteronuclear chemical shift correlation 2D-NMR experiments. Spin population inversion or SPI experiments employ the selective application of a 180° or π pulse to a $^{13}$C satellite line in the proton spectrum, producing a perturbation in the energy level diagram for the heteronuclear spin system and a resulting, very characteristic perturbation of the $^{13}$C spectrum.

Consider first the energy level diagram (ELD) for a simple AX heteronuclear spin system. A typical example would be chloroform. Populations of the energy levels are shown in Fig. 3-2. The application of a selective 180° pulse to the 2→4 transition gives populations of $\alpha_H\alpha_C$ and $\beta_H\alpha_C$ for levels 2 and 4, respectively. When the perturbed energy levels are sampled by the application of a 90° $^{13}$C observation pulse, the intensities of the transitions will reflect the perturbation just created. Based on the ratios of the gyromagnetic ratios, $\gamma$, let us set $\beta_H$ = +4 and $\alpha_H$ = -4 while $\beta_C$ = +1 and $\alpha_C$ = -1. Using these values in Fig. 3-2, the normal $^{13}$C transitions would each have an algebraic value of +2. In contrast, as a result of the perturbation produced by the selective 180° pulse applied to the 2→4 transition, the 1→2 $^{13}$C transition would assume a net value of +10 while the parallel 3→4 transition assumes a net value of -6. Thus, the proton coupled $^{13}$C spectrum of our AX spin system, rather than being represented by a doublet with limbs of positive and equal intensity, would be a doublet with intensities of +10 and -6.

To illustrate the results of the manipulation just described experimentally, let us consider the assignment of the $^{13}$C-NMR spectrum of the alkaloid oxaline (3-2) reported by Pachler and coworkers.[13] Inversion of the high field C8 proton transition perturbs the multiplets of the resonances at 157.3 and 101.6 ppm. The former may be assigned as C10, the latter as C2 as shown in Fig. 3-3. Inversion of the low field C15 proton transition affects the C12 and C13 multiplets in similar fashion. The same technique has been used for the assignment of the C7 and C11 resonances of aflatoxin B-1 (3-3).[14]

Two very important points are derived from the preceding paragraph and the data presented in Fig. 3-3. First, the SPI/SPT experiments provide the means for specifically identifying proton-carbon heteronuclear coupling pathways. Second, the transfer of magnetization via perturbations induced in the energy level diagram leads to substantial enhancements in the observed $^{13}$C-NMR spectrum. As we shall see in the succeeding sections of this chapter, both of these points are of critical importance to the operation of the proton-carbon heteronuclear chemical shift correlation experiments. However, we must also recall that the SPI/SPT experiments operate on a single coupling pathway, whereas we would prefer to be able to transfer magnetization simultaneously and nonselectively for all direct carbon-hydrogen pairings in a molecule. It is most convenient to introduce the features of the heteronuclear chemical shift correlation experiment in a stepwise fashion. Let us begin with the one-dimensional nonselective polarization transfer experiments.

## NONSELECTIVE POLARIZATION TRANSFER

To lay the foundation for the development of heteronuclear chemical shift correlation, let us first consider nonselective polarization transfer experiments. The first such experiment to be developed was the Insensitive Nucleus Enhancement by Polarization Transfer or INEPT experiment pioneered by Freeman and coworkers.[15] The prototypical

**3 - 2**                                  **3-3**

pulse sequence, which is now seldom used because of more recent refinements, is shown in Fig. 3-4a.

Consider an AX heteronuclear spin system. Transverse proton magnetization is created by the first 90°(x) proton pulse of the pulse sequence shown in Fig. 3-4a which is then allowed to evolve during the first τ interval. The proton 180° pulse will exchange the labels of the two vectors. Applied alone, the vectors would refocus during the second τ interval forming a spin echo after an elapsed time of 2τ, assuming that τ = 1/2($^1J_{CH}$). The 180° carbon pulse applied simultaneously with the proton 180° pulse in the sequence modulates the echo by the heteronuclear coupling. If, however, the interval τ = 1/4($^1J_{CH}$) then the two proton magnetization components will achieve a net phase error of only 180° during the 2τ period and will thus be antiphase along the +x- and -x-axes at the completion of this inteval. Application of the 90°(y) pulse will align the proton magnetization components along the ±z-axis. Note here the similarity developing between the INEPT experiment and the SPI/SPT experiments discussed above. The magnetization component aligned along the -z-axis is analogous to the selective application of a 180° pulse in the SPI/SPT experiments. There is, however, one very significant difference. Alignment of the magnetization component along the -z-axis achieved in the INEPT experiment occurs in a frequency independent manner, whereas the selective 180° pulse in the SPI/SPT experiment must be applied in a frequency dependent fashion. The $^{13}$C magnetization sampled with the final 90° $^{13}$C pulse in the sequence yields a spectrum reflecting the polarization transfer. If the polarization of $^{13}$C is initially negligible, then the $^{13}$C doublet for an AX spin system will have components that are antiphase but equal in intensity. Extending the AX case to its $AX_2$ and $AX_3$ analogs leads to the expectation of intensity ratios of 1:0:-1 and 1:1:-1:-1, respectively.[16]

(b)

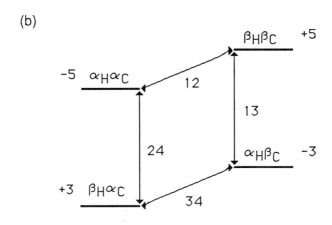

(a)

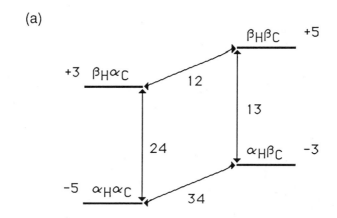

*Fig. 3-2.*    Energy level diagram for a simple heteronuclear AX spin system: (a) system at equilibrium; (b) system perturbed by the application of a selective 180° pulse to the 2→4 transition to invert the populations of levels 2 and 4.    Positive and negative integer values derive from the relationship $\gamma_H \doteq 4$ and $\gamma_C = 1$ making $\alpha_H = -4$, $\beta_H = +4$ and $\alpha_C = -1$, $\beta_C = +1$.    In the case of the unperturbed diagram (a), a 90° $^{13}C$ pulse would sample the magnetization and yield a spectrum containing two responses of equal, positive (+2) intensity.    In contrast, sampling the perturbed energy level diagram with the same 90° $^{13}C$ pulse would yield a spectrum containing two nonequivalent responses, one with an intensity of -6 the other +10.    Application of the selective 180° pulse to the 1→3 transition would ultimately yield nonequivalent responses with intensities of +10 and -6, the reverse of those obtained in the preceding case.

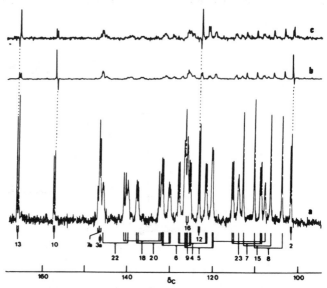

*Fig. 3-3.*    Low field region of the 25.2 MHz $^{13}$C-NMR spectrum of oxaline (**3-2**). (a) Normal proton coupled carbon spectrum. (b) SPI experiment, high field transition of the C8 proton selectively inverted using $T_\pi$ = 0.08 sec. (c) Same as (b) but $\pi$ pulse acting on the low-field transition of the C15 proton. (From D.W. Nagel, K.G.R. Pachler, P.S. Steyn, R. Vleggaar and P.L. Wessels, *Tetrahedron*, **32**, 2625 (1976) with permission.)

Antiphase peak intensities, plus the complexities engendered by multiple peaks for a given carbon, would clearly be unacceptable for heteronuclear correlation work, where there is substantial possibility of peak overlap and hence cancellation.  The peak intensity patterns described above also clearly preclude the use of proton broadband decoupling, which would cancel magnetization.  These difficulties are surmounted, however, by a later version of the INEPT experiment that adds the period 2Δ.[17,18] The modified pulse sequence is shown in Fig. 3-4b and also includes, midway through the 2Δ period, a pair of 180° pulses to suppress phase gradients.  Further, since the components of magnetization are coaligned at the end of the 2Δ period, broadband proton decoupling may be switched on during the acquisition period.

A final point about the INEPT experiment, the effect of the duration of Δ, will also be germane to our discussion of the heteronuclear chemical shift correlation experiment presented below. Importantly, there are differences between methine, methylene and methyl groups that emerge when the duration of 2Δ is considered. First, it should be realized that no single value for Δ will bring all of the components of a methyl quartet into phase.  The total signal intensity (I)

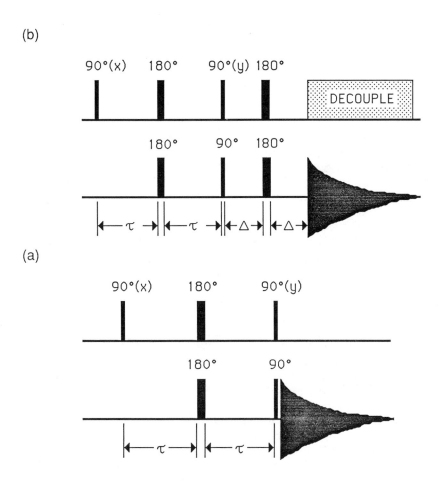

*Fig. 3-4.*   INEPT pulse sequences. (a) Five pulse INEPT sequence. Magnetization created by the simultaneous 90° pulses applied to proton and carbon creates antiphase magnetization, which is recorded directly without benefit of decoupling. (b) Seven pulse INEPT sequence. Antiphase magnetization created as in the previous sequence is refocused during the $2\Delta$ delay, which allows acquisition of the spectral data with the decoupler switched on.

obtained when proton decoupling is applied at the completion of the $2\Delta$ period is given by Eqns. [3-2] to [3-4]. Clearly, from Eqns. [3-2] to [3-4] it is possible to select a compromise value for $\Delta$ that gives good signal intensities for all multiplicity types. In principle this is generally not done with the INEPT experiment. Rather, the periodicity of the three expressions is used to sort carbon types according to their multiplicities. In contrast, for heteronuclear chemical shift correlation, the selection of a compromise value must be employed to produce optimal signal intensity for the various carbon types in the sample. The specifics of this type of compromise will be described in detail below.

CH $\qquad$ $I = \sin [2\pi \, {}^{1}J_{CH} \, \Delta]$ $\qquad\qquad\qquad\qquad$ [3-2]

CH$_2$ $\qquad$ $I = \sin [4\pi \, {}^{1}J_{CH} \, \Delta]$ $\qquad\qquad\qquad\qquad$ [3-3]

CH$_3$ $\qquad$ $I = 0.75 \{ \sin [2\pi \, {}^{1}J_{CH} \, \Delta] + \sin [6\pi \, {}^{1}J_{CH} \, \Delta]\}$ $\qquad$ [3-4]

Before proceeding into the heteronuclear chemical shift experiment itself, it should be noted that the INEPT experiment is not the only magnetization transfer experiment available. Doddrell and coworkers have devised an experiment known as Distortionless Enhancement Polarization Transfer (DEPT - see Fig. 3-21 below), which operates in a fashion quite similar to the INEPT experiment described above.[19,20] Excellent descriptions of the operation of both INEPT and DEPT experiments are found in a recent review[21] and in several recent chapters[22,23] to which the reader requiring more detail is referred.

## PROTON-CARBON HETERONUCLEAR
## CHEMICAL SHIFT CORRELATOIN

The basic proton-carbon heteronuclear chemical shift correlation experiment is reasonably complex in that it provides, in addition to heteronuclear chemical shift correlation, the means for heteronuclear decoupling in both frequency domains. To explain this experiment, we have found it best to develop the experiment in several steps, beginning with heteronuclear correlation itself and then adding heteronuclear decoupling in two subsequent steps.

The earliest work on the subject of heteronuclear chemical shift correlation was reported by Ernst and coworkers.[24] Their work differed from the now commonly employed heteronucleus detected experiment in that they utilized proton observation (indirect detection), which is technically much more difficult than its converse. Two additional proton detected experiments were described by Ernst and colleagues[25] before the work from Freeman's group, which provided the fundamental developments eventually resulting in the heteronucleus detected experiment as we now know it. We shall fully develop the heteronucleus detected experiment and its variants first before considering the more demanding indirect detection experiments pioneered by Ernst and coworkers.[24,25]

# Heteronuclear Chemical Shift Correlation:
# The Basic Experiment; Heteronuclear Spin
# Coupling on Both Frequency Axes

Scalar heteronuclear spin coupling provides the basis for heteronuclear chemical shift correlation in a manner analogous to that in which it is exploited to determine resonance multiplicities in the INEPT and DEPT experiments. Although Ernst and coworkers[24,25] had described several indirect detection schemes for performing heteronuclear chemical shift correlation, these experiments were much slower to gain favor than the heteronucleus detected experiments that evolved from the initial communication of Bodenhausen and Freeman.[26]

The pulse sequence for the experiment described by Bodenhausen and Freeman[26] is shown in Fig. 3-5. The first 90° proton pulse creates transverse proton magnetization in the same sense as in the INEPT experiment described above. If we consider, once again, a simple heteronuclear AX spin system, the components of proton magnetization will diverge through the evolution period, $t_1$, which follows the first proton pulse. This behavior is illustrated for two different evolution times in the top series of vector diagrams in Fig. 3-6 (a-d). At the end of the evolution period (Fig. 3-6c or 3-6d), the second $90°_x$ proton pulse is applied to the evolved proton magnetization. In the case of the evolved state of magnetization shown in Fig. 3-6c, the second $90°_x$ pulse will rotate component $M_{24}$ toward the +z-axis (equilibrium) while the $M_{13}$ component is rotated toward the -z-axis (non-equilibrium) as shown in Fig. 3-6e. The opposite orientation of the magnetization is achieved with a somewhat longer evolution period shown in Fig. 3-6d. The second $90°_x$ proton pulse operating on the magnetization shown in Fig. 3-6d will produce component orientations shown in Fig. 3-6f which are exactly opposite that shown in Fig. 3-6e. Clearly from Fig. 3-6, it can be seen that the systematic incrementation of the evolution period, $t_1$, will result in the encoding of a frequency dependent oscillation of the components of proton magnetization and the recorded signal from the heteronucleus.

Sampling of the evolved spin state by the application of the 90° $^{13}C$ pulse in the pulse sequence shown in Fig. 3-5 will afford perturbed, proton coupled $^{13}C$ akin to those of the INEPT experiment described above. In total, as shown by Bax[27] for chloroform in Fig. 3-7, the experiment will afford a proton-carbon heteronuclear chemical shift correlation spectrum containing a total of four responses. The total number of responses is due to heteronuclear spin coupling in both frequency domains, and intensities will be ±1 as before.[16] Clearly, such complexity for a molecule as simple as chloroform would render the spectrum of a much more compelx molecule hopelessly complicated. Realizing this, Bodenhausen and Freeman[26] managed to simplify their spectra somewhat by the application of coherent single frequency

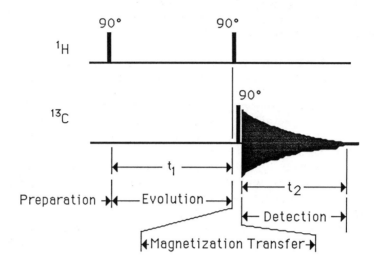

*Fig. 3-5.*   Basic pulse sequence for heteronuclear chemical shift correlation with heteronuclear spin coupling in both frequency domains. The first 90° proton pulse creates transverse proton magnetization, which evolves during the $t_1$ interval under the influence of the one bond heteronuclear spin coupling ($^1J_{CH}$). The second 90° proton pulse rotates antiphase components of magnetization (see Fig. 3-6) onto the z-axis creating a non-equilibrium condition analogous to that created in the INEPT sequence, which is sampled by the 90° carbon pulse.

decoupling during the acquisition period to partially collapse couplings. A small, residual coupling was allowed to remain of sufficient magnitude to prevent the mutual cancellation of signals, which would occur if full broadband homonuclear proton decoupling were applied. Equivalent simplification was achieved during the evolution period, $t_1$, by the introduction of a 180° $^{13}C$ refocusing pulse at time = $2t_1/3$. In a slightly later paper, Bodenhausen and Freeman[28] described basically the same experiment but deleted the carbon 180° refocusing pulse.

### Heteronuclear Chemical Shift Correlation: Heteronuclear Decoupling in the $F_2$ Frequency Domain

Initial work by Bodenhausen and Freeman[26,28] had demonstrated the feasibility of partial collapse of the multiplet structure in heteronuclear chemical shift correlation spectra. Only a partial collapse was permissible using the pulse sequence shown in Fig. 3-5, since broadband decoupling would lead to the mutual cancellation of the signal. Broadband proton decoupling may be introduced, however, if appropriate care is taken.

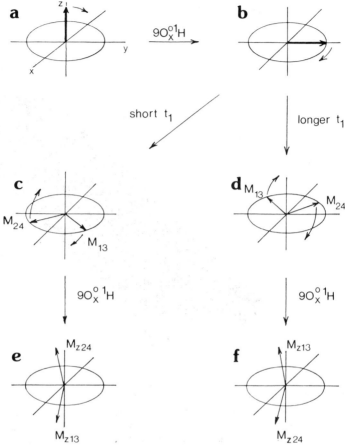

Fig. 3-6.   Vector diagram illustrating the pulse sequence shown in Fig. 3-5.   (a) Equilibrium proton magnetization is rotated by the first 90° pulse to the y-axis. (b) Precession begins to occur in the xy-plane.  In the case of an AX heteronuclear spin system, eg, chloroform, the proton vector ensemble will be comprised of two components, which will defocus as a result of the heteronuclear spin coupling.   (c) Possible orientation of the proton vectors $M_{13}$ and $M_{24}$ after a short evolution period.   (d) Plausible orientation of the proton vectors after longer evolution. Non-equilibrium proton magnetization caused by the action of the second 90° pulse is illustrated in (e) and (f) which are derived from orientations shown in (c) and (d), respectively.   In (e) component $M_{13}$ has experienced a rotation analogous to the 180° selective pulse applied to the energy level diagram shown in Fig. 3-2.   In (f) component $M_{24}$ has experienced the 180° rotation.   Nonequilibrium magnetization conditions in (e) and )f) are ready to be sampled by the 90° carbon pulse (Fig. 3-5) which will complete the magnetization transfer from proton to carbon. The spectrum of chloroform shown in Fig. 3-7 was recorded with the pulse sequence shown in Fig. 3-5.

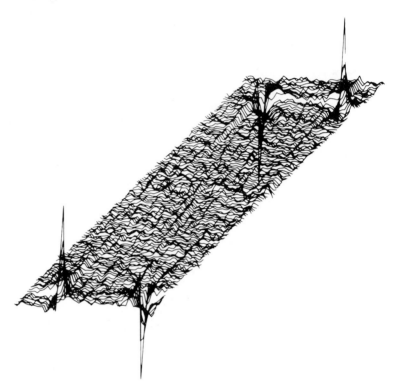

*Fig. 3-7.*  Heteronuclear proton-carbon chemical shift correlation spectrum of chloroform. The data is shown as a stack plot to illustrate the positive and negative phase of the responses. The spectrum was recorded using the pulse sequence shown in Fig. 3-5 at observation frequencies of 300.068 and 75.459 MHz for $^1H$ and $^{13}C$, respectively. Total acquisition time was about 30 minutes.

Enabling the decoupler immediately following the 90° carbon pulse would be experimentally fatal, since the carbon magnetization at that point in time is antiphase in character and would undergo mutual cancellation. However, by introducing a fixed duration delay, which we will label $\Delta_2$ (the delay $\Delta_1$ will be reserved for use in the following section), the components of antiphase carbon magnetization will come into phase after a period of time, $\Delta_2 = 1/2(^1J_{CH})$, in the case of a simple methine carbon "doublet." With the components of magnetization in phase, the decoupler may be gated on and the beneficial effects of broadband proton decoupling enjoyed. This modification of the pulse sequence, incorporating the $\Delta_2$ interval, is shown in Fig. 3-8. Experimentally, the result of broadband proton decoupling is to collapse the four responses contained in the spectrum of chloroform shown in Fig. 3-7 into the two responses shown in Fig. 3-9. As will be noted, the heteronuclear spin coupling has been removed from the $F_2$ axis and only the heteronuclear coupling in $F_1$ remains.

## Heteronuclear Chemical Shift Correlation: Broadband Heteronuclear Decoupling in the $F_1$ Frequency Domain

With the capability of broadband proton decoupling during the acquisition period, $t_2$, facilitated by the insertion of the fixed duration delay, $\Delta_2$, we may now turn our attention to the final stage of modification of the original pulse sequence, heteronuclear decoupling in the proton or $F_1$ frequency domain. Clearly, application of broadband $^{13}C$ decoupling is not feasible even at the present time, since only a small percentage of the spectrometers now in operation have X-band decoupling capabilities. The alternative is the insertion of a 180° pulse midway through the evolution period, $t_1$. Practically, a 180° pulse midway through $t_1$ cannot be added to the pulse sequence shown in Fig. 3-8 because complete refocusing would result in cancellation of the signal. For this reason, Bodenhausen and Freeman[26] initially inserted a 180° pulse displaced from the midpoint of the evolution period at time = $2t_1/3$, which gave only partial decoupling. The modification described by Bodenhausen and Freeman[26] affords a partial collapse of the heteronuclear spin coupling in the $F_1$ frequency domain. Freeman and Morris[29] solved the problem of cancellation with the modified pulse sequence shown in Fig. 3-10. A 180° refocusing pulse may be used at time = $t_1/2$ provided that the magnetization is allowed to defocus after evolution during a fixed duration delay, $\Delta_1$. During the evolution period, $t_1$, protons are labeled, as necessary, with their characteristic precession frequencies. The 180° $^{13}C$ pulse applied midway through $t_1$ decouples the heteronuclear spin coupling in the proton time domain ($t_1$), which will eventually become the frequency domain, $F_1$. Components of proton magnetization will be refocused at the completion of $t_1$, which would normally preclude magnetization transfer. However, during the fixed duration delay, $\Delta_1 = 1/2(^1J_{CH})$, defocusing occurs to provide a state of magnetization analogous to that portrayed in Fig. 3-6c and 3-6d. Subsequent application of the second 90° proton pulse transfers magnetization in the usual fashion. The 90° carbon pulse followed by the $\Delta_2$ delay samples the magnetization and brings it into phase for broadband decoupling during the acquisition period, $t_2$.

Experimentally, the results of the cumulative modifications developed by Freeman and his students through the series of papers just described[26,28,29] are shown in Fig. 3-11. Broadband heteronuclear decoupling has now been applied in both frequency domains and a single response, for chloroform, is observed at the intersection of the chemical shift frequencies. The pulse sequence shown in Fig. 3-10 is one of several now in usage for the acquisition of heteronuclear

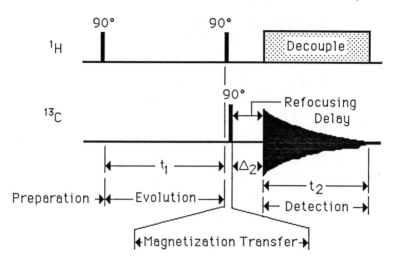

Fig. 3-8 . Modified heteronuclear chemical shift correlation pulse sequence to provide the capability of broadband proton decoupling in the carbon ($F_2$) frequency domain. The fixed duration delay, $\Delta_2$, is used to bring antiphase components of carbon magnetization back into phase so that the decoupler may be gated on. If the decoupler were gated on immediately after transfer, cancellation of the magnetization would occur because of its antiphase configuration. The experimental result of the modification shown in this pulse sequence, collapse of the heteronuclear spin coupling in the $F_2$ frequency domain, is shown for chloroform in Fig. 3-9.

chemical shift correlation spectra. The sequence devised by Freeman and Morris[29] is perhaps the most widely utilized, but this statement is conjectural as no definitive survey on relative usage of the various heteronuclear chemical shift correlation experiments has been reported.

## Phase Cycling

Phase cycling for the heteronuclear chemical shift correlation experiment described above is relatively simple and straightforward. Ideally we would like to be able to manipulate phases in such a way as to afford quadrature detection in both frequency domains. Bax and Morris[30] described a method of achieving this goal in 1981.

In the simplest form of the phase cycle, the four step cycle, the phases of the first proton pulse and both carbon pulses may be held constant while the phase of the second 90° proton pulse and the receiver are cycled in a four step cycle. The four step cycle of the receiver phase provides quadrature detection of the carbon spectrum ($F_2$) while the cycling of the second 90° proton pulse ultimately provides quadrature detection of the encoded proton frequency information. In addition, the phase cycling scheme just described and shown in tabular form in Table 3-1 also provides the means of suppressing undesirable

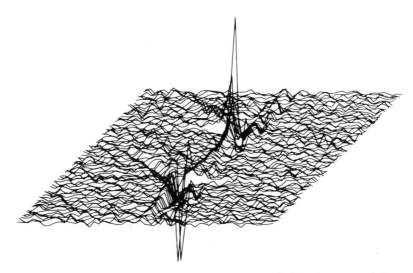

*Fig. 3-9.*    Heteronuclear chemical shift correlation spectrum of chloroform recorded
using the pulse sequence shown in Fig. 3-8. The duration of the fixed
duration refocusing delay, $\Delta_2$, was set to $1/2(^1J_{CH}) = 2.3$ msec to
accommodate the 216 Hz one bond coupling of chloroform. Note the
collapse of the heteronuclear spin coupling in the carbon or $F_2$ frequency
domain relative to the spectrum shown in Fig. 3-7.

axial responses. Finally, as will be noted from Table 3-1, the
investigator has a choice of receiver phase cycles. The receiver phase
cycle designated $\phi_A$ selects for the acquisition of the antiecho while the
phase cycle designated $\phi_E$ permits the acquisition of the coherence
transfer echo.

With regard to choosing which of the receiver phase cycles to
employ, the acquisition of the coherence transfer echo ($\phi_E$) is generally
preferred for reasons treated in detail by Bax.[31] For purposes of this
discussion, it is sufficient to note that the coherence transfer echo
generally affords somewhat higher sensitivity in the case of field
inhomogeneity. It should also be noted, however, that the acquisition of
the coherence transfer echo effectively rotates the proton spectrum
about the transmitter on the $F_1$ axis by 180°, thus necessitating a rotation
of the data matrix prior to plotting if proper orientation of the proton
spectrum is to be achieved. Practically, this last point may limit some
individuals to acquisition of the antiecho if their spectrometer software
does not provide a command for spectral rotation. This final point will be
discussed further below in the section describing data processing.

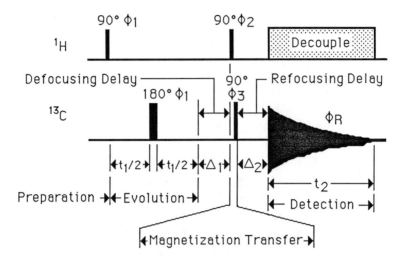

*Fig. 3-10.*    Contemporary heteronuclear chemical shift correlation pulse sequence providing broadband heteronuclear decoupling in both frequency domains. As explained in the text, the 180° pulse midway through the evolution period serves to refocus the proton magnetization vectors at time = $t_1$. Normally, this condition would preclude magnetization transfer. However, introduction of the fixed duration defocusing delay, $\Delta_1$, allows the vectors to assume an antiphase orientation required for transfer to occur. Since the $\Delta_1$ delay duration is constant, it effectively becomes transparent insofar as the spectral result is concerned. Phase cycling for the pulse sequence is specified in Table 3-1 and provides quadrature detection in both frequency domains.

We have found the eight step modification of the four step phase cycle shown in Table 3-1 to afford some improvement in the quality of the data. Phases of all of the pulses except for the carbon 90° pulse are cycled, again with the choice of detecting either the antiecho ($\phi_A$) or the coherence transfer echo ($\phi_E$). The eight step sequence could, in those cases where long data accumulations are necessary, be extended to a thirty-two step cycle by superimposing a cyclops cycle[32] over the eight step phase cycle shown.

## Parameter Selection

The number of parameters that must be selected for the two-dimensional heteronuclear chemical shift correlation experiment is relatively small. Pulse widths (whose calibration was described in Chapter 1) must be specified, a dwell time for the evolution period must be chosen and an appropriate level of digitization in $F_1/F_2$ decided, optimal values for the fixed delays, $\Delta_1$ and $\Delta_2$, must be selected and finally an interpulse delay of sufficient duration must be specified. Each

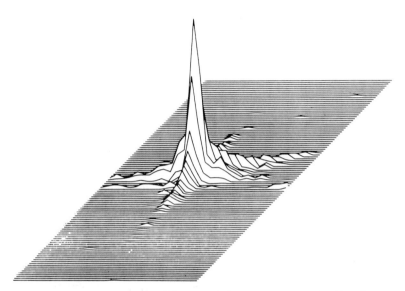

*Fig. 3-11.* Heteronuclear chemical shift correlation spectrum of chloroform recorded using the pulse sequence shown in Fig. 3-10 and the phase cycling shown in Table 3-1. Note that the heteronuclear spin coupling in the proton or $F_1$ frequency domain in the spectrum shown in Fig. 3-9 has been collapsed by the introduction of the 180° pulse midway through the evolution period and the fixed duration defocusing delay, $\Delta_1$. In more complex molecules, proton-proton homonuclear couplings will be preserved when the pulse sequence shown in Fig. 3-10 is employed. Broadband homonuclear proton decoupling can, however, be introduced by further modification of the pulse sequence as discussed below and illustrated in Figs. 3-25 and 3-30.

of these features of the experiment will be dealt with individually in the following subsections.

## Pulse Widths

The calibration of the individual pulses from the transmitter and the decoupler was described in detail in Chapter 1, to which the reader is referred for specifics.

## Dwell Time Selection

By examining the pulse sequence shown in Fig. 3-10, we note that the evolution time, $t_1$, is split into two segments by the inclusion of the 180° carbon pulse, which provides heteronuclear decoupling in the $F_1$ frequency domain (compare with Fig. 3-9). Thus, the acquisition of a

Table 3-1.    Phase cycling for the heteronuclear chemical shift correlation pulse sequence shown in Fig. 3-10.

| Acquisition | $\phi_1$ | $\phi_2$ | $\phi_3$ | Receiver Phases[a] | |
|---|---|---|---|---|---|
| | | | | $\phi_A$ | $\phi_E$ |
| **Four Step** | | | | | |
| 1 | 0 | 0 | 0 | 0 | 0 |
| 2 | 0 | 1 | 0 | 1 | 3 |
| 3 | 0 | 2 | 0 | 2 | 2 |
| 4 | 0 | 3 | 0 | 3 | 1 |
| **Eight Step** | | | | | |
| 1 | 0 | 0 | 0 | 0 | 0 |
| 2 | 2 | 1 | 0 | 1 | 3 |
| 3 | 0 | 2 | 0 | 2 | 2 |
| 4 | 2 | 3 | 0 | 3 | 1 |
| 5 | 2 | 2 | 0 | 0 | 0 |
| 6 | 0 | 3 | 0 | 1 | 3 |
| 7 | 2 | 0 | 0 | 2 | 2 |
| 8 | 0 | 1 | 0 | 3 | 1 |

[a] The receiver phase designated as $\phi_A$ allows the detection of the antiecho while $\phi_E$ allows detection of the coherence transfer echo.

proton reference spectrum will define the proton spectral width which is to be digitized. The dwell time (DW) derives from sampling theory and is defined quite simply by Eqn. [3-5]

$$DW = 1/SW \qquad [3-5]$$

where SW defines the total sweep width. To afford the same sweep width as that of the proton reference spectrum, the evolution period in heteronuclear chemical shift experiment (Fig. 3-10) must be incremented by a period equal to the proton dwell time in each successive block of data acquired. Further, since the evolution time, $t_1$, is split by the 180° carbon pulse, the individual segments must initiallly be set to one half the dwell time and also incremented by this period of time.

## Digitization in the $F_1$ Frequency Domain

The level of digitization of the proton or $F_1$ frequency domain is determined by the number of increments of the evolution time, $t_1$, which are to be performed. Generally, the factor that governs the digitization will be the density of responses in the proton spectrum. As the complexity and degree of overlap in the proton spectrum increase, the need for finer levels of digitization correspondingly increases. There are, however, finite limits, which are generally imposed by the quantity of material in the sample tube and the amount of time available for the acquisition of the experimental data. Hence, it is generally useful to record between 64 and 256 increments of the evolution period with zero filling prior to the second Fourier transformation, if disc storage capacity of the spectrometer permits it, to afford a final matrix digitized by 512 points in the $F_1$ frequency domain.

## Digitization of the $F_2$ Frequency Domain

Desirable levels of digitization in the carbon or $F_2$ frequency domain are variable and, once again, governed by the complexity of the spectrum. Generally, the goal is to have sufficient resolution to differentiate individual resonances from one another. We find it useful to record survey proton-carbon heteronuclear chemical shift correlation spectra using 1K-2K complex points to digitize $F_2$, which will afford 512 x 1K real points after Fourier transformation.

## Optimal Fixed Delay ($\Delta_1$ and $\Delta_2$) Durations

By virtue of the incorporation of the 180° carbon pulse midway through the evolution period, $t_1$, heteronuclear spin vectors are refocused at the end of the evolution period which would normally preclude magnetization transfer. To circumvent the adverse effect of refocusing, it is necessary to insert the fixed delay, $\Delta_1$, into the pulse sequence as shown in Fig. 3-10. If the duration of $\Delta_1$ is selected as a function of the heteronuclear spin coupling, a controlled degree of dephasing can be introduced. The optimal setting for $\Delta_1$ is given by Eqn. [3-6].

$$\Delta_1 = 1/2(^1J_{CH}) \qquad\qquad [3\text{-}6]$$

The duration of $\Delta_1$ defined by Eqn. [3-6] is optimal, since the heteronuclear magnetization vectors will attain a maximal phase error of 180° when the condition $1/2(^1J_{CH})$ is satisfied. Under this constraint, the transfer of magnetization from proton to carbon will correspondingly be maximized.

Obviously, some compromise value for $\Delta_1$ must be selected, since there will almost invariably be a range of one bond heteronuclear spin coupling constants encountered in any given sample. In the case of purely aliphatic compounds such as terpenes, coupling constants will range from a low of about 125 Hz to approximately 145 Hz for carbons bearing functional groups such as a hydroxyl. Assuming an average one bond coupling constant of 135 Hz in this case, the optimal setting for $\Delta_1$ would probably be in the vicinity of 3.7 msec. For compounds that also contain vinyl moieties whose one bond couplings range up to about 165 Hz, a somewhat smaller value of $\Delta_1$ would be desirable, perhaps about 3.45 msec, which corresponds to a 145 Hz average one bond coupling. In the case of purely aromatic compounds, where the average one bond couplings are large and generally range from 165 to 185 Hz, a shorter duration for $\Delta_1$, perhaps 2.85 msec for a 175 Hz average, would be optimal. In the case of an unknown structure, a useful survey setting for the $\Delta_1$ delay is about 3.2 msec which corresponds to an average coupling of 155 Hz, which is sufficient for anything from about 125 Hz on the low end to a heteroaromatic one bond coupling of up to 185 Hz on the high end.

Optimization of the $\Delta_2$ delay may be handled in a fashion at least somewhat similar to the considerations implicit in the optimization of the $\Delta_1$ interval. During the $\Delta_2$ interval, however, the multiplicity of the carbon comes into question and different optima arise for different degrees of protonation. As shown in Fig. 3-12, response intensity for a methine carbon is at its maximum when the duration of $\Delta_2 = 1/2(^1J_{CH})$. Unfortunately, under the condition just defined, it will be noted that the response intensity for both methylene and methyl carbons goes to zero, clearly an undesirable situation if a molecule such as a steroid is being studied. Fortunately, however, it will be noted from Fig. 3-12 that a compromise setting of $\Delta_2$ that affords reasonable signal intensity for all three possible species of protonated carbons is obtained when:

$$\Delta_2 = 1/3(^1J_{CH}) \qquad [3\text{-}7]$$

Hence for survey conditions, using the 155 Hz average coupling described for $\Delta_1$ above, the optimal setting for $\Delta_2$ will be approximately 2.1 msec. In contrast, for aromatic compounds that contain only quaternary and methine carbons, the duration of $\Delta_2$ should be set equal to that of $\Delta_1$, giving an optimal setting of 2.85 msec for an average couping of 175 Hz.

### Selection of an
### Interpulse Delay

One significant advantage of the heteronuclear chemical shift correlation experiment over conventional $^{13}C$ spectral acquisition is in the selection of the interpulse delay. Since magnetization is being transferred from $^1H$ to $^{13}C$, the optimal duration for the interpulse delay will be governed by the spin-lattice or longitudinal relaxation time of the proton(s) involved in the process. This is in contrast to conventional $^{13}C$ data acquisition, which must normally be cycled as a function of the generally longer inherent relaxation times of $^{13}C$. Normally an interpulse delay ranging from about 1.3 to 1.5 x $T_1(^1H)$ will be optimal.[33] In the absence of a determined proton relaxation time, we generally have found it convenient to recycle the experiment with a 1-1.5 sec interpulse delay plus the acquisition time.

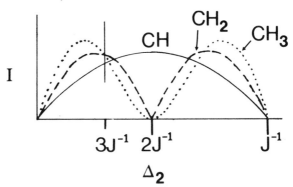

*Fig. 3-12.*    Graphical illustration of the dependence of response intensity, I, in a heteronuclear chemical shift correlation experiment as a function of the duration of the refocusing delay, $\Delta_2$, and the multiplicity of the carbon being observed. Methine carbons are shown by the solid line, methylene carbons by the dashed line and methyl carbons by the dotted line. Effects of Overhauser enhancement and relaxation are ignored in the computation of the curves.

## Data Processing

After the acquisition of a two-dimensional heteronuclear chemical shift correlation data set, attention must next be directed to processing the data for presentation. Many of the same constraints imposed on the processing of proton-proton connectivity experiments apply to heteronuclear correlation experiments. We will deal with the selection of processing parameters for each operation in turn below.

To illustrate the processing, we will utilize a heteronuclear chemical shift correlation data set that was acquired using the simple alkaloid norharmane (**3-4**) ($\beta$-carboline). This compound was selected because of its ready availability from many chemical suppliers and its good solubility and stability in $d_6$-dimethyl sulfoxide. The data whose processing is shown below was acquried using a 100 mg sample of **3-4** dissolved in 0.5 ml of $d_6$-dimethyl sulfoxide in a 5 mm sample tube.  A total of 96 blocks of data were acquired with 64 acquisitions/block in approximately 90 minutes.  The data was acquired using the eight step phase cycle shown in Table 3-1 with the receiver phase cycled to collect the coherence transfer echo.

**3-4**

## The First Fourier Transformation

Given the respective chemical shift ranges for $^1H$ and $^{13}C$, the sweep width necessitated for the latter will generally be several to many times that required for the former.  The difference in sweep widths hence necessitates somewhat different considerations in the selection of weighting parameters for the two Fourier transformations that must be performed.

To minimize the noise floor in the data matrix, it is generally possible to tolerate some broadening of the carbon resonances if the spectrum is reasonably well resolved.  One alternative would be to use an exponential multiplication that is the reciprocal of the acquisition time in $t_2$.  If somewhat better resolution is required, double exponential apodization represents a useful alternative.

Irrespective of the weighting used in performing the first Fourier transformation, a series of phase modulated spectra will result.  The series of phase modulated spectra arising from the first Fourier transformation of a heteronuclear chemical shift correlation data set collected for norharmane (**3-4**) is shown in Fig. 3-13.  A normal carbon reference spectrum is plotted beneath the stack plotted spectra to illustrate the location of the protonated carbon resonances relative to the quaternary carbons, which do not give a response in the heteronuclear correlation experiment when optimized for transfer via   direct heteronuclear coupling. The spectra plotted above the reference spectrum were processed using a double exponential apodization. Focusing our attention on the protonated carbon furthest downfield, which was intentionally phased to give a positive response in the first trace, it is quite easily to see the nature of the phase modulation occurring from the first evolution time to each successively incremented duration.  It should also be noted that the phase of the spectra at this stage in the data processing is completely arbitrary. The sole requirement is that a constant phase correction be applied to each spectrum generated by the first Fourier transformation.

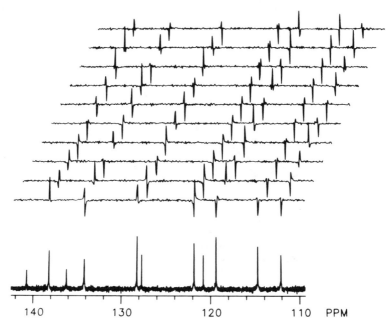

*Fig. 3-13.*    A carbon reference spectrum (bottom trace) of the alkaloid norharmane
(**3-4**) is shown with phase modulated traces taken from a series of
$S(t_1, F_2)$ files from a heteronuclear correlation data set obtained using
the pulse sequence shown in Fig. 3-10.  The phase of the first trace
above the reference spectrum was arbitrarily set to afford positive
peak intensity for the carbon resonating furthest downfield.  All other
traces have had the same phase correction applied.  It should also be
noted that the quaternary carbon resonances in the reference spectrum
do not produce any signal in the heteronuclear correlation experiment.

The oscillatory nature of the phase modulation of the resonance located furthest downfield is also shown in the stack plotted series of spectra presented in Fig. 3-14. This view gives us what is essentially a preview of what we will see in the interferogram after transposition. At this point, we are ready to transpose the data matrix to prepare for the second Fourier transformation.

## Transposition

The first Fourier transformation operates with respect to the $t_2$ time domain as shown in Eqn. [2-1]. After the first Fourier transformation, it is necessary to bring the data matrix into a form amenable to further processing.  Hence, it is necessary to transpose the data matrix as shown in Eqn. [2-2] to allow the computer to operate with respect to the $t_1$ time domain during the second Fourier transformation.

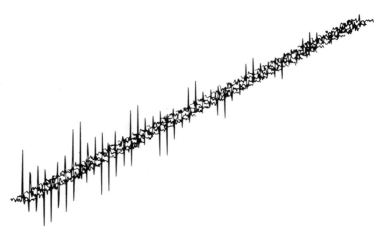

Fig. 3-14    Stack plotted series of spectra (48) of the protonated carbon resonating furthest downfield in the spectrum shown in Fig. 3-13 to illustrate the oscillatory nature of the resonance as a function of the incremented evolution time.    Proton frequency and coupling information is encoded in the oscillation in $t_1$ shown.

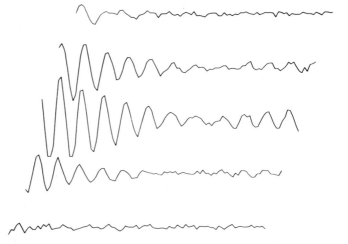

Fig. 3-15.    Interferograms extracted from the region of the downfield protonated resonance shown in Figs. 3-13 and 3-14.

The encoded proton frequency information contained in the series of spectra plotted in Fig. 3-14 results in the series of interferograms shown in Fig. 3-15 after transformation.    The interferograms shown again correspond to the protonated carbon resonating furthest downfield in the spectrum shown in Fig. 3-13.

## The Second Fourier Transformation

Following transposition the data are amenable to Fourier transformation as shown in Eqn. [2-3]. Fourier transformation of the interferograms in the region of those shown in Fig. 3-15 give rise to the single response shown as a white washed stack plot in Fig. 3-16. The interferograms were subjected to exponential apodization prior to Fourier transformation and a magnitude calculation prior to plotting.

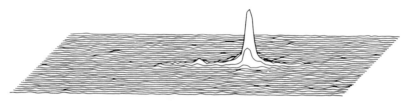

*Fig. 3-16.*    Whitewashed stack plot of the heteronuclear correlation response for the protonated carbon resonating furthest downfield in the reference spectrum shown in Fig. 3-13.

## Other Processing Considerations

As noted in the section above, which describes phase cycling considerations, cycling the phase of the receiver to collect the coherence transfer echo necessitates the rotation of the final data matrix about the $F_1 = 0$ axis. To illustrate this rather important point, the data matrix for norharmane (**3-4**) was first processed without the spectrum rotation. The data at this stage of the processing is shown in Fig. 3-17 as a contour plot. Horizontally beneath the contour plot the 0° projection of the data matrix that recovers the proton spectrum is shown. Compare the orientation of the spectrum to the normal orientation shown in the high resolution spectrum which is plotted horizontally above the contour plot. It will be noted that, as expected, the H1 singlet resonates furthest downfield in the normal spectrum. In contrast, the 0° projection shows the H1 resonance to be furthest upfield. This orientation represents a 180° rotation about the $F_1 = 0$ axis.

To reestablish the normal orientation of the proton or $F_1$ frequency domain, we must rotate the spectrum shown in Fig. 3-17 by 180°. Using the Nicolet spectrometer on which the data was collected and processed, this task is easily accomplished, and the result is shown in Fig. 3-18. As noted in the discussion of phase cycling above, instruments lacking the capability of the rotating the spectrum will be restricted to a receiver phase cycle in which the antiecho is collected. In the case of data collected with phase cycling to accumulate the antiecho, the result following the second Fourier transformation will be that shown in Fig. 3-18, the rotation, which is normally performed automatically following the second Fourier transformation, obviated.

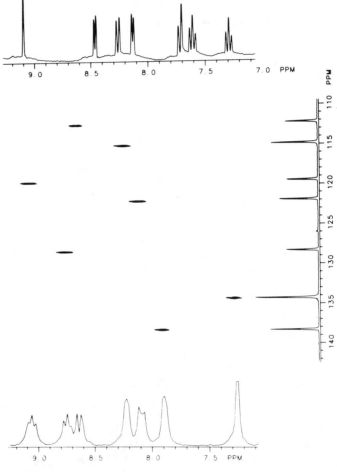

*Fig. 3-17.*   Heteronuclear correlation spectrum of **3-4** obtained by Fourier transformation of the data set without performing a spectrum rotation when the coherence transfer echo was recorded (see Table 3-1 and the discussion of phase cycling presented above). The spectrum is flanked above and below by the proton reference spectrum and the proton ($F_1$) projection, respectively. Note the rotation of the projected spectrum about the carrier relative to the reference spectrum. The projected carbon spectrum ($F_2$) is plotted vertically along the contour plot.

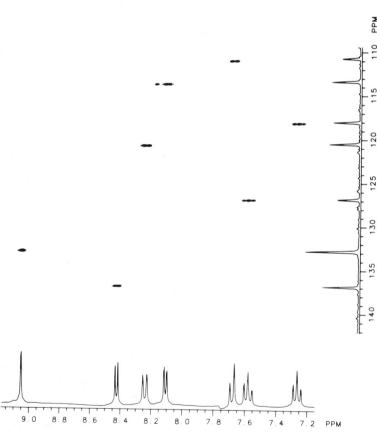

*Fig. 3-18.* Heteronuclear correlation spectrum of **3-4** recorded as above but processed using a spectrum rotation during the second Fourier transformation to reestablish the correct orientation of the proton ($F_1$) frequency domain.

## Artifacts and Spurious Responses

Heteronuclear chemical shift correlation spectra, as a class of experiments, are comparatively free of misleading responses. There are, however, some circumstances that can lead to the generation of spurious signals. Several categories of artifacts or unusual responses will be described in the following subsections. We shall deal first with responses arising due to heteronuclear "virtual coupling" which are the best studied and perhaps have the opportunity of being the most prevalent type of artifact in heteronuclear chemical shift correlation spectra.

## "Virtual Coupling" in Heteronuclear
## Chemical Shift Correlation Spectra

Perhaps the most likely source of artifacts in heteronuclear correlation spectra are those that can arise due to heteronuclear "virtual coupling." Generally, when heteronuclear chemical shift correlation spectra are recorded it is assumed that proton-proton coupling will be weak. In the weak coupling case, the 180° $^{13}$C pulse midway through the evolution time (see Fig. 3-10) exchanges the $\alpha$ and $\beta$ spin states, causing them to refocus at the end of the evolution period. In the weak coupling case, the defocusing interval that follows evolution, $\Delta_1$, serves to dephase proton magnetization by 180° to prepare it for magnetization transfer. Strong proton-proton coupling, in contrast, creates a very much different picture. The 180° $^{13}$C pulse does not simply exchange the spin labels. Rather, the mixing is more complicated, which brings us to a departure point for considering heteronuclear virtual coupling.

The problem of heteronuclear virtual coupling was first characterized by Morris and Smith.[34] Consider an ABX heteronuclear spin system in which the X spin is $^{13}$C and both A and B are $^1$H, the B spin associated with some carbon other than X. Couplings important to the development of this argument will include $J_{AX}$ and $J_{AB}$ where $J_{AX}$ >> $J_{AB}$. The long range heteronuclear coupling $J_{BX}$ may be ignored in the case to be developed. In the case of the vast majority of such ABX spin systems, the difference in the proton chemical shifts, $\Delta\nu_{AB}$, will be such that the system will be weakly coupled in a heteronuclear sense. The experimental result in the normal case will be that a response will be observed at $F_1 = \nu_A$, $F_2 = \nu_X$, which will appear as a doublet because of $J_{AB}$. In the special case where $\nu_{AB} = 1/2 J_{AX}$, a very much different situation develops. One of the $^{13}$C satellites of the A spin will be degenerate with the B spin while the other $^{13}$C satellite will be far removed and will have no interaction with the B spin. Hence, the satellite exhibiting the degeneracy will be strongly coupled with the B spin while the other satellite will be weakly coupled in the normal sense. In principle, due to the mixing of the proton states, the strongly coupled satellite will be indistinguishable from proton B magnetization. Consequently, the 180° pulse widway through the evolution period (see

Fig. 3-10) will mix proton A and B magnetization. Due to the mixing, magnetizations transferred between A and B will acquire a 90° phase difference even prior to the $\Delta_1$ delay. In this circumstance, some proton B magnetization will acquire a phase difference between $^{13}C$ $\alpha$ and $\beta$. This leads to the appearance of signal at $F_1 = \nu_B$, $F_2 = \nu_X$, which represents what is essentially heteronuclear relayed coherence transfer, which is described in detail in Chapter 4.

Morris and Smith[35] have reported a detailed analysis of the one bond virtual coupling effects. For present purposes, it is sufficient to note that the spurious response in the heteronuclear correlation spectrum may be confirmed as due to virtual couping by repeating the experiment with $\Delta_1 = 0$. Only heteronuclear spin systems exhibiting virtual coupling can give rise to signals when $\Delta_1 = 0$, since all other magnetization components will be in phase at the end of $t_1$ and hence incapable of polarization transfer.

An example of a spurious heteronuclear virtual coupling response is seen in the heteronuclear chemical shift correlation spectrum of the *seco*-gorgosterol analog 5α,6α-epoxy-3,11-dihydroxy-seco-5α-gorgostan-9-one 3,11-diacetate (**3-5**). The side chain proton resonances were assigned via a COSY spectrum as reported by Musmar, et al. [36] The heteronuclear chemical shift correlation spectrum of **3-5** is shown in Fig. 3-19. It will be noted that C22, although a methine carbon, exhibits signals in $F_1$ at the chemical shift of its directly attached proton ($\delta_H = 0.13$) as well as at the shift of one of the C30 cyclopropyl methylene protons, which resonates at $\delta_H = 0.44$ ppm as shown in Fig. 3-19.

**3-5**

Molecules that are prime candidates for the observation of heteronuclear one bond virtual coupling artifacts are those that inherently have highly congested proton spectra. Examples would include polynuclear aromatics, oligosaccharides and perhaps terpenes and some alkaloids. We have occasionally observed virtual coupling responses in the spectra of polynuclear aromatics and in the steroid (terpene) example just shown. It is probably only a matter of time before examples are reported for the other classes of compounds just cited as well as for others that meet the $\Delta v_{AB} = 1/2J_{AX}$ criterion.

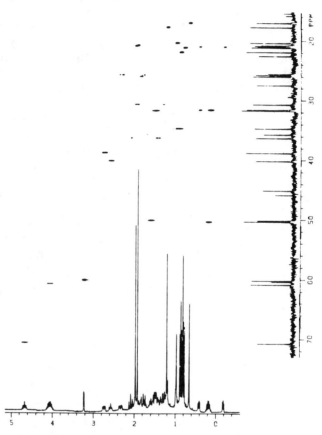

*Fig. 3-19.*    Heteronuclear chemical shift correlation spectrum of *seco*-gorgosterol analog 5α,6α-epoxy-3,11-dihydroxy-*seco*-5α-gorgostan-9-one 3,11-diacetate (**3-5**). The C22 resonance, although due to a methine carbon, exhibits a second response to one of the H30 protons that arises due to heteronuclear virtual coupling.

## Doublets in $F_1$ Rather Than Singlets as a Result of Isotopomeric Breaking of Proton chemical shift degeneracy

The appearance of a proton two spin system may run the gamut from AX when $\Delta v >> J_{AX}$ to a singlet when $\Delta v = 0$. In the latter case, we are observing, in essentially a 100 fold majority, molecules of the type shown by **3-6** in which both of the carbons in question are $^{12}C$. In the

$$^1H - {}^{12}C_A = {}^{12}C_B - {}^1H$$

**3-6**

$$^1H - {}^{13}C_6 = {}^{12}C_7 - {}^1H$$

**3-7**

$$^1H - {}^{12}C_6 = {}^{13}C_7 - {}^1H$$

**3-8**

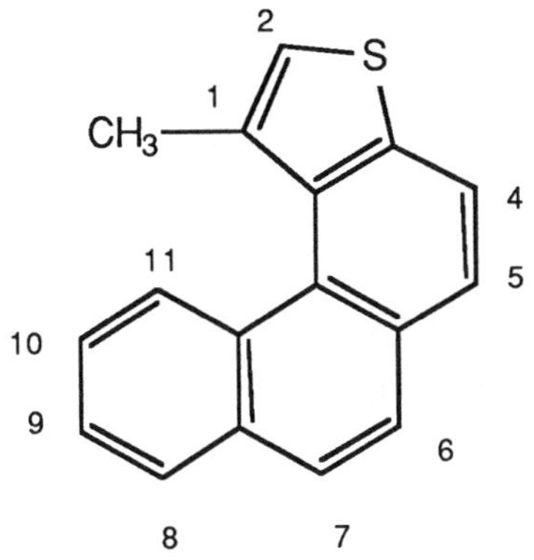

**3-9**

case of the heteronuclear chemical shift correlation spectrum, we are actually considering two different isotropically labeled molecules represented by **3-7** and **3-8**, whose carbons are numbered to correspond to 1-methylphenanthro[3,4-*b*]thiophene (**3-9**) described immediately below, in which the difference in chemical shifts of the $^{13}C$ in the two isotopomers breaks the degeneracy of the proton spin system. Consider as an example, 1-methylphenanthro[3,4-*b*]thiophene (**3-9**) in which the H6 and H7 proton resonances have identical chemical shifts and thus appear as a two proton singlet in the conventional proton NMR spectrum.[37] In the normal proton spectrum, we are observing the molecule as represented by **3-6**. In the heteronuclear chemical shift correlation spectrum of **3-9** shown in Fig. 3-20, it will immediately be noted that the projected proton spectrum contains a doublet resonating at 7.87 ppm (J = 8.5 Hz), which arises because the spectrum is actually recorded by observing the isotopomers represented by **3-7** and **3-8**. Similar effects have been described by Bolton.[38]

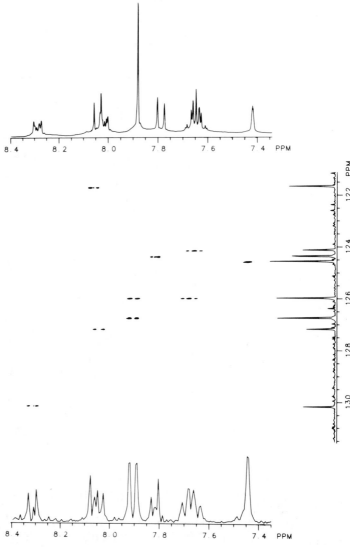

*Fig. 3-20.*   Heteronuclear chemical shift correlation spectrum of 1-methylphenan-
thro[3,4-*b*]thiophene (**3-9**) recorded in deuterochloroform. The one-
dimensional proton spectrum above the contour plot contains an
intense singlet at 7.87 ppm, which corresponds to the H6 and H7
resonances ($\Delta v_{AB} = 0$). The projected proton spectrum (below) shows
a doublet J = 8.5 Hz for the H6/H7 resonances, each of which appears
as a doublet in the contour plot, the chemical shift degeneracy broken
by the isotopomeric differences of the observed species represented
by **3-7** and **3-8**.

# DEPT BASED HETERONUCLEAR CHEMICAL SHIFT CORRELATION

The heteronuclear chemical shift correlation pulse sequence devised by Freeman and Morris[29] and shown in Fig. 3-10 above is derived, in principle, from the INEPT pulse sequence for polarization transfer.[15,17] In similar fashion, the DEPT pulse sequence, which is shown in Fig. 3-21,[19,20] has also spawned several pulse sequences for heteronuclear chemical shift correlation. Chronologically, the development of the DEPT based heteronuclear chemical shift correlation experiment is similar to that encountered for the INEPT derived experiment of Freeman and Morris.[29]

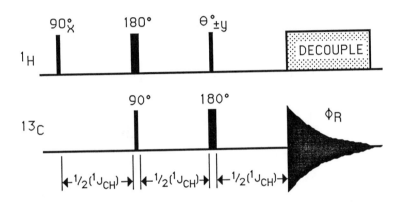

*Fig. 3-21.*    Distortionless Enhancement Polarization Transfer (DEPT) pulse sequence for establishing carbon resonance multiplicities.[19,20]

## Sequences Leading to DEPT Based Heteronuclear Chemical Shift Correlation

When both $^1H$ and $^{13}C$ are on-resonance, Pegg and coworkers[39] were able to demonstrate that the three pulse sequence shown in Fig. 3-22a gives rise to a $^{13}C$ spectrum that is invariant relative to the duration of the interval labeled $\Delta$. Vector components of $^{13}C$ magnetization, $C_\alpha$ and $C_\beta$, can easily be shown to be antiphase at the completion of the pulse sequence shown in Fig. 3-22a. Obviously, broadband homonuclear decoupling cannot be used in conjunction with this pulse sequence because cancellation of the $^{13}C$ magnetization components would result.

The next modification arose[40] with the addition of a second refocusing delay, $1/2(^1J_{CH})$. During the second refocusing interval, antiphase components of $^{13}C$ magnetization ($C_\alpha$ and $C_\beta$) will refocus, allowing the decoupler to be usefully gated on. Furthermore, in the case

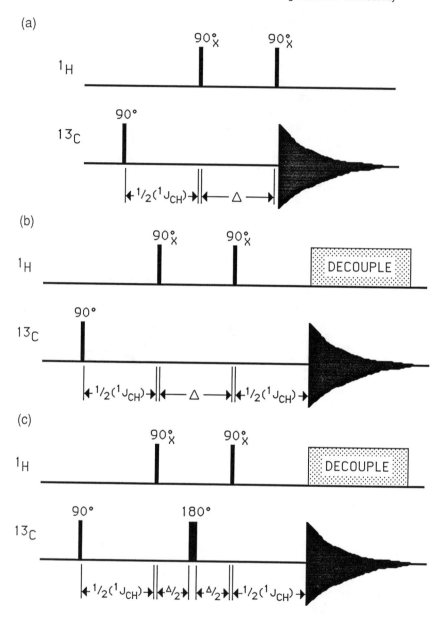

*Fig. 3-22.*     Pulse sequences leading to DEPT based chemical shift correlation. (a)
Three pulse sequence requiring that ¹H and ¹³C be on resonance.[39] (b)
Addition of the refocusing interval $1/2(^1J_{CH})$ eliminates the need for ¹H
to be on resonance and brings the ¹³C into phase, allowing the decoupler
to be used. Response intensity for ¹³C will depend on Δ and $\delta_H$, thus
providing the basis for heteronuclear shift correlation.[40] (c) Modified
sequence with a 180° ¹³C refocusing pulse midway through the Δ
interval, which removes the necessity for ¹³C to be on resonance.[40]

where the proton is moved off-resonance, the effect of the second 90° proton pulse and the response intensity of $^{13}C$ will exhibit a dependence on the length of $\Delta$. The worst case occurs when a 90° or 270° phase error of the proton is built up due to proton chemical shift effects where the second 90° proton pulse has no net effect on the $^1H$ spin system. Consequently, the $^{13}C$ magnetization will remain antiphase and will cancel when the decoupler is gated on. While seemingly non-productive, this observation is important in that it demonstrates the dependence of $^{13}C$ intensity on both $\Delta$ and $^1H$ chemical shift terms, which provide the basis for heteronuclear chemical shift correlation.

The third and final modification before actual DEPT based heteronuclear chemical shift correlation is shown in Fig. 3-22c.[40] Incorporation of a 180° carbon pulse midway through the $\Delta$ interval finally eliminates the need for $^{13}C$ to be on-resonance. The 180° $^{13}C$ pulse removes phase errors built up during both the $\Delta$ and $1/2(^1J_{CH})$ intervals. Spectra may be phase corrected after the first Fourier transformation using a constant phase correction. After the second Fourier transformation, either a 0° or 180° phase correction may be applied to the spectrum.

The experimental results obtained from the variant of the pulse sequence shown in Fig. 3-22c may now be described. Fourier transformation with respect to $\Delta$ gives a spectral result that is a function of the degree of protonation (multiplicity) of the carbon in question. Hence, methine carbons give rise to a single response at the proton chemical shift ($v_H$); methylene carbons give rise to two responses, one at the origin, the other at $2v_H$; finally, methyl carbons also give rise to two responses, one at $v_H$ the second at $3v_H$. In this sense, the spectral editing feature of the DEPT experiment is perhaps now more apparent. In addition, it should be noted that all protonated resonances exhibit their characteristic $^1H$-$^1H$ coupling patterns in the second frequency dimension when using the pulse sequence shown in Fig. 3-22c.

## Heteronuclear Chemical Shift Correlation Using the DEPT Sequence

Building on the pulse sequences shown in Fig. 3-22, Bendall and Pegg[41] next reported a DEPT based heteronuclear chemical shift correlation in the true sense, which is shown in Fig. 3-23. Several potential advantages were cited in the original report of this pulse sequence. First, given phase sensitive software, separate methine, methylene and methyl 2D-NMR subspectra could be computed. Second, since phase errors are fully refocused by the symmetrical DEPT sequence, the need for magnitude calculation is obviated, permitting phase sensitive heteronuclear chemical shift correlation spectra to be utilized.

For $CH_n$ species, the sequence shown in Eqn. [3-8] inserted between the second refocusing interval ($1/2(^1J_{CH})$) and the $\theta°$ $^1H$ pulse, provides the basis for heteronuclear chemical shift correlation, which is

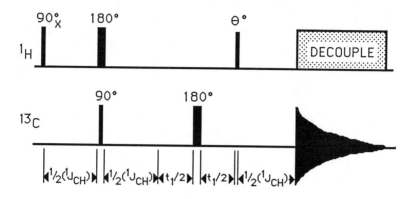

*Fig. 3-23.*    Heteronuclear chemical shift correlation pulse sequence employing DEPT
based polarization transfer.[41] The evolution period, $t_1/2$, is incremented
as in the INEPT based pulse sequence described by Freeman and Morris[29]
shown in Fig. 3-10 above.  The pulse sequence is shown without a
composite 180° pulse midway through the evolution period, although it
may be incorporated beneficially  at this point.

not found in the basic DEPT pulse sequence shown in Fig. 3-21.

$$t_1/2 - 180°(^{13}C) - t_1/2 \qquad\qquad [3\text{-}8]$$

As a result of the sequence shown in Eqn. [3-8], the $^{13}C$ signal is
modulated by the proton chemical shift frequency as a function of $t_1$.
The 180° $^{13}C$ pulse applied coincidentally with the $\theta°$ $^1H$ pulse in the
normal DEPT (Fig. 3-21) experiment is rendered redundant and may
therefore be omitted.

## Reorganization of the DEPT Based Heteronuclear Chemical Shift Correlation Pulse Sequence

Perhaps now the most commonly used variant of the basic DEPT
derived heteronuclear chemical shift correlation pulse sequence shown
in Fig. 3-23 entails the repositioning of the evolution time interval shown
in Eqn. [3-8] to a location immediately after the first 90° proton pulse
(Fig. 3-24) as described by Levitt and coworkers.[42] Reorganization  of
the pulse sequence in this fashion brings it to a configuration almost
identical to the more conventional sequence described by Freeman and
Morris[29] in which evolution is allowed to transpire followed
magnetization transfer and then detection.  It should also be noted that
this arrangement of the DEPT based heteronuclear chemical shift
correlation experiment serves as the basis for a number of the variants
described in a section below.

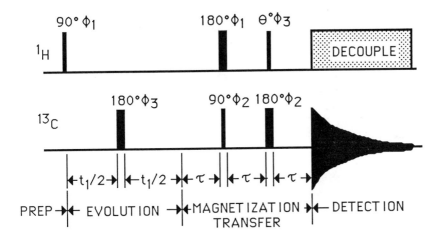

*Fig. 3-24.*    Rearranged variant of the DEPT based heteronuclear chemical shift correlation pulse sequence[42] in which evolution follows the first 90° pulse in a fashion analogous to the INEPT based experiment described by Freeman and Morris[29] and shown in Fig. 3-10 above. Phases are cycled according to Table 3-2. The variable flip angle pulse, $\theta$, may be used to spectrally edit the heteronuclear chemical shift correlation pulse sequence as described in the following section.

## Spectral Editing Using the DEPT Based Heteronuclear Chemical shift Correlation Experiment

When the interval $t_1 = 0$, the pulse sequences shown in Figs. 3-23 and 3-24 are essentially identical to the standard DEPT experiment and may be set up in like fashion.[19,20,43] The spectral editing feature of the one-dimensional DEPT experiment is preserved in its 2D analogs with the $\theta°$ [1]H pulse. If $\theta° = 135°$, data collected and processed in a phase sensitive manner will give a spectrum in which methine and methyl responses shown positive intensity while methylene responses exhibit negative intensity. Acquisition of half of the total accumulated transients for each increment of $t_1$ with $\theta° = 45°$ and the balance with $\theta° = 135°$ followed by processing to afford two phase sensitive spectra also provides an interesting result. Addition of the two spectra just described affords a spectrum containing only responses due to methine and methyl resonances. Subtraction of the two spectra gives a methylene-only spectrum. Separate methine and methyl spectra may be computed by acquisition of a further data matrix with $\theta° = 90°$. Thus, using the DEPT based experiment just described provides the means of spectrally editing two-dimensional heteronuclear chemical shift correlation spectra on the basis of multiplicity.

## Phase Cycling

Phase cycling to afford quadrature detection in the proton of $F_1$ frequency domain of the DEPT based experiment described above may be conducted in a manner analogous to that developed by Bax and Morris[30] for the standard Freeman-Morris[29] heteronuclear chemical shift correlation pulse sequence which was shown in Fig. 3-10. The simple four step phase cycle which is shown in Table 3-2 may be employed where the signal in the receiver is alternately added and subtracted while cycled in a normal four step (0123) quadrature cycle.

Table 3-2.    Phase cycling scheme for the DEPT based heteronuclear chemical shift correlation pulse sequence shown in Fig. 3-24.[44] An alternative phase cycling scheme is given in the original work of Sørensen, et al.[42]

| Acquisition | $\phi_1$ | $\phi_2$ | $\phi_3$ |
|:---:|:---:|:---:|:---:|
| 1 | 3 | 0 | 0 |
| 2 | 2 | 3 | 2 |
| 3 | 1 | 2 | 0 |
| 4 | 0 | 1 | 2 |

## Other DEPT Based Heteronuclear Chemical Shift Correlation Variants and Applications

A relatively wide variety of variants of the basic DEPT based heteronuclear chemical shift correlation experiment shown in Fig. 3-23 were described in a comprehensive paper by Pegg and Bendall.[45] A complete discussion of these techniques is beyond the scope of this chapter and the interested reader is referred to the very thorough paper just cited for additional details.

In addition to the variants of the basic DEPT based heteronuclear chemical shift correlation experiment described by its originators, Wong and his coworkers have reported several interesting variations. By incorporating a form of the bilinear rotation decoupling or BIRD pulse[46] to provide homonuclear decoupling, Wong and coworkers have examined proton chemical shifts and $^{19}F$ couplings in 9α-fluocortisol.[47] By incorporating a BIRD pulse selective for distant protons, Wong and Rutar[44] have reported a DEPT based sequence with broadband homonuclear decoupling analogous to the modification of the Freeman and Morris[29] sequence described by Bax.[48] The Wong and Rutar paper[44] also describes a sequence incorporating a BIRD pulse selective for directly attached protons that allows the determination of the one bond ($^1J_{CH}$) heteronuclear coupling constants. Wong[49] has also reported a DEPT based method for observing satellites of low abundance nuclei such as $^{207}Pb$ and $^{199}Hg$ in heteronuclear correlation spectra.  Finally, Batta and coworkers have reported the adjustment of the delays of the standard DEPT experiment for long range transfer of magnetization in the study of various oligosaccharides.[50,51]  The experiments described by Batta are discussed in more detail in the segment devoted to long range heteronuclear chemical shift correlation below.

## HETERONUCLEAR CHEMICAL SHIFT CORRELATION WITH BROADBAND HOMONUCLEAR PROTON DECOUPLING VIA BILINEAR ROTATIONAL DECOUPLING

### INEPT Based Polarization Transfer

Having established the means for heteronuclear chemical shift correlation in the preceding sections of this chapter, it is useful to next turn our attention to  proton multiplet structure and its impact on the overall sensitivity of the experiment.  Obviously, homonuclear spin coupling constant information is quite useful and may, in cases where there is sufficient digital resolution, be extracted from the heteronuclear chemical shift correlation spectrum.  More commonly, however, levels of digital resolution in the $F_1$ frequency domain will be insufficient to allow the accurate determination of proton homonuclear spin coupling constants.  Furthermore, as proton multiplet structures increase in complexity, the overall signal is spread across progressively greater numbers of responses, thereby diminishing response intensity.  In conjunction, the problem of insufficient digital resolution and diminished response intensity would make it desirable to eliminate proton-proton homonuclear couplings from the heteronuclear chemical shift correlation spectrum.

The mechanism for applying broadband homonuclear proton decoupling is provided by the work of Garbow, Weitekamp and Pines[46] in the form of a BIRD pulse as shown in Eqn. [3-9], whose function is shown using vector diagrams in Fig. 1-25.

$$^1H \qquad 90°(+x) - \tau - 180°(+x) - \tau - 90°(-x)$$

$$^{13}C \qquad 180°(+x) \qquad\qquad [3-9]$$

Bax[52] has demonstrated the utility of the BIRD pulse in broadband homonuclear proton decoupling by incorporating it in lieu  of the 180° $^{13}C$ pulse midway through the evolution period of the Freeman-Morris[29] pulse sequence (Fig. 3-10) to give the pulse sequence in Fig. 3-25.

The BIRD pulse shown in Eqn. [3-9] and located in the middle of the evolution period of the pulse sequence shown in Fig. 3-25 serves as a 180° pulse for those protons not directly coupled to $^{13}C$  while those protons directly bound to $^{13}C$ are left essentially unaffected.  The role of the BIRD pulse operator in providing broadband homonuclear proton-proton decoupling (which should be carefully distinguished from

broadband heteronuclear decoupling, which is provided in Fig. 3-10 by
the 180° carbon pulse located at time t = $t_1$/2) can be understood by
considering a hypothetical AMX heteronuclear spin system where
proton spin A is directly coupled to $^{13}$C spin X and vicinally coupled to
another proton spin M. As the AMX heteronuclear spin system evolves
during the first interval $t_1$/2, let us assume that the m spin evolves with a

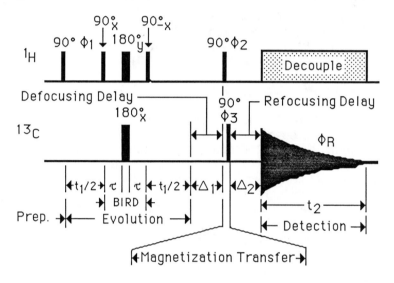

*Fig. 3-25.*    Pulse sequence for heteronuclear chemical shift correlation with
broadband homonuclear proton decoupled developed by Bax.[52] The
defocusing and refocusing delays, $\Delta_1$ and $\Delta_2$, respectively, are
optimized in a manner identical to the conventional heteronuclear
correlation experiment (see discussion pertaining to Fig. 3-10). Phases
are cycled according to the eight step cycle shown in Table 3-1. The $\tau$
interval in the BIRD pulse is normally set as a function of 1/2($^1J_{CH}$).

spin state m = 1/2. The BIRD pulse operator will flip the M spin
selectively while leaving the A spin transverse magnetization
unaffected. Hence, during the second interval $t_1$/2, the M spin will
evolve in spin state m = -1/2. Thus, at the end of the evolution period,
there will be no spin coupling between A and M, since the M spin
evolved for equal intervals in spin state m = 1/2 and m = -1/2.
Consequently, A magnetization will depend only on the chemical shift of
this proton, $\delta_A$, and the length of the evolution period, $t_1$. Magnetization
is then transferred to $^{13}$C and sampled in the usual fashion.

A shortcoming, in some senses, of the pulse sequence shown in
Fig. 3-25 is found in the case of nonequivalent geminal protons. Here,
there is no mechanism by which the BIRD pulse may selectively flip the
magnetization of one of the anisochronous protons to provide
decoupling. Hence, spectra of compounds containing anisochronous
methylene protons will exhibit homonuclear proton-proton coupling.

Examples of both the nominal operation of the experiment and its effect on nonequivalent geminal protons will be illustrated below.

## Phase Cycling

The phase cycling for the broadband homonuclear proton decoupled heteronuclear correlation experiment (Fig. 3-25) is identical to the eight step phase cycle contained in Table 3-1. The receiver phase may be cycled to collect either the antiecho or the coherence transfer echo. In the former, which is the preferred choice, it is necessary to perform a spectrum rotation following the second Fourier transformation. Users with instruments without this capacity will be restricted to the collection of the coherence transfer echo.

## Parameter Selection

The selection of virtually all of the parameters for the pulse sequence shown in Fig. 3-25 is identical to that described above with regard to Fig. 3-10 except for the duration of the $\tau$ interval, which is not used in the conventional Freeman-Morris[29] sequence. Normally, the duration of $\tau$ is optimized as a function of $1/2(^1J_{CH})$. In the case where there is a range of one bond couplings in the system being studied, an average value should be employed. However, it should be noted that Bax[52] has observed that for the BIRD pulse operator to function properly, the value of the coupling constant used to optimize the $\tau$ interval should be within $\pm 10\%$ of ideal. Furthermore, the width of the spin multiplet to be decoupled through the use of the BIRD pulse operator should also be $< 1/5$ of the $^1J_{CH}$ value used in optimizing $\tau$.[46]

## Data Processing

Data processing considerations for the broadband homonuclear proton decoupled heteronuclear correlation experiment shown in Fig. 3-25 are basically no different than those associated with the conventional Freeman-Morris[29] sequence shown in Fig. 3-10. In general, sensitivity will be somewhat better with the former, allowing perhaps more generous use of exponential broadening factors. Once again, it should also be recalled that if the antiecho is recorded it will be necessary to perform a spectrum rotation following the second Fourier transformation to reestablish normal orientation of the proton spectrum ($F_1$). To allow direct comparison, we will consider first the spectrum of norharmane (**3-4**) recorded using the pulse sequence shown in Fig. 3-25 above.

## Examples of the Application of Broadband Homonuclear Proton Decoupling in Heteronuclear Chemical Shift Correlation

To illustrate the capabilities of the BIRD pulse in broadband homonuclear decoupling we will consider several examples. First, we will utilize norharmane (**3-4**), which is used essentially as a benchmark reference compound because of its high solubility and spectral simplicity. Next we shall turn our attention to a more complex polynuclear aromatic compound, 9-methylphenanthro[4,3-*a*]dibenzo-thiophene (**3-10**), whose congested proton spectrum was considered in relation to the COSY experiment in Chapter 2. Finally, we will examine the problem of anisochronous geminal protons using strychnine (**3-11**).

## Norharmane

The spectrum of norharmane (**3-4**) recorded using the pulse sequence shown in Fig. 3-25 in conjunction with the eight step phase cycle specified in Table 3-1 is shown in Fig. 3-26. If we compare the projected proton spectrum of norharmane (**3-4**) from Fig. 3-18, which was obtained using the conventional Freeman-Morris[29] pulse sequence shown in Fig. 3-10, we will note that the proton resonances display multiplet structure even though the couplings are not well resolved for reasons of low digitization. In contrast, the projected proton spectrum plotted beneath the contour plot in Fig. 3-26 shows that all of the homonuclear proton-proton couplings have been collapsed and the projected proton spectrum is now comprised exclusively of singlets.

## 9-Methylphenanthro[4,3-*a*]dibenzothiophene.

The more complex helically distorted polynuclear aromatic 9-methylphenanthro[4,3-*a*]dibenzothiophene (**3-10**) provides an example of the ability of the broadband homonuclear proton decoupled heteronuclear chemical shift correlation experiment to extract proton chemical shift information from a highly congested proton spectrum.[53] If we examine the high resolution proton reference spectrum plotted beneath the two-dimensional contour plot shown in Fig. 3-27, it is difficult to determine that the region between approximately 7.8 and 8.0 ppm contains resonances for seven protons simply by inspection. Only the proton resonating furthest upfield is clearly resolved in this region of the 300 MHz spectrum. In contrast, seven identifiable signals are observed for the same region in the proton projection of the two-dimensional NMR spectrum of **3-10** acquired using the pulse sequence shown in Fig. 3-25. By extracting individual interferograms from the $S(F_2,t_1)$ data matrix and zero filling them to 2K points prior to the second Fourier transformation it was possible in this case to determine proton chemical shifts of **3-10** with an accuracy of ±0.01 ppm. Proton and carbon NMR chemical shift assignments for **3-10** are collected in Table 3-3.

## Strychnine

The final point to be made regarding the broadband homonuclear proton decoupling capabilities of the BIRD pulse contained in the pulse sequence shown in Fig. 3-25 concerns the lack of ability for the pulse to

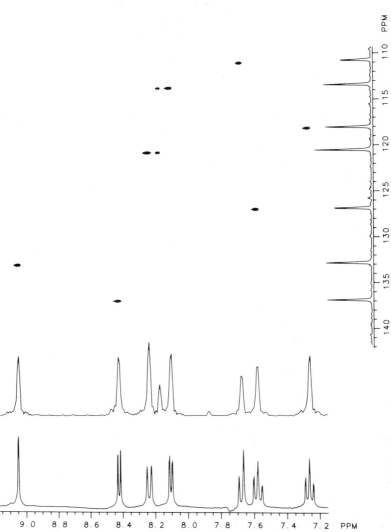

*Fig. 3-26.*    Broadband homonuclear proton decoupled heteronuclear correlation
spectrum of norharmane (**3-4**) recorded using the pulse sequence
shown in Fig. 3-25 and the phase cycling specified in Table 3-1. The
spectrum was acquired by taking 96 blocks of data with 32
acquisitions/block (overkill) in about 1.3 hours.

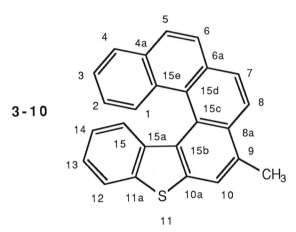

**3-10**

Table 3-3.    Proton and carbon resonance assignments of 9-methylphenanthro-[4,3-a]dibenzothiophene (**3-10**)[53] recorded in deuterochloroform at 30° C at observation frequencies of 300.068 and 75.459 MHz, respectively, for $^1$H and $^{13}$C.

| Resonance | $\delta_H$ | $\delta_C$ | Resonance | $\delta_H$ | $\delta_C$ |
|---|---|---|---|---|---|
| 1 | 7.97 | 128.80 | 10a | | 139.08 |
| 2 | 6.99 | 125.53 | 11 | | |
| 3 | 7.38 | 125.72 | 11a | | 138.04 |
| 4 | 7.94 | 127.69 | 12 | 7.80 | 122.10 |
| 4a | | 131.50 | 13 | 7.16 | 124.83 |
| 5 | 7.95 | 128.03 | 14 | 6.72 | 122.45 |
| 6 | 7.88 | 125.87 | 15 | 6.83 | 125.93 |
| 6a | | 131.71 | 15a | | 136.35 |
| 7 | 7.85 | 125.61 | 15b | | 130.23 |
| 8 | 8.13 | 123.35 | 15c | | 125.04 |
| 8a | | 130.99 | 15d | | 125.77 |
| 9 | | 133.19 | 15e | | 130.48 |
| 10 | 7.88 | 122.10 | -CH$_3$ | 2.88 | 19.67 |

collapse geminal couplings in the case of nonequivalent methylene protons. Strychnine (**3-11**) provides a useful model compound with which to illustrate the impact of geminal methylene proton nonequivalence, since every methylene geminal pair is nonequivalent to some degree.

The degree of geminal methylene proton nonequivalence for strychnine (**3-11**) runs the gamut, ranging from 0.01 ppm in the case of the H17a/H17b pair to nearly 1.0 ppm difference in the case of the H15a/H15b and H20a/H20b pairs.[54] Rigorous assignments for the proton nmr spectrum of strychnine (**3-11**) have been reported by Chazin, Colebrok and Edward[55] and are reflected in Chapter 2 (see Figs. 2-35, 2-36, 2-37 and 2-46).

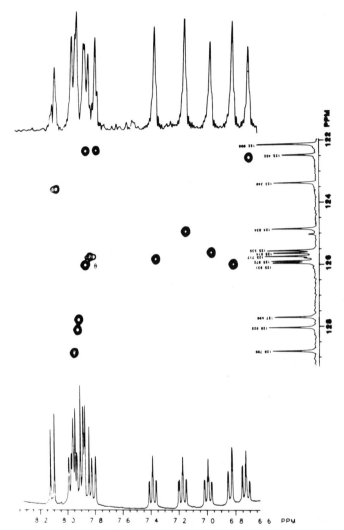

*Fig.  3-27.*  Broadband  homonuclear  proton  decoupled  heteronuclear  correlation
spectrum  of  35  mg  of  **3-10**  recorded  in  0.4  ml  deuterochloroform  at
30°C.  The  spectrum  was  acquired  by  taking  190  blocks  of  data  in  $F_1$
with  192  acquisitions/block.      The  spectrum  was  zero-filled  to  512
points  in  $F_1$  during  processing  to  give  the  512  x  512  point  contour  plot
shown.    The  high  resolution  reference  spectrum  (below)  was  recorded
using  4K  complex  points.  The  projected  proton  spectrum  is  plotted  above
the  data  matrix;  the  high  resolution  carbon  spectrum  is  plotted
vertically  beside  the  contour  plot.

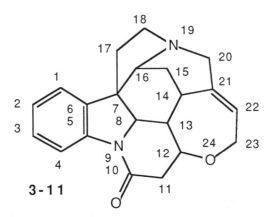

**3-11**

All of the methylene resonances of strychnine, with the exception of the C17 resonance demonstrate easily observed multiplets due to their respective geminal couplings (Fig. 3-28). In the case of H17a/b, the chemical shifts of these protons are so close that it would be impossible to resolve the geminal coupling in the heteronuclear correlation spectrum and it hence appears as a "singlet." The C15 resonance is located furthest upfield at 26.9δ; C11 may be unequivocally assigned as the resonance observed at 42.43δ, C17 at 42.9δ, C18 at 50.3δ and C20 at 52.7δ.

### DEPT Based Polarization Transfer

The heteronuclear chemical shift correlation experiment based on the reorganized DEPT experiment shown in Fig. 3-24 has also served as the basis for a broadband homonuclear proton decoupled heteronuclear chemical shift correlation experiment. Once again, decoupling has been provided by the replacement of the 180° refocusing pulse midway through the evolution period, $t_1$, by a BIRD pulse of the type used above the the pulse sequence described by Bax[52] and shown in Fig. 3-25.

In 1984 Wong and coworkers[56] first described a DEPT based heteronuclear chemical shift correlation experiment that incorporated a BIRD pulse midway through the evolution period. Their pulse sequence, which is shown in Fig. 3-29, represents an amalgamation of the experiments shown in Figs. 3-24 and 3-25. Results obtainable with the sequence of Wong and coworkers[56] are essentially identical to those illustrated by the spectra in Figs. 3-26, 3-27 and 3-28.

### Parameter Selection

Parameter selection for the DEPT based heteronuclear chemical shift correlation experiment with broadband decoupling is fundamentally no different from the conventional DEPT experiment. The duration of the

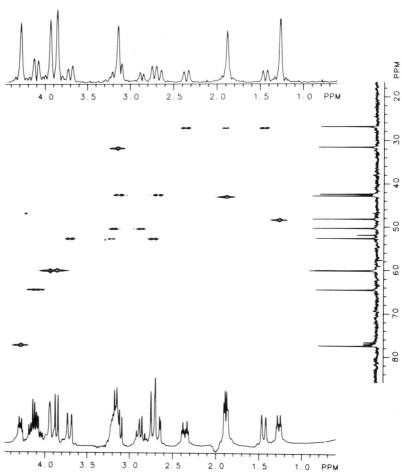

*Fig. 3-28.*     Heteronuclear chemical shift correlation spectrum of strychnine (**3-
11**) acquired using the pulse sequence shown in Fig. 3-25 to provide
semiselective broadband homonuclear proton decoupling.

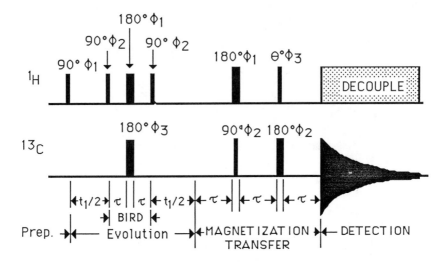

*Fig. 3-29.*        DEPT based heteronuclear chemical shift correlation experiment with broadband homonuclear proton decoupling described by Wong and coworkers.[56]

$\tau$ intervals is normally set to $1/2(^1J_{CH})$. The final proton pulse, $\theta$, may be set to 45° as a compromise in routine applications. Alternately, the final proton pulse may be set to 90° or 135°. The interested reader should see the discussion in the section above on heteronuclear DEPT correlation for a description of the effects of these choices.

## Phase Cycling

Phase cycling for the pulse sequence shown in Fig. 3-29 is rather straightforward, as shown in Table 3-4. The receiver phase is cycled in the normal fashion to afford quadrature detection with an alternate scans subtracted.

Table 3-4.        Phase cycle for DEPT based heteronuclear chemical shift correlation with broadband homonuclear proton decoupling.

| Acquisition | $\phi_1$ | $\phi_2$ | $\phi_3$ |
|:-----------:|:--------:|:--------:|:--------:|
| 1 | 3 | 0 | 0 |
| 2 | 2 | 3 | 2 |
| 3 | 1 | 2 | 0 |
| 4 | 0 | 1 | 2 |

## Variations

Wong and coworkers, in addition to the DEPT based heteronuclear correlation experiment with broadband homonuclear proton decoupling, have reported a variant in which the phases of the BIRD pulse are altered to selectively flip only the attached protons. Using this variant of the DEPT based heteronuclear correlation expeirment, Wong and coworkers[57,58] have been able to separately measure one bond coupling constants and geminal coupling constants, respectively.

## HETERONUCLEAR CHEMICAL SHIFT CORRELATION WITH BROADBAND PROTON HOMONUCLEAR DECOUPLING VIA A CONSTANT EVOLUTION TIME

The heteronuclear chemical shift correlation pulse sequence developed by Bax[52] and shown in Fig. 3-25, as noted in the case of the strychnine spectrum discussed above (Fig. 3-28), cannot collapse geminal couplings. This limitation of the experiment is, of course, a consequence of the use of the bilinear rotational decoupling or BIRD pulse to provide the decoupling. In applications where it is desirable or necessary to eliminate all homonuclear couplings, it is possible to accomplish this by the use of a constant duration evolution time as described by Reynolds and coworkers.[59]

Constant duration evolution periods have been extensively utilized in long range heteronuclear correlation experiments such as the Kessler-Griesinger COLOC sequence[60] and the Reynolds XCORFE sequence,[61] which are described below. Simultaneous 180° proton and carbon pulses are systematically repositioned through the evolution period, the location changed in each successive experiment. Homonuclear spin couplings, $^nJ_{HH}$, are insensitive to the mobile refocusing pulses and further, since the duration of the evolution period is fixed, they do not modulate the signal as a function of $t_1$ as in the conventional heteronuclear correlation experiments, in which the duration of the evolution period is systematically incremented. This feature of the experiment thus provides broadband homonuclear decoupling of all $^nJ_{HH}$ including geminal couplings. The pulse sequence devised by Reynolds and coworkers[59] is shown in Fig. 3-30 and has been christened QWIKCORR by Krishnamurthy and Casida.[62]

### Parameter Selection

The selection of parameters for use with the pulse sequence proposed by Reynolds and coworkers[59] for complete broadband homonuclear decoupling is somewhat restricted if optimal sensitivity is to be maintained. Maximum resolution is governed by $T = t_1(max)/2$, where $t_1(max)$ corresponds to the number of increments divided by the spectral width in $F_1$. Basically, Reynolds and coworkers recommend that T be restricted to either of two values: $0.012 \pm 0.002$ sec or $0.020 \pm 0.002$ sec, in which case the number of increments of the evolution time should be 24 or 16, respectively.

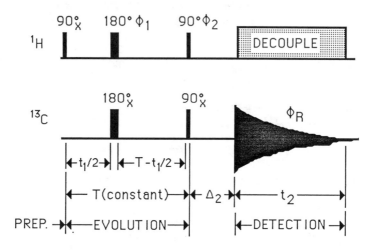

*Fig. 3-30.*    Pulse sequence for heteronuclear chemical shift correlation with complete broadband homonuclear decoupling (QWIKCORR) proposed by Reynolds and coworkers.[59]

Obviously, digital resolution in the $F_1$ frequency domain will be quite coarse because of the small number of $t_1$ increments. The coarse digitization in $F_1$ and the correspondingly low accuracy in the estimation of peak positions in $F_1$ may be overcome by extracting individual slices of the data matrix prior to the second Fourier transformation followed by extensive zero filling. While no new spectral information is gained in zero-filling beyond twice the number of points in $t_1$,[63] improved interpolation and hence estimation of peak positions does result.[64] A problem can occur, however, when there is a small chemical shift difference between geminal methylene protons. Specifically, the theoretical limit of resolution is the reciprocal of $t_1(max)$, which corresponds to about 50 Hz when T = 0.020 sec. Peaks closer than this will not be resolved by the experiment.

## Phase Cycling

The phase cycling of the pulse sequence shown in Fig. 3-30 requires a simple sixteen step recipe as shown in Table 3-5. Basically, the phase cycling breaks down to a four step phase cycle of the $^1H$ polarization transfer pulse and an independent four step cycle on the 180° proton pulse combined with a 180° phase shifts after each four acquistions to provide the equivalent of receiver phase alternation.

Table 3-5.        Phase cycling for the QWIKCORR pulse sequence shown in Fig. 3-30.

| $\phi_1$ | 0123 | 2301 | 0123 | 2301 |
| $\phi_2$ | 0000 | 1111 | 2222 | 3333 |
| $\phi_R$ | 0123 | 0123 | 0123 | 0123 |

## PROTON DETECTED (INVERSE) HETERONUCLEAR PROTON-CARBON CHEMICAL SHIFT CORRELATION

During each type of heteronuclear transfer of magnetization, whether we recognize or choose to ignore it, multiple quantum coherence is involved. Heteronuclear multiple quantum coherence can also be created and then indirectly observed by reconversion to single quantum coherence. The sole stipulation is that the involved spins, I and S in the general case ($^1$H and $^{13}$C, respectively, in the context that follows), must be weakly coupled.

### Introduction

It may surprise some to learn that heteronuclear multiple quantum coherence and indirect detected experiments were first experimentally demonstrated by Müller in 1979.[65] Applying the pulse sequence shown in Fig. 3-31 less the 180° refocusing pulse midway through $t_1$ to a sample of $^{13}$C enriched methyl iodide, Müller was able to demonstrate the indirect observation heteronuclear multiple quantum coherence. Methyl iodide has four zero and four double quantum transitions, which produce two triplets in $F_2$ at the methyl $^{13}$C shift. One triplet corresponds to heteronuclear zero quantum coherence (HNZQC), the other to heteronuclear double quantum coherence (HNDQC). Frequencies in $F_1$ may be defined by Eqns. [3-10] and [3-11]:

$$\omega_Z^0 = (\omega_I^0 - \omega_I) - (\omega_S^0 - \omega_S) \qquad [3\text{-}10]$$

and

$$\omega_D^0 = (\omega_I^0 - \omega_I) + (\omega_S^0 - \omega_S) \qquad [3\text{-}11]$$

where $\omega_I^0$ = proton Larmor frequence, $\omega_S^0$ = carbon Larmor frequency, $\omega_I$ = proton carrier frequency and $\omega_S$ = carbon carrier frequency.

As will be noted quickly from Fig. 3-31, the experiment just described is heteronucleus detected but the sensitivity advantage of proton detection has been foregone. It is interesting too in that the pulse sequence in Fig. 3-31 affords an alternative to the Freeman-Morris (Fig. 3-10) and DEPT based (Fig. 3-24) heteronuclear chemical shift correlation experiments. Indeed, Müller[65] did demonstrate this

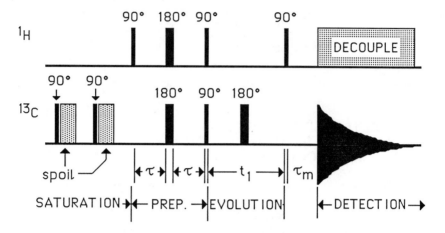

*Fig. 3-31.*    Multiple quantum based heteronuclear chemical shift correlation pulse sequence described by Müller.[65]

application of the experiment. To the best of our knowledge, however, this was the only such use of this experiment ever reported.

Homonuclear multiple quantum coherence can be generated by the application of two nonselective 90° pulses separated by the interval $\tau_p$ (see also Chapter 2). Heteronuclear multiple quantum coherence (HNMQC) can be created, in principle, by applying the sequence shown in Eqn. [3-12] to both spins simultaneously.

$$90° - \tau_p - 90° \qquad\qquad [3\text{-}12]$$

The utility of this approach is limited, however, by the difficulty in determining $\tau_p$ for the entire heteronuclear spin system. Considerable simplification is provided by the insertion of 180° "echo" pulses to both heteronuclear spins midway though the $\tau_p$ interval. Then, the generation of HNMQC depends only on $J_{IS}$, which will normally be $J_{CH}$ or $J_{NH}$.

The insertion of 180° pulses midway through the $\tau_p$ interval led very usefully to the proton detected experimental variants described by Müller.[65] The basic HNMQC pulse sequence with proton detection is shown in Fig. 3-32. On comparing the sequence shown in Fig. 3-32 to that in Fig. 3-31 we note that there has been only a very simple exchange of the nuclei to which the final two pulses in the sequence are applied. Müller proposed several variants of the pulse sequence shown in Fig. 3-32, which differed solely in the operation applied to $^{13}C$ during detection. Options ranged from no decoupling through full decoupling as shown.

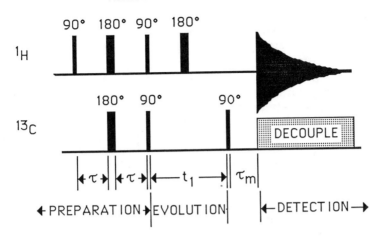

*Fig. 3-32.*    Heteronuclear multiple quantum coherence pulse sequence with proton detection described by Müller.[65]

## Suppression of the Signal for Protons Not Directly Bonded to $^{13}$C

Following Müller's[65] initial work, a number of other groups have described and/or utilized proton detected heteronuclear multiple quantum experiments.[66-80] A problem common to all of these experiments is the need to suppress the signal from the protons not directly bonded to $^{13}$C (which will be denoted as $^1$H$^{12}$C hereafter), combined with the dynamic range problem that would otherwise be associated with these protons.

There are a number of plausible approaches to the problem. In principle, it should be possible to saturate the protons followed by transfer of the now enhanced low gamma nuclide signal back to proton for detection.[81] Unfortunately, the sensitivity gained by proton detection is partially lost by this approach. The sequence described by Müller[65] and shown in Fig. 3-32 provides the basis for eliminating $^1$H$^{12}$C signals by dint of a simple phase shift of the second 90° proton pulse as described by Brühwiler and Wagner.[76] The sequence up to the point of the second 90° pulse creates a 180° phase error for magnetization arising from $^1$H$^{13}$C, leaving these components coaligned with the x'-axis while the undesired $^1$H$^{12}$C magnetization is left essentially along the y'-axis. As noted by Brühwiler and Wagner[76] applying a 90°$_{-x}$ pulse to the protons at this point will leave the desired $^1$H$^{13}$C component of magnetization unaffected while the undesired $^1$H$^{12}$C component is rotated back to the +z'-axis.

A second alternative described by Brühwiler and Wagner[76] utilizes a 90°$_y$ second proton pulse. This operation preferentially rotates the desired $^1$H$^{13}$C magnetization coaligned with the x'-axis to the z'-axis while leaving the undesired $^1$H$^{12}$C component unaffected. Next, a homogeneity spoiling pulse may be applied to destroy the undesirable proton magnetization. The subsequent application of a proton $90°_{-y}$ pulse restores the $^{1}H^{13}C$ component of magnetization to the x'-axis, whereby a 90° carbon pulse may create the desired HNMQC. This approach was found to be superior in the development of a proton detected heteronuclear relay sequence, while the preceeding approach was better for normal heteronuclear chemical shift correlation.

Another novel approach was described by Bax and Subramanian.[77] This approach takes advantage of the discriminatory capabilities of the BIRD pulse that allow protons not directly attached to $^{13}C$ ($^{1}H^{12}C$) to be inverted as shown using vector diagrams in Chapter 1. The pulse sequence of Bax and Subramanian[77] is shown in Fig. 3-33 below. Following inversion of $^{1}H^{12}C$ magnetization by the BIRD pulse, a fixed delay is inserted during which relaxation may occur. By proper choice of this interval, $^{1}H^{12}C$ magnetization will be saturated. In actual practice, the BIRD pulse is applied at a time, $\tau = T/2.7$, before the end of the delay period, where T is chosen to be approximately $1.3T_1$ of the most rapidly relaxing proton. Hence, $^{1}H^{12}C$ magnetization will be saturated when the first 90° $^{13}C$ pulse is applied to begin the creation of HNMQC.

The method of Bax and Subramanian[77] just described is quite effective and works well for small molecules where $\omega\tau_C < 1$. Unfortunately, the BIRD presaturation method cannot be used with larger molecules because the period between the BIRD pulse and the beginning of the actual sequence is of sufficient length for the negative nOe to attentuate the signals for protons attached to $^{13}C$. To circumvent this problem, another pulse sequence was reported by Sklenar and Bax[79] as shown below in Fig. 3-35.

## Chemical Shift Correlation via Heteronuclear Multiple Quantum Coherence: "Small" Molecules

With various means available to circumvent the problem of undesired $^{1}H^{12}C$ magnetization, it is appropriate to direct our attention to the actual creation of heteronuclear multiple quantum coherence (HNMQC). A density based treatment of the creation of HNMQC is given by Bax, Griffey and Hawkins.[67] An alternative treatment using a Heisenberg vector picture has been proposed by Bendall and co-workers.[69] Regardless of which approach the reader may favor, a rigorous treatment is beyond the scope of this chapter. However, it is useful to note that following the first pulse applied to one of the two weakly coupled heteronuclear spins, a fixed delay, $\Delta$, may be allowed to transpire, during which the vector components from $J_{IS}$ develop an antiphase relationship. Immediately prior to the application of the 90° pulse to the other heteronucleus, the density matrix will contain only single quantum terms. Following the 90° pulse application to the second heteronucleus, the density matrix will contain only terms for zero and double quantum coherence; hence all transverse magnetization has been converted to HNMQC.

Let us consider the heteronuclear multiple quantum pulse sequence described by Bax and Subramanian[77] and shown in Fig. 3-33. As described above, the BIRD pulse eliminates the undesired $^1H^{12}C$ magnetization, leaving it saturated at the beginning of the generation of HNMQC. The first 90° $^{13}C$ pulse serves to eliminate from the spectrum artifacts that would otherwise arise from longitudinal $^{13}C$ magnetization present at the end of the preparation period. Hence, the first pulse actually involved in the generation of HNMQC is the $90°_x$ proton pulse. The fixed duration delay, $\Delta$, is set as a function of $1/2(^1J_{CH})$ and HNMQC is actually created by the 90° $^{13}C$ pulse applied at the end of this period. HNMQC is allowed to evolve during the evolution period, $t_1$. A 180° $^1H$ pulse is applied midway through this interval to refocus $^1H$ chemical shift evolution so that the signals are labeled with the carbon frequency during $t_1$. The evolved HNMQC is reconverted to observable single quantum coherence by the final $90°_x$ $^{13}C$ pulse applied in the sequence. When recreated, the single quantum coherence is antiphase and a second fixed delay, $\Delta$, is necessary before $^{13}C$ decoupling and detection is initiated. The duration of the $\tau$ interval and the optimization of the time interval, T, were discussed above.

### Phase Cycling

The phase cycling prescription for the Bax/ Subramanian[77] pulse sequence shown in Fig. 3-33 is relatively simple, as shown in Table 3-6. The procedure requires storage of odd-and-even numbered scans in separate locations in memory for separate processing to provide a phase sensitive spectrum.

### Parameter Selection

The determination of the pulse lengths and selection of parameters has effectively been described in preceding sections and chapters. Perhaps the sole considerations that remain are the pulse power of the X-band or $^{13}C$ decoupler used as a pulse source for the heteronucleus and the digitization of the carbon frequency domain.

At high observation frequencies with broad sweep widths, it may be desirable to split the $^{13}C$ frequency range in half if pulse lengths for carbon are too long. This has been done by Bax and coworkers[82-85] with studies at 500 MHz using long range heteronuclear multiple quantum coherence experiments, which are described below, although it was in their case done to prevent an amplifier from going CW. Regardless, it is important to note that ineffective flip angles engendered by long pulses and high offsets may lead to inefficient excitation of HNMQC with a corresponding increase in the number of artifacts that will be observed.

The second factor, digitization of the carbon frequency domain, $F_1$, differs substantially from the considerations implicit in the

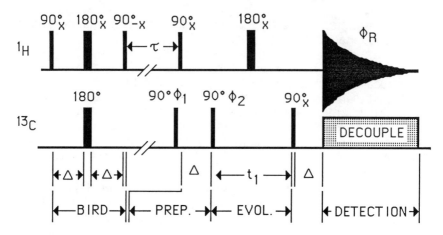

Fig. 3-33.    Proton detected heteronuclear multiple quantum chemical shift
correlation pulse sequence devised by Bax and Subramanian.[77] The
duration of the delay, $\Delta$, is set to $1/2(^1J_{CH})$. The interval $\tau$ = T/2.7,
where T is chosen to be about 1.3 x $T_1$ of the fastest relaxing proton in
the molecule.

Table 3-6.    Phase cycle for the Bax/Subramanian[77] inverse detected heteronuclear chemical shift correlation pulse sequence shown in Fig. 3-33.

| | | | $\phi_R$ | |
| --- | --- | --- | --- | --- |
| Scan | $\phi_1$ | $\phi_2$ | Odd | Even |
| 1 | 0 | 0 | 0 | |
| 2 | 0 | 1 | | 0 |
| 3 | 0 | 2 | 2 | |
| 4 | 0 | 3 | | 2 |
| 5 | 2 | 0 | 0 | |
| 6 | 2 | 1 | | 0 |
| 7 | 2 | 2 | 2 | |
| 8 | 2 | 3 | | 2 |

heteronucleus detected experiments. In reverse detected experiments,
the carbon frequency range must be digitized *via* the incrementation of
the evolution period. Hence, it would be desirable in most instances to
collect 300 or more increments of the evolution period if time and disc
storage capacity permit. Indeed, in a long range study recently reported
by Kessler and coworkers[86] a total of 2K experiments were performed to
ensure adequate digitization of the $F_1$ frequency domain.

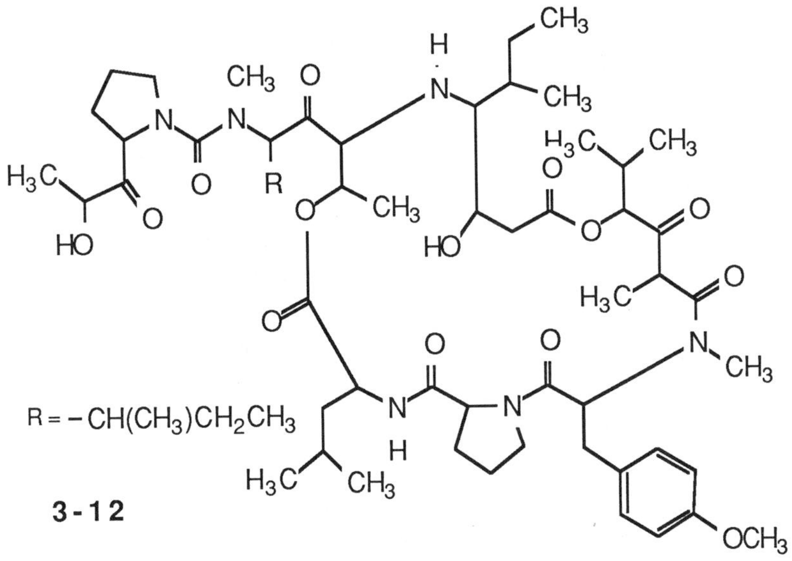

**3-12**

$R = -CH(CH_3)CH_2CH_3$

## Application to the Cyclic Depsipeptide Didemnim-B

The cyclic depsipeptide, didemnim-B (**3-12**), while a large molecule by some standards, falls under the category of small molecules in so far as biopolymers are concerned. The heteronuclear correlation spectrum of didemnim-B (**3-12**) shown in Fig. 3-34 was acquired overnight using the pulse sequence of Bax and Subramanian[77] shown in Fig. 3-33 on a 25 mg sample dissolved in 0.4 ml of deuterochloroform.

## Chemical Shift Correlation via Heteronuclear Multiple Quantum Coherence: Larger Molecules

For larger molecules ($\omega\tau_C < 1$), the BIRD pulse in the sequence presented in Fig. 3-33 cannot be used because the negative nOe would attenuate signals arising from protons attached to $^{13}C$. In this type of application, Sklenar and Bax[79] have reported that the pulse sequence shown in Fig. 3-35.

The application of the pulse sequence shown in Fig. 3-35 to the protein hen egg white lysozyme ($M_r$ 14,400 daltons) has been reported by Sklenar and Bax.[79] Brühwiler and Wagner[76] have used a variant of the pulse sequence shown in Fig. 3-32 to acquire a proton carbon correlation spectrum of bovine pancreatic trypsin inhibitor ($M_r$ = 6500 daltons). Where the line of demarcation between the use of these pulse sequences occurs at which one becomes more effective than the other remains to be determined. What is important is that heteronuclear

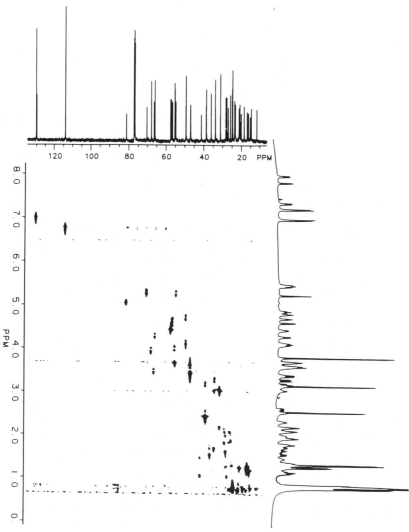

*Fig. 3-34.*    Proton-carbon heteronuclear correlation spectrum of didemnim-B (**3-12**) acquired on a 25 mg sample using the proton detected heteronuclear multiple quantum correlation experiment of Bax and Subramanian[77] shown in Fig. 3-33. The data was acquired at 500 MHz and the matrix consisted of 350 x 2K points.

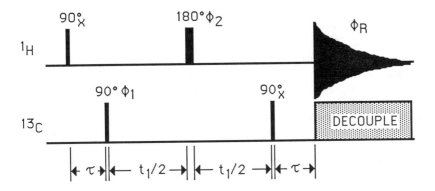

*Fig. 3-35.* Proton detected heteronuclear multiple quantum pulse sequence suggested by Sklenar and Bax[79] for use in obtaining spectra of larger molecules.

chemical shift correlation experiments may now be extended to much larger molecules than previously could be accomodated.

## A Brief Survey of Applications of Chemical Shift Correlation via Heteronuclear Multiple Quantum Coherence

A number of applications of reverse detected chemical shift correlation via heteronuclear multiple quantum coherence have already appeared. As the power of this method becomes more widely recognized and the number of spectrometers capable of performing such experiments grows, there will probably be burgeoning numbers of applications of these techniques. The approach has been demonstrated for several isotopically labeled proteins[78,87,88] and in two cases at natural abundance.[76,79] Several reported applications to nucleic acids have also appeared.[80,89,90]

## LONG RANGE HETERONUCLEAR CHEMICAL SHIFT CORRELATION

Heteronuclear chemical shift correlation experiments, whether utilizing INEPT or DEPT as the basis for polarization transfer, exploit the heteronuclear scalar coupling to permit the actual transfer to take place. In the examples described thus far, we have exclusively utilized the large direct (one bond) heteronuclear spin coupling constant, which nominally range from 125 to 250 Hz. More recently, a growing number of studies have utilized longer range heteronuclear coupling pathways ($^nJ_{CH}$, $n$ = 2-4). The following sections of this chapter will consider in detail the now relatively large number of experiments designed for long range heteronuclear chemical shift correlation.

From a careful survey of the literature, the earliest suggestion of using longer range heteronuclear scalar coupling pathways as a mechanism for chemical shift correlation is found in a communication by Hallenga and Van Binst[91] that appeared in 1980. Hallenga and Van Binst suggested that it would be possible to use heteronuclear chemical shift correlation to assign carbonyl resonances in peptides by utilizing the long range coupling to the $\alpha$ proton, thereby sequencing the peptide. Unfortunately, Hallenga and Binst were forced to use one dimensional spin population transfer (SPT), which was described in a section above, for reasons of sensitivity. It was not until 1984 that the first papers describing applications of long range heteronuclear chemical shift correlation began to appear in the literature.

In addition to the obvious application of the long range heteronuclear correlation experiment to tasks such as determining peptide sequences, it also rather conveniently provides the means of establishing scalar coupling connectivities across heteroatoms that cannot be determined by any other two-dimensional NMR experiment presently available. Furthermore, the long range heteronuclear correlation experiment also offers a substantial gain in sensitivity relative to the carbon-carbon double quantum INADEQUATE experiment, which is described in Chapter 5. For these reasons, currently available long range correlation experiments are described individually in the following sections.

## Long Range Optimization of the Conventional Heteronuclear Chemical Shift Correlation Experiment

The simplest and perhaps most direct approach to determining long range heteronuclear connectivities is through the reoptimization of the conventional heteronuclear chemical shift correlation experiments described above for smaller, long range couplings in lieu of the much larger direct or one bond couplings. In applying this strategy, two alternatives are available: long range polarization transfer via the INEPT based pulse sequence shown in Fig. 3-10 originally described by Freeman and Morris[29]; long range transfer using the DEPT based experiment of Levitt and coworkers[42] shown in Fig. 3-24. Both approaches to the problem have been used with success. For whatever reasons, however, the vast majority of reports now contained in the literature utilize the former approach, which we shall describe first.

## General Considerations of Long Range Heteronuclear Coupling Constants

Before one may contemplate long range heteronuclear correlation experiments, a knowledge of the size of the various long range heteronuclear coupling constants must be available. Here again, several options are available. A proton coupled carbon spectrum can, of course, be obtained and may or may not resolve the coupling constants of interest. Unfortunately, there will frequently be so many long range couplings into some carbons, especially when dealing with terpenoids, that it will be impossible to resolve the individual coupling constants even at the highest available observation frequencies.

An alternative if couplings cannot be directly measured is to consult one of the various review articles or monographs that treat the topic of heteronuclear spin coupling. Unfortunately, compilations of heteronuclear coupling constant information are far fewer in number than those pertaining to homonuclear spin coupling. An excellent review of heteronuclear couplings in aromatic systems is, however, contained in reviews by Hansen.[92, 93] Additional useful information on aromatic systems is also to be found in the monograph by Memory and Wilson.[94] Valuable tabulations of a more general nature are found in the monograph by Breitmaier and Voelter.[95] Perhaps the most general source, however, it the monograph specifically devoted to carbon-carbon and carbon-proton couplings by Marshall.[96]

A final alternative to direct measurement of the actual couplings is to employ a set of survey parameters. This approach was suggested in the early work of Reynolds and coworkers[97] and more recently by Wernly and Lauterwein.[98] Unfortunately, the success of this approach may be varied because of the problems associated with one bond modulation of long range response intensities which was first described by Bauer, Freeman and Wimperis[99] and more recently has been the subject of intensive investigation by groups led by Martin,[100-104] Reynolds[105] and Batta.[106] These problems are largely ameliorated, however, by some of the more sophisticated long range heteronuclear correlation experiments described in later sections.

## Long Range Optimization of the Heteronuclear Chemical Shift Correlation Sequence of Freeman and Morris

The simplest approach to establishing long range correlations is to modify the duration of the fixed defocusing ($\Delta_1$) and refocusing ($\Delta_2$) delays of the conventional heteronuclear correlation sequence of Freeman and Morris[29] shown in Fig. 3-10. In essence, this was the approach first suggested by Hallenga and Van Binst.[91] The approach has the advantage of being directly implemented without the need of programming new pulse sequences, which may be an impediment for some users.

Reoptimization of the duration of the de/refocusing ($\Delta_1/\Delta_2$) delays for a long range heteronuclear coupling rather than the direct or one bond coupling constant affords a vector picture quite similar to that shown in Fig. 3-6. The initial 90° proton pulse, since it is nonselective, excites both direct and long range components of magnetization as previously. In the direct correlation experiment described above we were able to ignore the long range components of magnetization since they never defocus to a sufficient extent during the $\Delta_1$ interval when optimized for $^1J_{CH}$ to afford any magnetization transfer. In contrast, the duration of the $\Delta_1$ interval when optimized for a long range coupling constant, i. e., 10 Hz, ensures that the long range components of magnetization will attain an antiphase orientation and will thereby engage in magnetization transfer. The inherently higher frequency of the components associated with $^1J_{CH}$ will carry them in and out of focus several times during the $\Delta_1$ interval and may or may not leave them in a state amenable to magnetization transfer.[99]

Following transfer of magnetization from a remote proton, $H_R$, ($^nJ_{CH}$ where n ≠ 1) to carbon, components of $^{13}C$ magnetization may be represented by two pairs of vectors when a given carbon is coupled to a directly attached proton ($H_D$) and the remote proton. Since $H_R$ was responsible for the transfer, the components of magnetization for this heteronuclear coupling will obviously begin to evolve during the $\Delta_2$ interval from an antiphase orientation. Assuming that the couplings due to $H_D$ were refocused at the terminus of $\Delta_1$, coupling due to $H_D$ will be "passive" and the evolution of these vector components during the $\Delta_2$ interval will commence from an in-phase orientation.[99]

During the $\Delta_2$ interval, which has been optimized for the long range coupling pathway, the initially antiphase components of magnetization due to $H_R$ will be refocused at the end of interval. The components due to $H_D$, in contrast, will evolve at their inherently higher frequency, going in and out of phase a number of times during the $\Delta_2$ interval. If the duration of the $\Delta_2$ interval unfortuitously happens to correspond to one of the antiphase alignments of the components of magnetization due to $H_D$, very little signal will be detected during the subsequent acquisition period. The $H_D$ components of magnetization in this case will have adversely affected the detection of the components of magnetization due to $H_R$, and the anticipated long range coupling information will have been lost.

**Selection of optimal delay intervals for long range heteronuclear correlation**. Clearly from the discussion contained in the preceding section, there are potential problems associated with the choice of optimal $\Delta_1$ and $\Delta_2$ delay intervals when the pulse sequence shown in Fig. 3-10 is to be employed. Nevertheless, long range optimization of the heteronuclear correlation experiment described by Freeman and Morris[29] has been very successfully employed by a relatively large number of research groups beginning with papers that first appeared in 1984. Applications of the long range optimization technique are reviewed below. Suffice it to say, however, that several recommendations appeared quite early on regarding the optimal selection of fixed delay durations.

Reynolds and coworkers[97] reported that the best results had been obtained for a terpenoid when the experiment was optimized for

10 Hz. Wernly and Lauterwein[98] also reported having obtained optimal results with β-carotene by using a 10 Hz optimization of the $\Delta_1$ and $\Delta_2$ intervals. Hence, these suggestions would give 50 and 33 msec durations for $\Delta_1$ and $\Delta_2$, respectively, when varying degrees of protonation are taken into account as in the conventional heteronuclear correlation experiment. Van Binst and colleagues[107] reported using a "compromise" value of essentially 7 Hz (70 msec for both $\Delta_1$ and $\Delta_2$) in sequencing a small peptide. While the suggested optimal settings were apparently based upon several trials in the preceding studies,[97,98,107] the rationale behind choices made in several other papers that appeared in 1984 is perhaps less clear. Lallemand and coworkers[108] reported using a 12 Hz optimization (41.67 msec for both $\Delta_1$ and $\Delta_2$) with the ionophorous antibiotic X-14547 A. Connolly, Fakunle and Rycroft, in assigning ester linkages in a terpenoidal compound, reported optimizing the $\Delta_1$ delay for 6 Hz (83.4 msec) while $\Delta_2$ was optimized for 40 msec.[109] Shoolery[110] reported using a 7 Hz optimization for a reasonably complex spiro-indole compound. Finally, in the last paper to appear in 1984 that we are aware of, Kintzinger et al. reported using optimizations of 12.5 and 20 Hz for a model pyridine compound, while only the former was employed for a 1,3-diazapyrene analog.[111]

In general, we have found that 10 Hz optimization of the fixed delays is a reasonable starting point when the pulse sequence shown in Fig. 3-10 is to be employed. Assuming that $\Delta_1$ is optimized as a function of $1/2(^1J_{CH})$, 10 Hz optimization will thus entail a 50 msec delay. Because of the relatively long periods for the $\Delta_1$ and $\Delta_2$ delays that "small" optimizations (in Hz) mandate, we have found it to be preferable to optimize $\Delta_2$ as a function of $1/3(^1J_{CH})$ in all cases. Prudent users of the experiment, spectrometer time permitting, would also be well advised to back up the data from the 10 Hz optimized experiment with that from a second experiment performed with different optimal settings because of the potential loss of signals due to the modulation by the direct coupling, which was alluded to above and will be discussed in some detail below.

**An example of long range heteronuclear correlation data obtained with the simple alkaloid norharmane**. To illustrate both the nature of the data and the potential problems associated with the manipulation of magnetization components in the generation of the data, let us consider the long range heteronuclear correlation spectrum of the simple alkaloid norharmane (**3-4**), which is shown in Fig. 3-36.[102] The data were collected using a 10.4 Hz optimization of the $\Delta_1$ and $\Delta_2$ delays giving durations of 40.1 and 32.1 msec, respectively. The total data accumulation time was about 90 minutes when 128 increments of $t_1$ were recorded. Key long range couplings in the pyridine ring of norharmane (**3-4**) are shown in Fig. 3-37.

For the protonated carbon resonances of the pyridine portion of the molecule, it will be noted that C1 exhibits a reasonably intense response for the directly attached proton and a much weaker response for the H3 doublet to which it is long range coupled. Continuing, the C3 resonance also exhibits a weak long range response for coupling to the H1 resonance despite the fact that the 10.4 Hz optimization chosen compares favorably with the actual coupling of 11.1 Hz for $^3J_{C3H1}$. There is a specific reason for the preceding point, which we shall elaborate further in the following section. Finally, C4 exhibits a response with good intensity for the two bond coupling to H3 in addition to a response for the direct coupling. The quaternary carbon resonances common to the pyridine/pyrrole ring fusion, C9a and C4a, also exhibit responses in the spectrum shown in Fig. 3-36. C4a shows usable three bond couplings to the H1 and H3 resonances. The two bond coupling to the H4 resonance is absent. The C9a resonance, in contrast, exhibits a two bond coupling response with H1 that is considerably more intense than the response correlating C9a with H4 via three bonds.

Clearly, there is a mixture of two and three bond coupling information conveyed by the long range optimized spectrum shown in Fig. 3-36, and care must be taken in interpreting data from long range correlation experiments. As mentioned in the analysis of the data just presented, there are also numerous direct correlation responses contained in the spectrum. Direct responses are undesirable and needlessly complicate the spectrum. Usefully, the means are available for removing direct responses and will be discussed below (see section on low pass J filters). By far the most important point to be made regarding the data shown in Fig. 3-36, however, is the weak intensity of the three bond responses correlating H1 with C3 and H3 with C1 across the pyridine nitrogen atom. The low intensity of these responses has nothing at all to do with the presence of the annular nitrogen atom. Rather, the intensity of the C3H1 and C1H3 responses has been adversely affected by the one bond couplings of H1 to C1 and H3 to C3, which was mentioned as a complicating factor above. Indeed, the 10.4 Hz optimization was specifically chosen to highlight this point, and a detailed explanation of the effects of modulation due to the directly attached proton is found in the following section.

**One bond modulation effects in the long range optimization of the Freeman-Morris heteronuclear correlation experiment.** Having encountered the potential problems associated with modulation induced by the direct ($^1J_{CH}$) coupling, it is now appropriate to consider in more detail the origins of the problem. The first quantitative point of departure beyond the initial description of it by Freeman and coworkers[99] is contained in the study of one bond modulation of the long range INEPT experiment by Schenker and von Philipsborn.[112] Using expressions contained in that work, Martin and coworkers[100] first derived expressions that accurately predicted one bond modulations for an experiment derived from the pulse sequence of Freeman and Morris[29] in which a TANGO pulse operator is substituted in place of the second 90° proton pulse. The modified pulse sequence

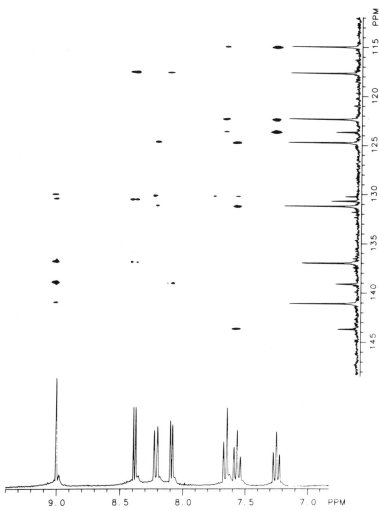

Fig. 3-36.   Long range heteronuclear chemical shift correlation spectrum of
norharmane (3-4) obtained using the pulse sequence of Freeman and
Morris[29] shown in Fig. 3-10.   Observation frequencies were 300.068
and 75.459 MHz for $^1$H and $^{13}$C, respectively. Pulse widths were 15.4
µsec for $^{13}$C and 34 µsec for $^1$H from the decoupler coils. The duration of
the $\Delta_1$ and $\Delta_2$ delays was optimized for 10.4 Hz giving durations of 40.1
and 32.1 msec, respectively.   A total of 128 increments of the evolution
time were performed, with 48 acquisitions taken per increment with an
interpulse delay of 1 sec giving an acquistion time of 90 minutes.   The
data was processed in a manner identical to that described for Fig. 3-18.
Reference proton and carbon spectra are plotted flanking the contour
plot.

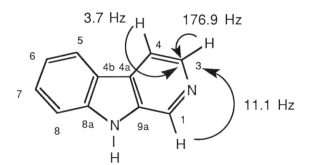

*Fig. 3-37.*   Selected long range heteronuclear coupling pathways in the pyridine ring of the alkaloid norharmane (**3-4**) determined from a gated decoupling $^{13}C$ experiment.

is described in a section below. For purposes of the present discussion, it is useful to note that the experiment presently under consideration behaves in an analogous fashion, as shown more recently by Zektzer, John and Martin.[102]

In the simplest case, where we are dealing with an isolated long range coupling into a carbon, for example a long range coupling into a quaternary carbon, response intensity may be described by the expression:

$$I = \sin [ \pi \Delta_1 \, ^{LR}J_{CH}] \sin [ \pi \Delta_2 \, ^{LR}J_{CH}] \qquad [3\text{-}13]$$

where $^{LR}J_{CH}$ represents the long range coupling being observed and where $\Delta_1$ and $\Delta_2$ are the defocusing and refocusing delays, respectively (see Fig. 3-10). The calculated response intensity over the optimization range from 2 to 15 Hz is shown in Fig. 3-38.

Response intensity (I) calculated using Eqn. [3-13] which is shown in Fig. 3-38, illustrates the fact that even where there is only a single isolated coupling to deal with, the long range response will not be observed under certain optimization conditions. Furthermore, it should also be noted that although good response intensity is predicted at some lower (in Hz) optimization values by the curve shown in Fig. 3-38, it is impractical to optimize the experiment for very small couplings because of the losses of signal that may occur during the lengthy defocusing and refocusing delays entailed by small optimization values. Generally, we have not found it useful to employ optimizations much below about 5 Hz (100 msec) in situations where there are multiple long range couplings into a given carbon. As with any generality, however, there will always be exceptions to be noted. Nishida, Morris and Enzell[113] for example, have reported successfully using a 3.85 Hz (130 msec) delay duration to establish long range couplings into acetate carbonyl resonances of sucrose octaacetate. In any case, other factors

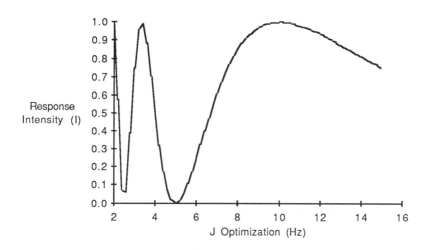

*Fig. 3-38.*     Response intensity (I) calculated for an isolated 10 Hz coupling as a
                 function of the optimization of the defocusing and refocusing delays
                 ($\Delta_1$ and $\Delta_2$, respectively) using Eqn. [3-13]. The fixed delays were of
                 equal duration. Relaxation effects were not considered in the
                 calculation.

below should be taken into consideration before making the choice of
optimal defocusing and refocusing delay durations.

The simple equation shown above (Eqn. [3-13]) becomes more
complex when the effects of additional couplings must be factored in.
Consider the response intensity for the H1-C3 coupling pathway in the
pyridine ring shown in Fig. 3-37. In addition to the three bond coupling
to actually be observed, $^3J_{C3H1}$ = 11.1 Hz, we must now consider the
effects of the direct coupling, $^1J_{C3H3}$ = 176.9 Hz as well as the two bond
coupling, $^2J_{C3H4}$ = 3.7 Hz. While the three bond coupling will require
terms in an expression identical to that shown in Eqn. [3-13], the one
bond and two bond couplings must be factored in as cosine terms in a
product operator ($\prod$) as shown by Eqn. [3-14]:

$$I = \sin [\pi \, \Delta_1 \, ^{LR}J_{CH}] \sin [\pi \, \Delta_2 \, ^{LR}J_{CH}] \times$$

$$\prod \cos[\pi \, \Delta_2 \, ^1J_{CH}] \cos [\pi \, \Delta_2 \, ^{LR'}J_{CH}] \qquad [3\text{-}14]$$

where the term $^1J_{CH}$ represents the one bond coupling and $^{LR'}J_{CH}$
represents the two bond coupling with all other terms defined as
above.[100, 102]

The response intensity curve calculated using Eqn. [3-14] and the coupling constants of the pyridine ring shown in Fig. 3-37 is shown in Fig. 3-39. Experimentally determined response intensities for a series of optimizations of the defocusing and refocusing delay ($\Delta_1$ and $\Delta_2$ in Fig. 3-10) are plotted as individual points superimposed over the calculated curve.[102]

The rapid modulation of the response curve shown in Fig. 3-39 is induced by the cosine term for the direct coupling, $\cos [\pi \Delta_2 \, ^1J_{CH}]$. From the curve it should be noted that the precipitous valleys in the response curve may lead to losses of long range response intensity despite optimization at values quite near the actual coupling constant. In the case above, an 11.1 Hz optimization gives excellent signal intensity while optimizations of 10.4 and 11.8 Hz give only minimal response intensity. Hence, long range optimization of the pulse sequence shown in Fig. 3-10 should be used with caution. We will see below that it is possible to suppress the rapid modulation caused by the direct coupling by a relatively simple modification of the heteronuclear correlation pulse sequence.

The modulation problem inherent to the methine carbon case just described is by no means unique. Similar problems will plague any protonated carbon to which magnetization is transferred. In the case of a methylene carbon, the term $(\{1 + \cos [2\pi \Delta_2 \, ^1J_{CH}]\}/2)$ may be substituted for the term $\cos [\pi \Delta_2 \, ^1J_{CH}]$ contained in Eqn. [3-14 to give Eqn. [3-15]. The modification of the response intensity equation in Eqn. [3-15] describes the modulation of a methylene carbon center.

$$I = \sin [\pi \Delta_1 \, ^{LR}J_{CH}] \sin [\pi \Delta_2 \, ^{LR}J_{CH}] \times$$

$$\prod (\{ 1 + \cos [2\pi \Delta_2 \, ^1J_{CH}]\}/2) \cos [\pi \Delta_2 \, ^{LR'}J_{CH}] \quad [3\text{-}15]$$

A similar expression may be written for a methyl carbon by substituting the term $(\{ 3 \cos[\pi \Delta_2 \, ^1J_{CH}] + \cos [3\pi \Delta_2 \, ^1J_{CH}]\}/4)$ for the term $(\{1 + \cos [2\pi \Delta_2 \, ^1J_{CH}]\}/2)$ contained in Eqn. [3-15] to give Eqn. [3-16].

$$I = \sin [\pi \Delta_1 \, ^{LR}J_{CH}] \sin [\pi \Delta_2 \, ^{LR}J_{CH}] \times$$

$$\prod (\{ 3 \cos[\pi \Delta_2 \, ^1J_{CH}] + \cos [3\pi \Delta_2 \, ^1J_{CH}]\}/4) \cos [\pi \Delta_2 \, ^{LR'}J_{CH}] \quad [3\text{-}16]$$

The important feature to be noted regarding these euqations ([3-15] and [3-16]) is that the valleys created in the response curve by the modulation induced by the directly attached protons get progressively broader as the degree of protonation increases. This point has been experimentally demonstated by Salazar, Zektzer and Martin[103] for 6,7-dimethoxy-3,4-dihydroisoquinoline (**3-13**) whose coupling pathways are shown in Fig. 3-40, by examining the transfer from H1 into the C3 methylene carbon. Similar results have also been noted by Reynolds and coworkers[105] for methylene and methyl carbons of 2-pentanone.

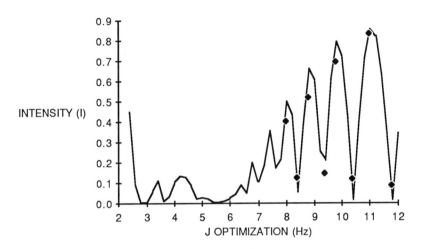

*Fig. 3-39.* Response intensity versus optimization curve calculated using Eqn. [3-14] for optimizations ranging from 2 to 12 Hz and the coupling constants shown in Fig. 3-37. The duration of the $\Delta_2$ delay was set equal to $\Delta_1$, since magnetization was being transferred into a methine carbon. Experimental points were determined from individual long range optimized heteronuclear chemical shift correlation spectra at the optimizations indicated by the solid diamonds. Optima were selected on the basis of the calculated curve shown.

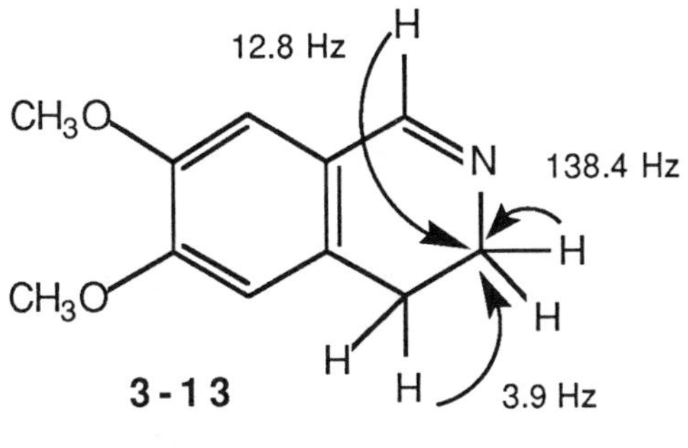

*Fig. 3-40* Selected heteronuclear coupling pathways of 6,7-dimethoxy-3,4-dihydroisoquinoline (**3-13**).

The important point to be noted from the response curve calculated for the C3 methylene carbon of **3-13** using Eqn. [3-15] and shown in Fig. 3-41 is that only very minimal intensity would be obtained for the long range response using the 10 Hz optimum suggested by the early studies of Reynolds and coworkers[97] and Wernly and Lauterwein.[98] Hence, it must be stressed that long range optimization of the pulse sequence developed by Freeman and Morris[29] must be employed with caution. Once again, the modulation effects described by Eqns. [3-15] and [3-16] may be eliminated by using the modified pulse sequences described below.

In summary, modulations induced by the directly attached proton are a critical concern when magnetization is being transferred to a protonated carbon. Valleys in the response curve broaden with successively higher degrees of protonation. Practically, the problem of modulation of methyl carbon long range response intensity, which may be described by Eqn. [3-16], will be of less concern than that of a methylene carbon, since information gleaned from the transfer of methyl protons to other carbons will normally be much more important than transfer from a proton into a methyl carbon.

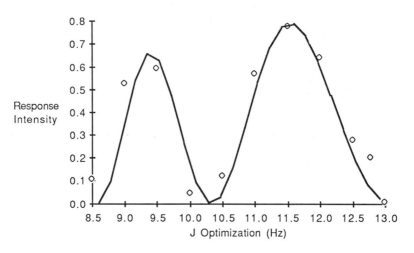

*Fig. 3-41.*     Response intensity curve for the $^3J_{C3H1}$ coupling pathway of 6,7-dimethoxy-3,4-dihydroisoquinoline (**3-13**) calculated using Eqn. [3-15] and the coupling constants shown in Fig. 3-40. Experimentally determined points superimposed over the calculated curve (open circles) were determined using the pulse sequence devised by Freeman and Morris[29] shown in Fig. 3-10 in which $\Delta_2 = 0.667\Delta_1$.

**A brief survey of applications of the long range optimized heteronuclear chemical shift correlation experiment developed by Freeman and Morris.**[29]     Of the several long range heteronuclear chemical shift correlation pulse sequences presently available, by far the largest number of applications must be credited to the pulse sequence devised by Freeman and Morris[29] as of the time that this monograph was written. Doubtless this distribution will change as more studies are performed, in part because of the modulation problems associated with the pulse sequence. In any case, there are numerous, interesting applications that are worth briefly reviewing at this point.

Reports of several applications in the assignment of small peptides have indicated varying degrees of success.[107,114] There have been single applications to a complex ionophorous antibiotic,[108] a corrinoid[115] and an oligosaccharide.[113] Several applications to alkaloids have also appeared.[99,102-104,116-119] A substantial number of applications of long range heteronuclear chemical shift correlation using the pulse sequence of Freeman and Morris[29] have been reported for polynuclear aromatic systems.[111,120-125] Finally, by far the most numerous applications have been in the area of natural products, many of which have been terpenoid in nature and/or of marine origin.[97,98,109,110,126-139]

## Long Range Optimized DEPT

Unlike the long range optimized experiment developed by Freeman and Morris,[29] which has been used numerous times in spectral assignment and structure elucidation studies, the long range optimized DEPT experiment has been used only a relatively few times. In all instances, the variant of the DEPT heteronuclear correlation experiment selected for long range optimization has been that developed by Levitt and coworkers,[42] which is shown in Fig. 3-24.

Virtually all of the work reported using the long range optimized DEPT heteronuclear correlation experiment has emanated from the laboratory of Batta. The problem of establishing long range heteronuclear couplings across the ether linkages of oligosaccharides necessary for sequencing was first addressed by Batta and Liptak[50] using long range optimized DEPT in 1985. More recently, Somogyi, Herczegh and Batta have utilized the long range optimized DEPT experiment in the assignment of a thiazole containing product of the anomalous Wittig reaction of aldonic thioamide derivatives.[140] The first study of factors entailed in the optimization of the long range DEPT experiment was reported only recently by Batta and Köver.[51]

Signal intensity in the long range DEPT experiment is proportional to

$$\sin [\pi \tau \, {}^{LR}J_{CH}] \exp(-\tau/T_{2C}) \qquad [3\text{-}17]$$

where the term $\tau = 1/2({}^{LR}J_{CH})$ and where the term $T_{2C}$ is determined by the $T_2$ relaxation time of the $^{13}C$ nuclei. Taking into account relaxation processes, Batta and Köver[51] have shown that the final signal intensity, $M_x$, is defined by Eqn. [3-18] for a heteronuclear AX spin system

$$M_x = \exp (-2\tau/T_{2H}) \exp (-\tau/T_{2C}) \sin^2 [\pi \, {}^{LR}J_{CH}] \sin \theta \qquad [3\text{-}18]$$

which allows the determination of the relationship between the detected signal intensity and the optimal choice of $\tau$, which is given by Eqn. [3-19].

$$\tau_{opt} = \text{arc tan} \left[ 2\pi \, ^{LR}J_{CH} \left( T_{2H} T_{2C} / 2T_{2C} + T_{2H} \right) \right] \left( 1/\pi \, ^{LR}J_{CH} \right) \quad [3-19]$$

There have been no studies of the modulation effects in the long range optimized DEPT heteronuclear chemical shift correlation experiment.

## Long Range Heteronuclear Correlation Using the Fully Coupled (FUCOUP) Technique

While the final version of the heteronuclear chemical shift correlation technique devised by Freeman and Morris[29] uses fixed delays of specific duration to defocus and refocus specific heteronuclear couplings, the much earlier and simpler variant of the pulse sequence shown in Fig. 3-5 does not employ any fixed delays.[26]  Rather, the experiment, as shown in Fig. 3-7, affords a spectrum in which full heteronuclear coupling is retained in both frequency domains.  By virtue of the fact that specific couplings are not manipulated by the inclusion of fixed delays, the pulse sequence shown in Fig. 3-5 will put all couplings into the spectrum whether direct or long range.

The pulse sequence shown in Fig. 3-5, although supplanted by the one shown in Fig. 3-10  for direct heteronuclear correlation, has been utilized for long range correlation studies by a number of authors. In this application, the experiment has been occasionally referred to using the acronym FUCOUP. While the pulse sequence does have the advantage of being free from the need of selecting an optimal duration for the fixed delays and the inherent problems of modulation, which were discussed above with regard to the long range optimization of the pulse sequence shown in Fig. 3-10, it does have detractors.  First, the sensitivity of the experiment is inherently lower, since responses are distributed over all of the components of the multiplet.  Second, the experiment is complicated by virtue of containing responses for both the long range and direct responses, the latter superfluous if a conventional heteronuclear chemical shift correlation experiment is acquired.

## Parameter Selection

The FUCOUP experiment, in terms of parameters that must be selected, is probably the simplest possible long range heteronuclear correlation experiment.  Only the pulse widths, transmitter frequencies, dwell times and levels of digitization and sweep widths in $F_1$ and $F_2$ must be selected.  The reader is therefore directed to the appropriate sections where these considerations are described for the conventional heteronuclear chemical shift correlation experiment above for all but the level of digitization in $F_1$.  Here another concern must be addressed. Specifically, if the $F_1$ frequency domain is underdigitized, the long range components of magnetization will never attain a sufficiently antiphase orientation to allow efficient magnetization transfer. Thus, a sufficient number of blocks of data in $F_1$ should be selected to allow the evolution time, $t_1$, to equal $1/2(^{LR}J_{CH})$, thus ensuring efficient transfer for the long range couplings of interest.

## Phase Cycling

Phases of the pulses and the receiver in the FUCOUP experiment may be cycled according to Table 3-2. Thus, the first 90° proton pulse utilizes phase $\phi_1$; the second 90° proton pulse is cycled according to $\phi_2$.; the 90° $^{13}$C pulse is cycled according to $\phi_3$. The receiver may be cycled to acquire either the coherence transfer echo ($\phi_E$) or the antiecho ($\phi_A$). An alternative phase cycling scheme for the FUCOUP sequence, which has been employed by Waterhouse, Holden and Casida[141] is found in the work of Bleich et al.[142] Very recently, Bain[143] has described a phase sensitive variant of the experiment. Interestingly, Bain also suggested in that work that quadrature detection in $F_1$ may afford no real advantage over single phase detection, although this remains to be confirmed.

## Data Processing

Data processing considerations for the FUCOUP experiment are also largely analogous to those of the conventional direct correlation experiment. It may, however, be useful to be less generous with line broadening in the case of the FUCOUP experiment, since the responses in the spectrum will be much more numerous than in the conventional spectrum of any molecule being studied. It should also be noted that if the coherence transfer echo is collected ($\phi_E$ in Table 3-1), the spectrum must be rotated following the second Fourier transformation.

To provide a basis of comparison for the reader, the FUCOUP spectrum of norharmane (**3-4**) is shown in Fig. 3-42 and may be compared to the spectrum generated using long range optimization of the conventional heteronuclear correlation pulse sequence shown in Fig. 3-10.

## A Brief Survey of Applications
## of the FUCOUP Experiment

Despite the problems of sensitivity and congestion, the FUCOUP experiment has been employed successfully for both structure elucidation and spectral assignment. Waterhouse, Holden and Casida[141] have utilized FUCOUP in the total assignment of the $^{13}$C-NMR spectrum of the complex insecticide ryanodine. Gould and coworkers have used the FUCOUP experiment in the elucidation of the structure of the antibiotic kinamycin D[144] and murayaquinone, a phenanthraquinone metabolite of *Streptomyces murayamaensis*.[145] Representing what is thus far the most complex molecule to which FUCOUP has been applied, Floss et al. [146] successfully assigned the $^{13}$C-NMR spectra of

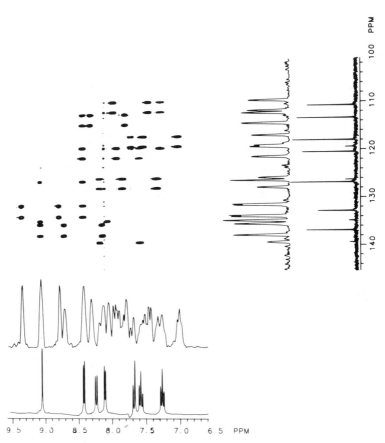

*Fig. 3-42.*   Long range heteronuclear correlation spectrum of norharmane (**3-4**)
acquired using the FUCOUP technique employing the pulse sequence
shown in Fig. 3-5.   The receiver phase was cycled to record the
coherence transfer echo.

the antibiotic boromycin. Kashman and colleagues have utilized the
technique in a study of the rearrangement and ring opening of the
complex macrolide latrunculin-B[147] and in the determination of the
structures of several novel imidazole alkaloids.[148] Kubo and
coworkers[149] have applied the technique to the *ent*-kaurene diterpene,
adenanthin.   Kawasaki and colleagues[150] have used FUCOUP in
determining the structure of a new labdane diterpene, scoparic acid.
Hochlowski and coworkers[151] have used the technique in elucidating
the structure of the tiacumicins, a complex of novel 18-membered
macrolide antibiotics.   Halterman and coworkers[152] have applied to the
technique to synthetic products and finally, Waterhouse[153] has used the

coupling information afforded by the experiment for conformational analysis.

## A Modification of the FUCOUP
## Experiment: Introduction
## of $F_1$ Decoupling.

Recognizing the sensitivity and congestion problems inherent to the FUCOUP approach to long range heteronuclear correlation, Wernly and Lauterwein[154] employed the pulse sequence shown in Fig. 3-8, which collapses the heteronuclear couplings in the $F_2$ frequency domain as shown in Fig. 3-9, for long range heteronuclear chemical shift in a study of β-carotene. This approach has the beneficial effect of reducing the total number of responses in the spectrum and thereby congestion. Furthermore, partial collapse of the heteronuclear multiplet structure improves sensitivity. However, the introduction of the $\Delta_1$ delay adds the necessity of optimization considerations and reintroduces the problem of one bond modulation found with the full sequence of Freeman and Morris[29] shown in Fig. 3-10, thereby offsetting any gains. It is perhaps for this reason that the single report of Wernly and Lauterwein[154] is the only appearance of this approach to long range heteronuclear chemical shift correlation that we are aware of.

**Phase cycling and parameter selection**. Phase cycling considerations implicit with the pulse sequence shown in Fig. 3-8 are identical to those discussed above with the FUCOUP approach to long range correlation.

Parameter selection for $F_1$ decoupled FUCOUP is only minimally more complex than for the FUCOUP experiment itself. In addition to pulse widths, transmitter frequencies and levels of digitization in $F_1$ and $F_2$, it is necessary to select the duration of the refocusing delay, $\Delta_2$. Considerations that enter into the selection of an appropriate duration for $\Delta_2$ in this case are identical to those discussed above with regard to the long range optimized conventional experiment.

**Data processing**. Processing considerations, practically, are identical to those inherent to FUCOUP.

## Modifications Intended to
## Suppress Direct Responses

Direct responses in long range heteronuclear chemical shift correlation spectra are generally regarded as a nuisance since they may obscure information from long range coupling pathways in some instances. Thus, eliminating direct responses is desirable. Basically, there are two ways in which direct respones may be eliminated. First, by using a tailored pulse sequence such as a TANGO pulse,[155] which is intended to excite only protons distantly coupled to carbon atoms it is in principle possible to leave directly attached protons unexcited. Alternately, components of magnetization due to the directly attached proton may be eliminated by means of a low pass J filter[156] or though the use of semiselective refocusing via a BIRD pulse.[157] These approaches to the problem of removing direct responses will be discussed in turn.

### Replacing Proton 90° Pulses by TANGO Pulses: The CSCMLR Experiment.

Basically, there are two sites in which the pulse sequence shown in Fig. 3-10 may be modified by the incorporation of a TANGO pulse. Either of the proton 90° pulses may be replaced. By replacing the first 90° pulse, only those protons long range coupled to carbon will be excited and their magnetization allowed to evolve. Alternately, the second 90° pulse may be replaced by a TANGO pulse which, in principle, should not transfer components of magnetization associated with directly attached protons to carbon for subsequent refocusing. Thus far, in the case of variants of the Freeman-Morris sequence, only the latter approach has been reported by Zektzer et al.[158]

Two variants of the TANGO pulse operator are described in the original report of Wimperis and Freeman.[155] Using the form shown in Eqn. [3-I] a 360° pulse is applied to remote (long range coupled) protons while the directly attached protons experience a 90° pulse when $\tau = 1/2(^1J_{CH})$.

$^1H$ $\qquad$ $135°(+x) - \tau - 180°(+x) - \tau - 45°(+x)$

$^{13}C$ $\qquad\qquad\qquad\qquad$ $180°$ $\qquad\qquad\qquad\qquad$ [3-20]

A more appropriate form of the TANGO pulse for long range heteronuclear correlation requires the replacement of the initial 135° proton pulse by a 45° proton pulse. This modification, shown by Eqn. [3-21], with the $\tau$ interval defined as above, applies a 90° pulse to remote protons while selectively inverting directly attached protons.

$^1H$ $\qquad$ $45°(+x) - \tau - 180°(+y) - \tau - 45°(-x)$

$^{13}C$ $\qquad\qquad\qquad\qquad$ $180°$ $\qquad\qquad\qquad\qquad$ [3-21]

Vector presentations of the operation of both of these sequences are found in Chapter 1 (see Figs. 1-27 and 1-28).

Zektzer et al.[158] incorporated the TANGO pulse operator shown in Eqn. [3-21] in place of the second 90° proton pulse of the Freeman and Morris[29] sequence shown in Fig. 3-10 to give the modified pulse sequence shown in Fig. 3-43.

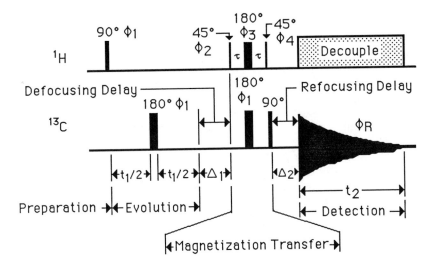

*Fig. 3-43.* CSCMLR pulse sequence[158] incorporating a TANGO pulse in place of the second proton 90° pulse with the intent of selectively transferring magnetization from remote protons to carbon.

**Phase cycling.** Phases in the CSCMLR pulse sequence may be cycled in a relatively simple eight step cycle, which is quite similar to that used for the conventional heteronuclear chemical shift correlation experiment shown in Table 3-1. The sole difference in Table 3-7 is the addition of the necessary phase statements to accommodate the phase cycling of the TANGO pulse. It should also be noted that the receiver phase in Table 3-7 is cycled for the coherence transfer echo, which will necessitate a spectrum rotation following the second Fourier transformation. Individuals using spectrometers without the capability of performing a spectrum rotation should cycle the receiver phase in the

Table 3-7. Pulse and receiver phases for the CSCMLR pulse sequence shown in Fig. 3-43.[a]

| Acquisition | $\phi_1$ | $\phi_2$ | $\phi_3$ | $\phi_4$ | $\phi_R$ |
|-------------|----------|----------|----------|----------|----------|
| 1 | 0 | 0 | 1 | 2 | 0 |
| 2 | 2 | 1 | 2 | 3 | 3 |
| 3 | 0 | 2 | 3 | 0 | 2 |
| 4 | 2 | 3 | 0 | 1 | 1 |
| 5 | 2 | 2 | 3 | 0 | 0 |
| 6 | 0 | 3 | 0 | 1 | 3 |
| 7 | 2 | 0 | 1 | 2 | 2 |
| 8 | 0 | 1 | 2 | 3 | 1 |

[a] The phase cycling shown can be extended to a 32 step cycle by superimposing a cyclops cycle over the phase of the 90° carbon pulse.

opposite sense (0123 0123) to record the antiecho, thus obviating the need for spectrum rotation during processing.

**Applications of the CSCMLR experiment and parameter selection.** The sequence has been successfully applied to the total assignment of a model heteroaromatic system. In appearance the spectrum resembles those acquired by simple long range optimization of the sequence devised by Freeman and Morris[29] discussed in considerable detail above. The number of direct responses observed using the CSCMLR pulse sequence shown in Fig. 3-43 was about half that observed with long range optimization of the pulse sequence shown in Fig. 3-10. In addition, it was also noted that the intensity of the direct responses observed was lower than the corresponding responses in the conventional experiment. Thus, even though the range of one bond couplings encountered for a heteroaromatic compound is considerably more restricted than for a natural product, the replacement of the second 90° pulse by a TANGO pulse does not successfully eliminate all of the direct responses in the spectrum.

The results obtained using the CSCMLR pulse sequence with norharmane (**3-4**) are shown in Fig. 3-44. The fixed delays for defocusing ($\Delta_1$) and refocusing ($\Delta_2$) may be optimized in a fashion analogous to the conventional heteronuclear chemical shift correlation (Fig. 3-10), considerations pertaining to the degree of protonation apply normally. In the case of the example spectrum shown, since magnetization was being transferred exclusively to quaternary and protonated methine carbons, the duration of the delays was set equal, both optimized for the $^3J_{C3H1} = 11.1$ Hz coupling. The duration of the $\tau$ interval in the TANGO pulse operator was optimized as a function of the average one bond coupling $[1/2(^1J_{CH}) = 2.9$ msec for 170 Hz].

**Modulation effects in the CSCMLR experiment**. The first of the long range experiments actually studied in terms of one bond modulation effects was the CSCMLR experiment. Working with 1-benzothieno[2,3-c]pyridine (**3-14**) Martin and coworkers[100] experimentally verified the Eqn. [3-14] describing the effects of one bond modulation. Response curves showing modulation effects analogous to that shown in Fig. 3-39 were reported.

## Using Low Pass J-filters
## to Remove Direct Responses

In the course of early work with the heteronuclear relayed coherence transfer experiment (see Chapter 4), Ernst and coworkers demonstrated a simple and effective means of eliminating direct responses from the relay spectrum, the low pass J filter.[156] The low pass J-filter is easily implemented and consists of a single 90° pulse applied to $^{13}C$ a period of time, $\tau$, after the first proton pulse where $\tau = 1/2(^1J_{CH})$ in the simplest case. The 90° $^{13}C$ pulse serves to convert the

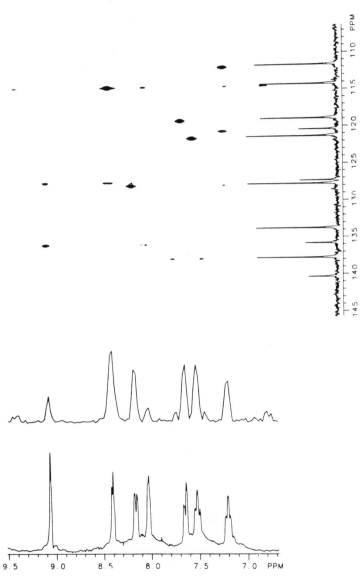

*Fig. 3-44 .*   Long range heteronuclear correlation spectrum of norharmane (**3-4**) recorded using the CSCMLR pulse sequence shown in Fig. 3-43. The spectrum was acquired as 106 x 1K complex points and was zero filled during processing to give a final data matrtix consisting of 256 x 512 points.

**3-14**

component of magnetization defocused by the direct heteronuclear into unobservable heteronuclear multiple quantum coherence. Magnetization associated with $^nJ_{CH}$ where n = 2-4 suffers only minimally, provided the duration of $\tau$ is short in comparison with the reciprocal of the long range coupling. Thus, in principle, low pass J filtration provides an excellent means of eliminating direct responses from long range heteronuclear chemical shift correlation spectra. An alternative approach utilizing semiselective refocusing provided by a BIRD pulse has been described by Bolton[157] but will not be discussed here.

Although Kessler and coworkers[159] suggested the incorporation of low pass J filtration in one of their first papers describing the COLOC long range experiment, the first reported application was by Salazar.[104,160] Using the conventional heteronuclear chemical shift correlation pulse sequence shown in Fig. 3-10, a low pass J filter was incorporated by the insertion of a 90° $^{13}C$ pulse a fixed interval, $\tau$, after the first 90° proton pulse as shown in Fig. 3-45.

**Phase cycling.** The phase cycling for the pulse sequence shown in Fig. 3-45 is only minimally different from that of the conventional long range optimized experiment. The sole difference is the alternation of the phase of the first 90° $^{13}C$ pulse along the ±x-axis as shown in Table 3-8. Once again, the receiver phase may be selected to collect either the antiecho or the coherence transfer echo, the latter shown in the table.

**Parameter selection**. Pulse widths, transmitter frequencies, dwell times and levels of digitization and sweep widths in $F_1$ and $F_2$ are selected in the normal fashion. The duration of the fixed delay, $\tau$, in the low pass J-filter may be set equal to $1/2(^1J_{CH})$. Selection of optimal values for $\Delta_1$ and $\Delta_2$ will be subject to the same considerations as were discussed for long range optimization of the conventional sequence of Freeman and Morris[29] presented above.

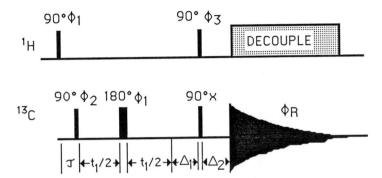

*Fig. 3-45.*   Pulse sequence for long range heteronuclear chemical shift correlation derived from the Freeman and Morris[29] sequence shown in Fig. 3-10 incorporating a low pass J-filter to eliminate responses due to directly attached protons.[104,160]

Table 3-8.   Phase Cycling Scheme for the Long Range Correlation Experiment with a Single Low Pass J-Filter shown in Fig. 3-45.[a]

| Acquisition | $\phi_1$ | $\phi_2$ | $\phi_3$ | $\phi_R$ |
|:---:|:---:|:---:|:---:|:---:|
| 1 | 0 | 0 | 0 | 0 |
| 2 | 2 | 2 | 1 | 3 |
| 3 | 0 | 0 | 2 | 2 |
| 4 | 2 | 2 | 3 | 1 |
| 5 | 2 | 0 | 2 | 0 |
| 6 | 0 | 2 | 3 | 3 |
| 7 | 2 | 0 | 1 | 2 |
| 8 | 0 | 2 | 0 | 1 |

[a] The phase cycle could be extended to 32 steps by superimposing a cyclops cycle over the second 90° $^{13}$C pulse.

### Data processing and results obtained.

Processing data accumulated using the pulse sequence shown in Fig. 3-45 is basically the same as the other long range heteronuclear correlation experiments described above. The spectrum of norharmane (**3-4**) recorded using the pulse sequence described above (Fig. 3-45) is shown in Fig. 3-46. The spectrum was recorded with 96 increments of the evolution time with 48 acquisitions taken/increment. The low pass J-filter was optimized for a 170 Hz average one bond coupling constant and $\Delta_2$ was set equal to $\Delta_1$, which was optimized for 10.4 Hz.

The long range spectrum of norharmane (**3-4**) employing long range optimization of the Freeman and Morris[29] sequence (Fig. 3-10) shown above in Fig. 3-36 may be compared with the low pass J-filtered spectrum shown in Fig. 3-46. These spectra were acquired and processed identically, the sole difference being the low pass J-filtering of the latter. The most important feature is the direct responses contained in both spectra. In the former, direct responses were observed for C1H1, C4H4, C5H5, C6H6, C7H7 and C8H8. In contrast, in the spectrum shown in Fig. 3-46, only the C4H4 direct response is visible in the contour plot.

It may be concluded that the single step, low pass J filtered pulse sequence as shown in Fig. 3-45 effectively removes direct responses over the limited range of one bond couplings (ca. 165-180 Hz) for the norharmane model system. Direct responses occurring when a broader range of one bond couplings is encountered may presumably be filtered out by resorting to multiple step J filtration as suggested by Ernst and coworkers.[156] An example of a pulse sequence containing a three step J filter in conjunction with a BIRD pulse midway through the $\Delta_2$ refocusing interval to provide decoupling of one bond modulation effects is described below (see Fig. 3-51 and 3-54).

**Artifacts and effects on one bond modulations**. It will be noted from the spectrum shown in Fig. 3-46 that a spurious response occurs at the chemical shift of the C6 resonance downfield of the H1 resonance in $F_1$, which does not correspond to any proton chemical shift. It has been suggested[104,160] that such responses may be due to the unintended reconversion of multiple quantum coherence back into observable single quantum coherence. Similar observations have been made by Brown and Bremer.[161] While the single spurious response observed in Fig. 3-46 is by no means unmanageable, much more numerous spurious responses have been observed when using the three step J filter in conjunction with the BIRD pulse for modulation decoupling.[104]

A rather interesting influence of the single step, low pass J filter on modulations caused by the one bond coupling was also noted by Salazar.[104,160] In a study of the effect of varying the optimization of the $\Delta_1$ and $\Delta_2$ delays Salazar noted that there was partial suppression of the one bond modulation over a reasonably broad range of optima. There was also a low frequency modulation of the response intensity that could not be accounted for by other heteronuclear couplings exhibited by the molecule. The "normal" one bond modulation data and the response intensities obtained over the range from 7 to 12 Hz using the pulse sequence shown in Fig. 3-45 is presented in Fig. 3-47.

## Decoupling One Bond Modulations in Long Range Heteronuclear Shift Correlation Experiments

Passive heteronuclear coupling involving the directly attached proton following polarization transfer has been shown to lead to pronounced modulation of the long range response intensity (see Fig. 3-39). Ideally, we would like to be able to "decouple" one bond modulations without adversely affecting long range responses. Several pulse sequences have been reported that are capable of disrupting one bond modulations. All of the experiments that fall into this category

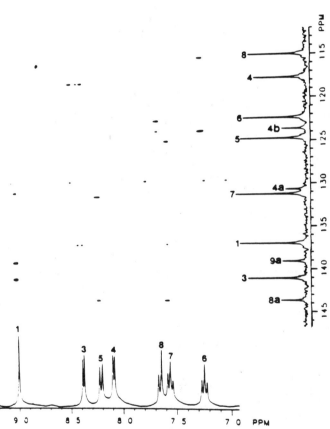

*Fig. 3-46.* Long range heteronuclear correlation spectrum of norharmane (**3-4**) recorded using the pulse sequence shown in Fig. 3-45 which incorporates a low pass J filter to remove direct responses.

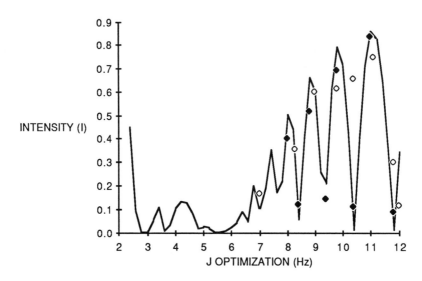

*Fig. 3-47.* Response intensity curve for the $^3J_{C3H1}$ coupling pathway of norharmane (**3-4**) calculated using Eqn. [3-14]. Solid diamonds represent experimentally determined response intensities obtained by long range optimization of the Freeman and Morris[29] sequence (see also Fig. 3-39). Open diamonds represent experimentally determined response intensities obtained using the one step low pass J-filtered pulse sequence shown in Fig. 3-44. Data from the low pass J-filtered experiment produces a result which is comparable to that obtained using modulation decoupling techniques which are discussed below (compare with Fig. 3-51).

employ bilinear rotational decoupling (BIRD) pulses, which take advantage of the significant difference in the size of the direct coupling relative to the long range coupling. These experiments may be further subdivided on the basis of how the evolution interval in the pulse sequence is handled. Thus, we will first describe the experiments in which the evolution time is incremented in the now familiar fashion. We will then consider experiments in which the duration of the evolution period is constant in the next section of this chapter.

## Suppression of One Bond Modulation in the CSCMLR Sequence by Incorporation of a BIRD Pulse

Passive coupling of the $^{13}C$ magnetization to the directly attached proton ultimately leads to losses of signals when the direct component of the $^{13}C$ magnetization attains an antiphase orientation at the end of the $\Delta_2$ interval. Practically, we may prevent the magnetization component from the directly attached protons from attaining an antiphase orientation at the end of the $\Delta_2$ interval by inserting a bilinear rotational decoupling or BIRD pulse[46] midway through the interval. Specifically, we would like to apply a 0° flip to the directly attached protons while both the remote protons and the $^{13}C$ nuclei experience a 180° flip. This may be achieved by employing a BIRD pulse in the form shown by Eqn. [3-22], which is described by vector diagrams in Fig. 1-25.

$$^1H \qquad 90°(+x) - \tau - 180°(+x) - \tau - 90°(-x)$$

$$^{13}C \qquad\qquad\qquad 180°(+x) \qquad\qquad\qquad [3\text{-}22]$$

The net result of this modification is to refocus all components of magnetization at the end of the $\Delta_2$ interval. Refocusing prevents the direct components of magnetization from attaining an antiphase orientation in which they may adversely affect long range response intensity.

Several research groups have utilized the principle just described in modifying heteronuclear correlation pulse sequence. The first such modification to be described was in the work of Bauer, Freeman and Wimperis.[99] More recently, Reynolds and coworkers[61] reported a similar sequence, which was given the acronym XCORFE. Both of these experiments employ constant duration evolution times and are discussed in detail below. The first experiment utilizing an incremented evolution period to incorporate a BIRD pulse midway through the $\Delta_2$ delay was the modified version of the CSCMLR sequence[158] (Fig. 3-43) reported by Zektzer, John, Castle and Martin,[101] which is shown in Fig. 3-48.

**Phase cycling**. Phases for the modified experiment may be cycled in eight steps using the scheme shown in Table 3-7. The phase cycle could be extended to 32 steps by imposing an additional cyclops cycle over either the 90° $^{13}C$ pulse or the BIRD pulse; alternately, a 128 step cycle could be generated by imposing separate cyclops cycles over both the 90° $^{13}C$ pulse and the BIRD pulse. We note, however, that completely satisfactory results have been obtained with the minimal eight step phase cycle specified in Table 3-7.

**Parameter selection**. Pulse widths, transmitter frequencies, sweep widths and levels of digitization in $F_1$ and $F_2$ are chosen in a fashion identical to any other heteronuclear chemical shift correlation experiment. The $\Delta_1$ interval is nominally set to $1/2(^nJ_{CH})$. To minimize losses due to relaxation during the longer delays necessary to accommodate long range couplings, we generally have found it advisable to optimize $\Delta_1$ for 10 Hz (50 msec) unless there is some compelling reason to optimize for a smaller coupling, which necessitates a correspondingly longer delay. For transfer to a methine

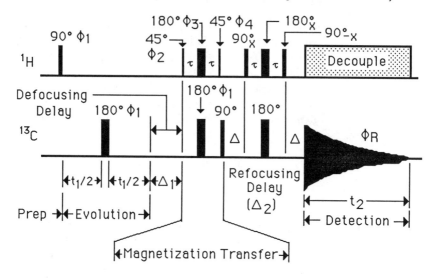

*Fig. 3-48.*     CSCMLR pulse sequence (Fig. 3-43) modified by incorporating a BIRD sequence midway through the refocusing delay ($\Delta_2$) to suppress one bond modulations. The defocusing interval, $\Delta_1$, is optimized as a function of $1/2(^nJ_{CH})$ in the usual sense. The refocusing interval, $\Delta_2$, will have a duration that may be equal to $\Delta_1$ in the case of a methine carbon or $0.667\Delta_1$ for general transfer. The $\tau$ interval is optimized as a function of $1/2(^1J_{CH})$; $\Delta = (\Delta_2 - 2\tau)/2$.

carbon, the refocusing delay, $\Delta_2$, should equal $\Delta_1$. In the general case, $\Delta_2$ may be set to $0.667\Delta_1$, thus affording optimal transfer to methine, methylene and methyl carbons. The $\tau$ interval in both the TANGO and BIRD pulses is normally set to $1/2(^1J_{CH})$, where $^1J_{CH}$ is taken as the average one bond coupling. The $\Delta$ interval of the refocusing delay is normally determined as a function of $(\Delta_2 - 2\tau/2)$. Lengthening the $\Delta_2$ interval should generally be avoided as corresponding decreases in observed signal intensity are obtained.

    **Modulation decoupling**. Response intensity for the $^3J_{C2H4}$ coupling pathway of 1-benzothieno[2,3-c]pyridine (**3-14**) was used to evaluate the ability of the BIRD pulse inserted into the refocusing delay to decouple one bond modulation effects. Using the coupling constants shown in Fig. 3-49, the response curve was calculated using Eqn. [3-14]. The calculated response curve and the experimentally determined response intensity over the range from 2 to 12 Hz are shown in Fig. 3-50a and are analogous to the results obtained using the conventional heteronuclear chemical shift correlation pulse sequence (Fig. 3-10) optimized for long range transfer.

    The behavior of the spins under the influence of the BIRD pulse located midway through the refocusing delay, $\Delta_2$, was described above.

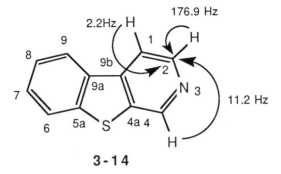

**3-14**

*Fig. 3-49.* Heteronuclear couplings in the pyridine portion of 1-benzothieno-[2,3-c]pyridine (**3-10**) employed as a model compound to evaluate the modulation decoupling capabilities of the pulse sequence shown above (Fig. 3-48).

The practical result of that discussion, mathematically, is to simplify Eqn. [3-14] to give Eqn. [3-23] by eliminating the first cosine term which describes the contribution of $^1J_{CH}$.

$$I = \sin [\pi \Delta_1 \, ^{LR}J_{CH}] \sin [\pi \Delta_2 \, ^{LR}J_{CH}] \prod \cos [\pi \Delta_2 \, ^{LR'}J_{CH}] \qquad [3\text{-}23]$$

Thus, we would expect the long range response curve for the coupling $^{LR}J_{CH}$ to be an essentially smooth curve exhibiting only the slow modulation contributed by $^{LR'}J_{CH}$. The response curve calculated for $^3J_{C2H4}$ using the coupling constants shown in Fig. 3-49 is presented in Fig. 3-50b. Experimentally determined points are superimposed over the calculated response curve confirming the "decoupling" of the one bond modulation of the long range coupling pathway.

The practical consequence of the experimental results just described is to eliminate the concern for losses of long range responses resulting from an inopportune choice for the optimization of the fixed delays in the long range heteronuclear correlation experiment! As will be noted from Fig. 3-50a, valleys in the response curve may be rather precipitous. For example, 10.4 and 10.8 Hz give good response intensity (> 0.75). The intermediary 10.6 Hz optimization gives only minimal intensity (0.10). In contrast, using the pulse sequence shown in Fig. 3-48 we obtain response intensity >0.8 when the experiment is optimized at 10.6 Hz. The benefits of the modified pulse sequence shown in Fig. 3-48 are obvious: durations of the fixed delays may be chosen with minimal concern for the loss of long range responses.

(b)

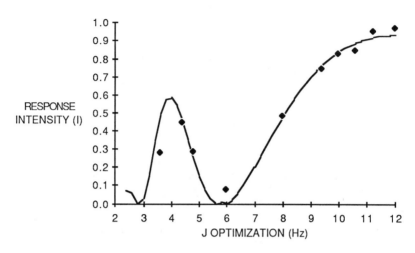

(a)

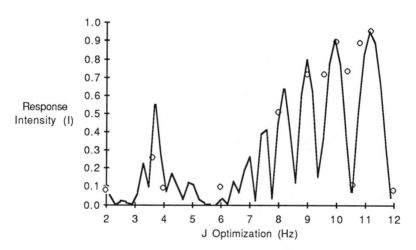

*Fig. 3-50.*    Calculated long range response curves for the $^3J_{C2H4}$ coupling of 1-
benzothieno[2,3-*c*]pyridine (**3-14**).  Couplings used in the computation
are shown in Fig. 3-49.   (a) One bond modulated response curve
calculated using Eqn. [3-14]. (b) Modulation decoupled response curve
calculated using Eqn. [3-23].   Both response curves are confirmed by
experimentally determined points with optima ranging from 2 to 12 Hz.

## Suppression of One Bond Modulation in the Freeman-Morris Sequence by Insertion of a BIRD Pulse

Following the reported modification of the CSCMLR pulse sequence just described, the conventional Freeman-Morris[29] sequence (Fig. 3-10) was also modified by the incorporation of a BIRD pulse midway through the refocusing delay ($\Delta_2$) to give the pulse sequence shown in Fig. 3-51.[102]

Using norharmane (3-4) as a model compound, modulation observed in the case of long range optimization of the Freeman-Morris sequence (Fig. 3-10) is shown in Fig. 3-39. With the modified sequence shown in Fig. 3-51, the response curve shown in Fig. 3-52 was obtained. Slices taken from the conventional experiment and that performed using the pulse sequence shown in Fig. 3-51 are shown for comparison in Fig. 3-53. Traces for the $^3J_{C3H1}$ and $^1J_{C1H1}$ responses are shown. Both spectra were recorded using a 10.4 Hz optimization, which is near one of the minima in response curve shown in Fig. 3-39. In the case of the conventional long range optimized experiment, the undesired direct response is shown in trace A and exhibited a signal-to-noise ratio (S/N) of 9.7:1. In contrast, the long range response shown in trace B gave a much weaker response (S/N = 4.9:1). A much different picture is obtained with the data acquired using the modified pulse sequence shown in Fig. 3-51. The undesired direct response is essentially absent, as seen from trace C. The long range coupling sought in the experiment appears in trace D with considerably enhanced intensity (S:N = 18.5:1). The information pertaining to the long range responses contained in traces B and D reflects the data presented in the curves shown in Figs. 3-39 and 3-52.

**Phase cycling**. Phases may be cycled for the pulse sequence shown in Fig. 3-51 using a steps shown in the eight step cycle specified in Table 3-7. The first 90° proton pulse is cycled according to $\phi_1$; the second proton 90° pulse is cycled according to $\phi_2$; the receiver phase is cycled using $\phi_R$, which selects for the coherence transfer echo. The antiecho may acquired by using 0123 0123 in lieu of the phase cycle contained in the table. The phase cycle may be extended to 32 or 128 steps by inserting additional cyclops cycles. Excellent results are, however, obtained using the minimum phase cycle contained in the table.

**Parameter selection**. Pulse widths, transmitter frequencies, sweep widths and levels of digitization in $F_1$ and $F_2$ are chosen in a fashion identical to any other heteronuclear chemical shift correlation experiment. The $\Delta_1$ interval is nominally set to $1/2(^nJ_{CH})$. To minimize losses due to relaxation during the longer delays necessary to accomodate long range couplings, we generally have found it

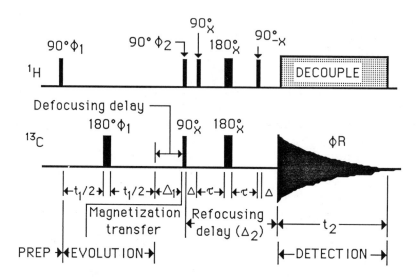

*Fig. 3-51.*    Modified Freeman-Morris[29] sequence incorporating a BIRD pulse midway
through the refocusing delay, $\Delta_2$.[102] The duration of $\Delta_1$ is optimized as a
function of $1/2(^nJ_{CH})$ in the normal fashion. For methine carbons, the
duration of $\Delta_2$ may be set equal $\Delta_1$. For general usage, $\Delta_2$ should be
$0.667\Delta_1$. The $\tau$ interval in the BIRD pulse is optimized as a function of
$1/2(^1J_{CH})$. The $\Delta$ interval $= (\Delta_2 - 2\tau/2)$.

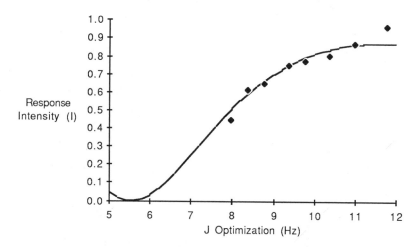

*Fig. 3-52.*    Response intensity curve for the $^3J_{C3H1}$ coupling pathway of norharmane
(**3-4**) calculated using Eqn. [3-23] and the couplings shown in Fig. 3-37.
Experimentally determined points were measured using the pulse
sequence shown in Fig. 3-51.

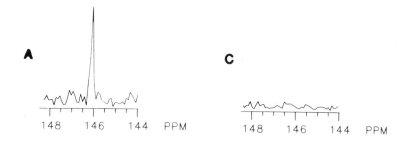

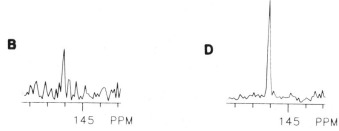

Fig. 3-53.    Slices taken from long range heteronuclear correlation spectra of norharmane (**3-4**).    Slices a and b were taken from a spectrum recorded by long range optimization of the Freeman-Morris[29] sequence (see Figs. 3-10 and 3-36).   Slices c and d were taken from a spectrum recorded using the pulse sequence shown in Fig. 3-51.    Both experiments were optimized for 10.4 Hz, which is near a minimum of the response curve for the $^3J_{C3H1}$ coupling pathway shown in Fig. 3-39.  Slice A shows the $^1J_{C1H1}$ direct response (S/N = 9.7:1).   Slice B shows the $^3J_{C3H1}$ long range response sought (S/N = 4.9:1).   Slice C shows the region of the direct response obtained form the spectrum recorded with the pulse sequence shown in Fig. 3-51.   Slice D shows the desired long range response with much improved intensity (S/N = 18.5:1).

preferable to optimize $\Delta_1$ for 10 Hz (50 msec) unless there is some compelling reason to optimize for a smaller coupling, which necessitates a correspondingly longer delay.  For transfer to a methine carbon, the refocusing delay, $\Delta_2$, should equal $\Delta_1$.  In the general case, $\Delta_2$ may be set to $0.667\Delta_1$, thus affording optimal transfer to methine, methylene and methyl carbons.  The $\tau$ interval in the BIRD pulses is normally set to $1/2(^1J_{CH})$ where $^1J_{CH}$ is taken as the average one bond coupling.  The $\Delta$ interval of the refocusing delay is normally determined as a function of $(\Delta_2 - 2\tau/2)$.  Lengthening the $\Delta_2$ interval should generally be avoided as corresponding decreases in observed signal intensity are obtained.

## Constant Evolution Time
## Long Range Heteronuclear
## Correlation Experiments

Long range heteronuclear correlation experiments based on the Freeman-Morris[29] sequence shown in Fig. 3-10 or the DEPT heteronuclear correlation sequence shown in Fig. 3-24 are exceedingly useful for structure elucidation and/or spectral assignment. However, the length of the fixed delays coupled with the incremented duration of the evolution interval may in some cases lead to substantial losses of signal and difficulties in optimizing the experiment. These potential difficulties have led to the development of several experiments that utilize constant evolution times. The first of these experiments were conceived by Bauer, Freeman and Wimperis[99] and by Kessler and coworkers[60,162] almost simultaneously. More recently by Reynolds et al.[61] developed another significant variant. The experiments are discussed in the following sections of this chapter and will be compared to the incremented duration evolution time experiments discussed above.

## The Bauer, Freeman and Wimperis
## Constant Evolution Time, Long Range
## Heteronuclear Correlation Experiment

One of the first experiments to be reported that employed a constant evolution time was the complex pulse sequence of Bauer, Wimperis and Freeman[99] shown in Fig. 3-54.

The pulse sequence described by Bauer, Freeman and Wimperis[99] was quite innovative, introducing features incorporated into a number of later pulse sequences by other workers. The first novel feature was the replacement of the initial 90° proton pulse of the Freeman-Morris sequence (Fig. 3-10)[29] by a TANGO pulse, which selects for protons remotely coupled to $^{13}C$. Next, the experiment incorporated a constant evolution time (T) with a 180° proton pulse, which was systematically relocated in successive experiments, thereby decoupling homonuclear proton-proton scalar couplings. Finally, after magnetization transfer was completed, a BIRD pulse was inserted midway through the refocusing interval that decouples modulations due to the one bond heteronuclear couplings. It should also be noted here that this last modification provided the impetus for what is now a growing number of studies of one bond modulations by a number of other research groups.[100-105,163]

Unfortunately, the pulse sequence shown in Fig. 3-54 has not been utilized other than in the initial report by its originators.[99] In part, the fact that this experiment has not been used may account for the lack of appreciation of many researchers for the problems engendered by one bond modulation of long range response intensity. Nevertheless,

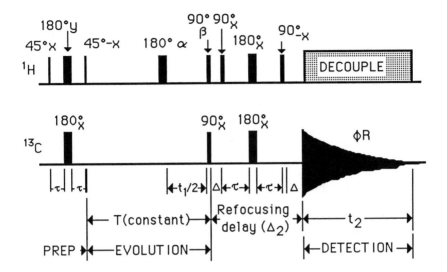

*Fig. 3-54.*    Long range heteronuclear chemical shift correlation pulse sequence
developed by Bauer, Freeman and Wimperis.[99] The initial TANGO pulse
selects for remote protons.  The 180° proton pulse is systematically
repositioned and decouples homonuclear proton scalar couplings.  The
BIRD pulse decouples one bond modulations.

pulse sequences developed by others have incorporated virtually all of
the novel features of the sequence shown in Fig. 3-54.  Sequences
having their origins rooted in the work of Bauer, Freeman and
Wimperis[99] include the CSCMLR sequence (Fig. 3-43) described
above,[158] the modified sequences for one bond modulation decoupling
(Figs. 3-48 and 3-51),[101,102,104] and two sequences described below:
the COLOC-S sequence (Fig. 3-58) of Krishnamurthy and Casida[163]
and the XCORFE sequence devised by Reynolds and coworkers (Fig. 3-
59).[61]

## The Kessler-Griesinger Correlation by Long Range Coupling (COLOC) Constant Evolution Time Pulse Sequence

The long range heteronuclear correlation sequence reported by
Kessler and coworkers[60,162] in 1984 utilizes a constant evolution time
analogous to that in the work of Bauer, Freeman and Wimperis.[99] The
COLOC sequence is shown in Fig. 3-55.  Once again, broadband
homonuclear decoupling is provided by the constant evolution time
through which simultaneous proton and carbon 180° pulses are
systematically repositioned in successive blocks of the experiment.
Unlike the preceding pulse sequence, numerous applications of the
COLOC experiment have appeared, and the experiment will therefore
be considered in greater detail.

Following the initial 90° proton pulse, the constant duration evolution period (T) has simultaneously applied proton and carbon 180° pulses, which are systematically relocated through the evolution period as successive blocks of data are taken. The $\Delta_1$ interval is incorporated into the evolution period and is governed by Eqn. [3-24]. The optimization of the $\Delta_1$ and $\Delta_2$ intervals is described in the following section on parameter selection.

**Parameter selection..** The COLOC experiment, as with any other experiment, requires selection of appropriate transmitter frequencies and a knowledge of the pulse widths for both proton and the heteronucleus. The sweep width and digitization in the $F_2$ frequency domain may also be freely selected. In regard to the $F_1$ domain, a somewhat different situation arises and it may not be possible in all cases to employ the desired sweep width and/or level of digitization.

The COLOC experiment may be optimized as desired provided that half of the evolution time is smaller than the $\Delta_1$ interval. Quite simply, this requirement reduces to:

$$t_1 max/2 = B \times IN \leq \Delta_1 \qquad [3-24]$$

where B is the number of increments (blocks) of the evolution time, which controls digitization in $F_1$, and IN is the $t_1$ increment time (1/2 x SW in $F_1$). To illustrate the impact of Eqn. [3-24] on the parameters that may be employed in setting up a COLOC experiment, let us consider as an example the simple alkaloid norharmane (**3-4**). Given a sweep width of 660 Hz (±330 Hz) in $F_1$ and a desire to digitize $F_1$ by taking 256 increments of $t_1$ and an optimization of $\Delta_1$ at 10 Hz (1/2($^nJ_{CH}$) = 50 msec), it is not possible to satisfy Eqn. [3-24]. If we desire to maintain the optimization of $\Delta_1$ at 10 Hz, we must either increase the sweep width and/or decrease the number of blocks of data to be taken in order to satisfy Eqn. [3-24]. In the COLOC spectrum of norharmane (**3-4**) shown in Fig. 3-57, below, $\Delta_1$ was optimized at 10.4 Hz (48.1 msec), the sweep width was 1000 (±500) Hz, giving IN = 500 μsec, and 96 blocks of data were taken, which barely satisfies Eqn. [3-24]. From the parameters utilized, it should be apparent that it was necessary to utilize a much wider sweep width than would normally be necessary to accomodate the requirements of Eqn. [3-24].

For optimization of the COLOC experiment, Kessler and colleagues[60] have suggested that it is possible to optimize the delays by acquiring a one-dimensional INEPT experiment to test the effects of the optimization. Reynolds and coworkers[105] have reported that while the INEPT experiment does provide an effective means of optimizing the COLOC experiment for nonprotonated carbon resonances, it is difficult to independently and reliably optimize the $\Delta_1$ and $\Delta_2$ intervals by means of an INEPT experiment. The difficulties between nonprotonated and protonated carbons arise from the fact that the INEPT experiment does

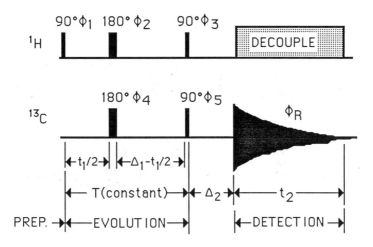

*Fig. 3-55.*    The COLOC constant evolution time experiment developed by Kessler and coworkers.[60,162] The constant evolution time is defined by Eqn. [3-24]. The refocusing interval, $\Delta_2$, may be optimized as discussed below. Phase cycling is given in Table 3-9.

not discriminate between the desired $^nJ_{CH}$, $n \geq 2$, and the undesired polarization transfer arising where $n = 1$ in the case of the protonated carbon resonances. In contrast, transfer for nonprotonated carbons is restricted to the desired long range couplings ($n \geq 2$). The solution to the problem of calibrating the COLOC experiment proposed by Reynolds et al.[105] was to devise a separate optimization sequence which they have termed CINCH-C (Calibrate INdirect Carbon Hydrogen correlation spectra - COLOC) which is shown in Fig. 3-56. The rationale behind the CINCH-C spectrum is to suppress polarization transfer via $^1J_{CH}$ coupling. The CINCH-C sequence achieves this task by replacing the first 90° proton pulse by a TANGO pulse[155] selective for protons remotely coupled to $^{13}C$ (see Eqn. [3-21]). Additionally, the final 180° pulse of the INEPT sequence has been replaced by the BIRD pulse. Data shown by Reynolds and coworkers[105] suggest that it is much easier to independently optimize the $\Delta_1$ and $\Delta_2$ intervals of the COLOC sequence by using the CINCH-C sequence.

Because of the restrictions imposed on optimization by Eqn. [3-24], the COLOC experiment may be somewhat restricted in utility relative to other long range heteronuclear correlation experiments when it becomes necessary to optimize the experiment for larger long range couplings, eg, 10 Hz, in conjunction with a restricted chemical shift range such as that encountered with polynuclear aromatics and and oligosaccharides. Despite these limitations, applications of the COLOC experiment to both of these classes of molecules have appeared and are surveyed below.

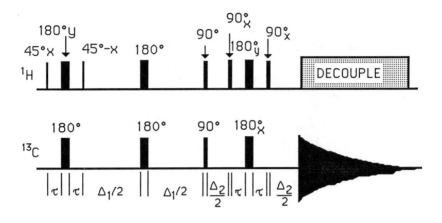

*Fig. 3-56.*     The CINCH-C pulse sequence for independent optimization of the $\Delta_1$ and $\Delta_2$ delays of the COLOC sequence.

**Phase cycling.** A number of phase cycles ranging from 16 to 256 steps are possible for the COLOC pulse sequence shown in Fig. 3-55.[60] The minimum 16 step cycle may be obtained by cycling only the phases of the two 90° proton pulses. Generally, we have found it necessary to use sufficient numbers of acquisitions to allow the use of the 64 step phase cycle shown in Table 3-9. As an alternative, the 32 step cycle shown in Table 3-9 may also be employed with both the COLOC pulse sequence shown in Fig. 3-55 and the newly reported COLOC-S pulse sequence, which is shown in Fig. 3-58[163] and discussed below.

**Data processing.** Considerations implicit in the processing of a COLOC data set differ only minimally from other long range experiments. The data prior to the first Fourier transformation may be massaged essentially to the taste of the user. Generally, we have found it preferable to use sufficient exponential line broadening to give good signal-to-noise ratios without sacrificing resolution. After transposition, the resonances in the $F_1$ domain will be rather coarsely digitized in cases where restricted $F_1$ sweep windows are utilized. Hence, it is usually advisable to zero-fill prior to the second Fourier transformation and to employ mild resolution enhancement. In addition, it should also be noted that the 64 step phase cycle given Table 3-9 requires a spectrum rotation to restore the proper orientation of the proton spectrum.

**Application of the COLOC experiment to the simple alkaloid norharmane (3-4).** The COLOC spectrum of norharmane (**3-4**) is shown in Fig. 3-57. The spectrum was acquired using a 10.4 Hz optimization to allow comparison to spectra acquired using other methods, which are described both above and below. To meet the

Table 3-9.      Phase cycling schemes for the COLOC pulse sequence in Fig. 3-55.

| Thirty-two Step Phase Cycle | | | | |
|---|---|---|---|---|
| $\phi_1$ | 00000000 | 00000000 | 00000000 | 00000000 |
| $\phi_2$ | 00112233 | 00112233 | 00112233 | 00112233 |
| $\phi_3 = \phi_5$ | 00000000 | 11111111 | 22222222 | 33333333 |
| $\phi_4$ | 13131313 | 13131313 | 13131313 | 13131313 |
| $\phi_R$ | 00220022 | 22002200 | 00220022 | 22002200 |

| Sixty-four Step Phase Cycle | | | | |
|---|---|---|---|---|
| $\phi_1$ | 00000000 | 00000000 | 11111111 | 11111111 |
|  | 22222222 | 22222222 | 33333333 | 33333333 |
| $\phi_2 = \phi_4$ | 00000000 | 00000000 | 00000000 | 00000000 |
|  | 00000000 | 00000000 | 00000000 | 00000000 |
| $\phi_3$ | 01010101 | 23232323 | 01010101 | 23232323 |
|  | 01010101 | 23232323 | 01010101 | 23232323 |
| $\phi_5$ | 01231230 | 23013012 | 12302301 | 30120123 |
|  | 23013012 | 01231230 | 30120123 | 12302301 |
| $\phi_R$ | 00220022 | 11331133 | 00220022 | 11331133 |
|  | 00220022 | 11331133 | 00220022 | 11331133 |

requirements imposed by Eqn. [3-24], it was necessary to use a 1000 Hz spectral width in $F_1$ digitized by 96 increments of the evolution time, $t_1$. The data were processed using an exponential multiplication prior to the first Fourier transformation and double exponential apodization with zero filling to 256 points prior to the second Fourier transformation.

Comparison of the spectrum shown in Fig. 3-57 to that shown in Fig. 3-36 recorded using long range optimization of the Freeman-Morris[29] sequence shows some minor differences in the responses observed. For example, the H1/C3 and H3/C4a and H7/C8a responses are missing from the COLOC spectrum but are observed in the conventional long range correlation experiment. The H4/C4 response in the COLOC spectrum shown in Fig. 3-57 is also somewhat weaker than in the conventional spectrum shown in Fig. 3-36, although this is not necessarily undesirable.

**Modulation effects in the COLOC experiment**. Bax and Summers[164] were the first to note the potential of the COLOC experiment to exhibit one bond modulation effects analogous to those discussed above when the Freeman-Morris[29] sequence is optimized for long range couplings. In addition, as a result of the constant evolution time, long range responses of protonated carbons may also be modulated by homonuclear proton-proton couplings according to Eqn. [3-25]:

$$I = \sin [\pi \, \Delta_1 \, {}^{LR}J_{CH}] \sin [\pi \, \Delta_2 \, {}^{LR}J_{CH}] \prod \cos [\pi \, \Delta_2 \, {}^1J_{CH}] \; x$$

$$\cos [\pi \, \Delta_2 \, {}^{LR'}J_{CH}] \cos [\pi \, \Delta_1 \, J_{HH}] \qquad\qquad [3\text{-}25]$$

where ${}^{LR}J_{CH}$ represents the long range coupling pathway being examined; ${}^1J_{CH}$ corresponds to the direct coupling of the protonated carbon to which magnetization is being transferred; ${}^{LR'}J_{CH}$ represents other long range couplings to the protonated carbon; and $J_{HH}$ corresponds to the homonuclear coupling of the directly attached proton. Equation [3-25] ignores relaxation effects which may be generally assumed to be minimal for small molecules such as norharmane (**3-4**).

**Sensitivity**. In general, we have found the COLOC experiment to require more acquisitions to attain a given signal-to-noise ratio than with the pulse sequences derived from the Freeman-Morris[29] sequence. It should be noted, however, that this comment pertains to our experience in the application of the experiment only to the spectra of polynuclear aromatic compounds. Bax and Summers[164] have also generally alluded to the sensitivity problems associated with the COLOC experiment. In contrast, Enzell and coworkers[165] have reported that under favorable circumstances COLOC spectra can be obtained with relatively small quantities of material. Overall, the utility of the experiment is probably dependent on the nature of the compound to which it is being applied and the coupling pathways being examined. It will probably take a comparison of the COLOC experiment with other experiments under rigorously controlled conditions with several model compounds before a well informed comment regarding the overall and relative sensitivity of the compound can be made.

**A brief survey of applications of the COLOC experiment**. Probably the most numerous applications of the COLOC experiment are found in the elucidation of peptide structures. In their first paper, Kessler and coworkers[60] applied the COLOC technique to *trans*-cyclo[-Pro-Phe-D-Trp-Lys-Thr-Gly-]. Other applications to cyclic peptides have followed[166,167] - perhaps the most notable usage is a sutdy of the important immunosuppressive cyclosporin-A.[159] Applications to glycopeptides have also been reported,[168] and the use of the COLOC experiment has been discussed in several chapters that treat the analysis of NMR data of peptides.[169,170] Kessler, Griesinger and Wagner[171] have also considered the role of the COLOC experiment in establishing the side chain confirmations of peptides. Finally, in another demonstration, Kessler and coworkers have also reported the application of the COLOC experiment with a synthetic compound with many similarities to simple natural products.[162] Müller[172] has utilized COLOC in conjunction with long range optimization of the Freeman-Morris[29] sequence in the assignment of bilirubin and its methyl ester. Wernly and Lauterwein[173] have applied the COLOC sequence to the assignment of olefinic carbons of β-carotene in the presence of chemical shift degeneracy, and Enzell et al.,[165] have applied the technique to the assignment of a sucrose ester. Jans and coworkers[174] have utilized the COLOC sequence in the assignment of the spectra of cyclopenta-[cd]pyrene. In the area of antibiotics chemistry, Müller and Jeffs[175] have utilized the technique to deduce the primary structure of the aridicin aglycone and Hashimoto and coworkers[176] have used the experiment is determining the structure of a novel antibiotic related to mitomycin. Tringali and coworkers[177] have determined structures of p-terphenyl derivatives with antibiotic properties. Fattorusso and coworkers[178] have applied the experiment in determining the structure of a novel β-carboline from a marine hydroid. In a pair of related papers, Yasumoto and colleagues[179,180] have used the COLOC technique in determining the structure of complex polyether shellfish poisoning toxins. Kernan and Faulkner[181] have employed the COLOC experiment in determining the structure of an antifungal macrolide from a sponge. Finally, Kusumi and coworkers[182] have used the COLOC experiment in conjunction with long range proton detected methods in determining the structure of a cyclic peptide toxin from a blue-green alga.

## COLOC-S: A Modified COLOC Experiment Incorporating TANGO and BIRD Pulses

Recognizing the potential problems associated with modulation due to the one bond couping when magnetization is transferred long range to a protonated carbon resonance, Krishnamurthy and Cassida[163] have reported a modified version of the COLOC experiment which they refer to as COLOC-S (COLOC-selective), shown in Fig. 3-58. The COLOC-S sequence employs a TANGO pulse[155] in place of the initial 90° proton pulse of the conventional COLOC sequence shown in Fig. 3-55. The TANGO pulse selectively excites only those protons which are long range coupled to $^{13}C$. The sequence also utilizes a BIRD pulse midway through the $\Delta_2$ interval to achieve one bond modulation decoupling analogous to that furnished by the pulse sequences shown in Figs. 3-48 and 3-51 described by Martin and coworkers.[101,102] The ability of the sequence shown in Fig. 3-58 to decouple one bond modulations has not been experimentally confirmed, but results obtainable with it should be analogous to those already reported for similar modifications.[101-104] Phases in the COLOC-S sequence are cycled according to the 32 step phase cycle presented in Table 3-10.

## The XCORFE Constant Evolution Time Experiment Devised by Reynolds and Coworkers

Following the leads of Bauer, Freeman and Wimperis[99] and the Kessler-Griesinger COLOC[60] sequence, Reynolds and coworkers[61] in 1985 described another long range heteronuclear correlation

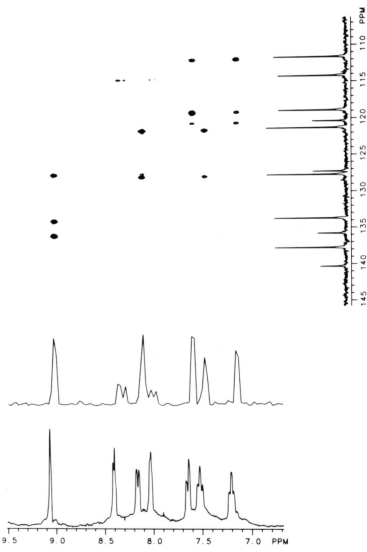

*Fig. 3-57.*     COLOC spectrum of norharmane (**3-4**). The spectrum was recorded using a sweep width of ±500 Hz in $F_1$ with 96 increments of $t_1$ to accommodate the 10.4 Hz optimization of $\Delta_1$ according to Eqn. [3-24]. A total of 256 acquisitions were taken/block and the data set was zero-filled to 256 points prior to the second Fourier transformation.

experiment that incorporated a constant evolution time and was given the acronym XCORFE. Interestingly, the XCORFE pulse sequence employs two BIRD pulses. As shown in Fig. 3-59, the first BIRD pulse is systematically repositioned through the evolution period in successive experiments in a fashion analogous to the COLOC sequence shown in Fig. 3-55. The second BIRD pulse is located midway through the refocusing delay and serves to decouple one bond modulation of long range response in a manner analogous to that for the modified Freeman-Morris sequences shown in Figs. 3-48 and 3-51.

While the XCORFE pulse sequence is quite similar to the COLOC and COLOC-S pulse sequences shown in Figs. 3-55 and 3-58, respectively, differences should be noted. Probably the most noteworthy differences are in the appearance of the various types of response contained in the data matrix. First, the movable BIRD pulse in the evolution period provides variable refocusing of certain parameters, which leads to modulation of polarization transfer and detection in the $F_1$ frequency domain. In contrast, other variables are not refocused and thus evolve over a fixed duration interval and are thus not detected in $F_1$. To illustrate these points, consider vicinal proton-proton couplings. The BIRD pulse relocated sequentially through the evolution period selectively flips all protons not directly bonded to $^{13}C$ ($^nJ_{CH}$ $n \geq 2$) while leaving directly attached protons ($n = 1$) unaffected. This differential treatment of protons leads to a modulation of indirect polarization transfer (via $^nJ_{CH}$, $n \geq 2$) by $^3J_{HH}$, which leads to a response in $F_1$ split by $^3J_{HH}$. Provided that sufficient digital resolution is available to digitize the $^3J_{HH}$ coupling, this feeature of the XCORFE pulse sequence is important in that it provides the means for differentiating $^2J_{CH}$ from $^3J_{CH}$ coupling pathways. The $^3J_{CH}$ responses will appear as singlets in $F_1$. The only exceptions to this behavior will be quaternary carbons, in which case both $^2J_{CH}$ and $^3J_{CH}$ responses will appear as singlets in $F_1$.

The major artifact in XCORFE spectra is likely to be the direct response ($^1J_{CH}$) that will appear as a doublet at $F_1 = 0 \pm {}^1J_{CH}/2$ with each limb exhibiting further splitting due to $^3J_{HH}$ if the requisite protons are present in the structure. Generally, direct responses are inefficiently detected if the duration of the $\tau$ interval in the BIRD pulse corresponds favorably to the actual one bond coupling.

**Parameter selection**. The selection of the parameters for the XCORFE pulse sequence is largely the same as for the COLOC experiment discussed above. Similar constraints in regard to the constant duration evolution time apply to both experiments. The impact of various parameter choices on the results obtained with the XCORFE pulse sequence has been rigorously examined in a recent publication by Reynolds and coworkers.[105] This same paper also describes a one dimensional pulse sequence, CINCH-X, shown in Fig. 3-60 which may used to optimize parameters for the XCORFE sequence.

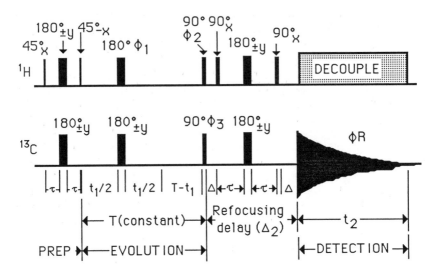

*Fig. 3-58.* The COLOC-S pulse sequence developed by Krishnamurthy and Casida[163] to suppress modulations of long range response intensity induced by one bond couplings in a manner analogous to that encountered with sequences derived from the original Freeman-Morris[29] sequence.

The CINCH-X pulse sequence shown in Fig. 3-60 has some similarities to the CINCH-C pulse sequence shown in Fig. 3-56. The TANGO pulse of the latter is replaced by a 90° proton pulse. Next the 180° proton pulse applied during the evolution period of the CINCH-C pulse sequence is also replaced by a BIRD pulse in the CINCH-X sequence. The latter change has the effect of mimicking the XCORFE sequence (Fig. 3-59) with respect to $^nJ_{HH}$ coupling vectors in addition to suppressing $^1J_{CH}$ polarization transfer; the CINCH-X sequence refocuses $^1J_{CH}$ coupling vectors, since the BIRD pulse is flanked by equal $\Delta_1/2$ intervals. Reynolds and coworkers[105] have further demonstrated that the XCORFE sequence is less sensitive to the choice of $\Delta_1$ than the COLOC sequence (Fig. 3-55).

**Phase cycling.** The phase cycling requirements of the XCORFE pulse sequence shown in Fig. 3-59 are relatively simple. The 90° proton pulse used to effect polarization transfer ($\phi_4$) is cycled in four steps to provide quadrature detection in $F_1$ as shown in Table 3-11. The BIRD pulse located in the evolution period is subjected to an independent phase cycle ($\phi_2$), the last pulse of the complex ($\phi_3$) cycled 180° ahead of the other pulses. Likewise, the BIRD pulse located midway through $\Delta_2$ is also phase cycled ($\phi_1$) the final pulse again cycled 180° ahead ($\phi_5$) of the other pulses in the cluster. Finally, the receiver is cycled as shown in Table 3-11 to complete the sixteen step phase cycle.

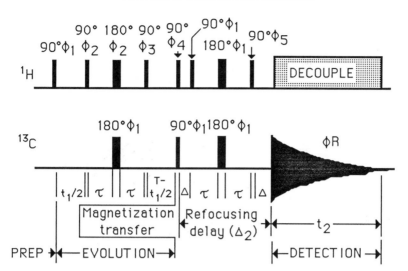

Fig. 3-59.    The XCORFE seqeuence developed by Reynolds et al.[61] The BIRD pulse
located in the constant length evolution period is systematically
repositioned in successive experiments as in the COLOC experiment.
The BIRD pulse located midway through the refocusing delay serves to
suppress one bond modulations of long range response intensity.   Phases
are cycled according to Table 3-11.

Table 3-10.    Phase cycling for the COLOC-S pulse sequence shown in Fig. 3-58.

| $\phi_1$ | 00112233 | 00112233 | 00112233 | 00112233 |
|---|---|---|---|---|
| $\phi_2$ | 00000000 | 11111111 | 22222222 | 33333333 |
| $\phi_R$ | 00220022 | 22002200 | 22002200 | 00220022 |

Table 3-11.    Phase cycling for the XCORFE pulse sequence shown in Fig. 3-59.

| $\phi_1$ | 0000 | 2222 | 0000 | 2222 |
|---|---|---|---|---|
| $\phi_2$ | 0000 | 1111 | 2222 | 3333 |
| $\phi_3$ | 2222 | 3333 | 0000 | 1111 |
| $\phi_4$ | 0321 | 0321 | 0321 | 0321 |
| $\phi_5$ | 2222 | 0000 | 2222 | 0000 |
| $\phi_R$ | 0123 | 0123 | 0123 | 0123 |

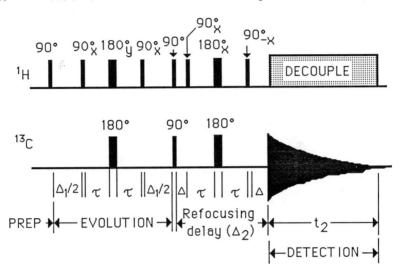

*Fig. 3-60.*    The CINCH-X pulse sequence developed by Reynolds et al.,[105] for independent optimization of the $\Delta_1$ and $\Delta_2$ delays of the XCORFE pulse sequence shown in Fig. 3-59.

**Modulation effects**.  Since the XCORFE pulse sequence contains a BIRD pulse located midway through the refocusing delay ($\Delta_2$) it would not be expected to suffer from the vagaries of one bond modulation of response intensity.  To confirm this point, Reynolds and coworkers[105] have examined response intensities for both methylene and methyl carbons of 2-pentanone recorded with and without the BIRD pulse normally contained in the refocusing delay.  As anticipated, when the BIRD pulse was in place, response intensities did not show the adverse effects of modulation, while when the BIRD pulse was intentionally removed, the response intensity was modulated in a fashion analogous to that described above for both derivatives of the Freeman-Morris sequence and the Kessler-Griesinger COLOC sequence.

**A brief survey of applications of the XCORFE pulse sequence**.  Only a relatively few applications of the XCORFE pulse sequence have appeared in the literature.  However, in the opinion of the authors, the ability to differentiate two bond from three bond long range couplings should lead to an increased usage of the experiment as more research groups appreciate its capabilities.

Ireland and coworkers have employed the XCORFE pulse sequence in establishing the structure of a novel tricyclic spiro ketal, muamvatin, from a marine sponge.[183]  It should also be noted that considerable usage was made of the abilities of the XCORFE pulse sequence to differentiate two and three bond heteronuclear spin couplings in establishing the structure of muamvatin.  In a completely different vein, Smith[184] has utilized the XCORFE pulse sequence to assign the spectra of highly substituted biphenyls and *m*-quaterphenyls. Reynolds and coworkers have employed their technique to assign the structcures of a series of natural products that have included: constituents from *Hortia regia*[185]; guyanin, a novel tetranortriterpenoid[186]; a group of isomeric triterpenols[187]; and the structure of a reaction product of the alkaloid aspidocarpine.[188] Finally, Petit and coworkers have employed the XCORFE experiment to assign the quaternary and aromatic carbons of a saccharid bearing aromatic substituents.[189]

## Heteronuclear Multiple Quantum Coherence Long Range Chemical Shift Correlation Experiments

Long range heteronucleus detected heteronuclear chemical shift correlation experiments, like any other group of experiments, will eventually suffer from sensitivity limitations due to either molecular complexity or solubility constraints. A substantial gain in sensitivity may be realized, however, by resorting to proton detection as in the case of the direct correlation experiments described above. Bax and coworkers[82] have recently described a long range proton detected chemical shift correlation experiment based on heteronuclear multiple quantum coherence. The particulars of the creation of heteronuclear multiple quantum coherence have been discussed above in the section dealing with proton detected direct heteronuclear correlation and will not be treated further here. There are, however, some differences in the pulse sequences for direct versus long range heteronuclear correlation because of the different objectives of the experiment. Thus, the pulse sequence constituents are considered in the following section.

## The Bax and Summers Long Range Heteronuclear Multiple Quantum Coherence Experiment

The pulse sequence devised by Bax and Summers[82] is shown in Fig. 3-61 and represents a derivative of earlier work on the detection of $^{15}N$ by heteronuclear multiple quantum coherence.[67] The first pair of pulses applied to proton and carbon serve as a low pass J filter[104,156] when the duration of the delay, $\Delta_1$, is adjusted to $1/2(^1J_{CH})$. Long range magnetization components are unaffected by the J-filter and thus continue to evolve through the $\Delta_2$ interval that follows. Heteronuclear multiple quantum coherence (HNMQC) is created for the long range components of magnetization by the application of the second 90° pulse applied to $^{13}C$. During evolution, proton chemical shift terms are refocused by the 180° proton pulse so that the signals are labeled with $^{13}C$ chemical shift during $t_1$. Consequently, the final 90° $^{13}C$ pulse recreates observable single quantum coherence from the evolved HNMQC. Signals from protons that originate from HNMQC are modulated by $^{13}$C chemical shifts and homonuclear proton spin couplings. Signals from protons that do not have a long range coupling to $^{13}$C are removed by the phase cycling of the second 90° $^{13}$C pulse.

Phase modulation of the detected proton signal by the homonuclear scalar coupling precludes the acquisition of phase sensitive spectra. Frey and co-workers[73] have suggested the use of z'-filters and purge pulses for the acquisition of phase sensitive $^{1}$H-$^{113}$Cd shift correlation spectra, but Bax and Summers[82] state that these modifications degrade sensitivity unacceptably and make the suppression of signals due to protons not long range coupled to $^{13}$C more difficult.

**Phase cycling**. The phase cycling prescription for the Bax and Summers experiment is a relatively simple eight step phase cycle shown in Table 3-12. We have found that when larger numbers of acquisitions are to be taken per $t_1$ increment it is useful to extend the phase cycle to 32 steps by superimposing a separate cyclops cycle over the low pass J filter pulse ($\phi_1$) with the other phases incremented in 90° steps accordingly. Recently, Cavanaugh and Keeler have proposed a modified phase cycling scheme reported to afford better suppression of $^{1}$H-$^{12}$C resonances.[190]

**Parameter selection**. The determination of pulse widths for $^{13}$C from the decoupler must be performed carefully as described in Chapter 1. The proton pulse should also be accurately calibrated. The duration of the $\Delta_1$ interval is a function of $1/2(^1J_{CH})$ with 3.0 msec

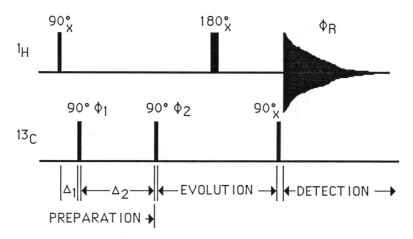

Fig. 3-61     Long range heteronuclear multiple quantum coherence based heteronuclear chemical shift correlation pulse sequence described by Bax and Summers.[82] The interval $\Delta_1$ serves as a low pass J filter. Long range heteronuclear multiple quantum coherence is created by the application of the 90° $^{13}$C pulse following the $\Delta_2$ interval. Phases are cycled according to Table 3-12.

Table 3-12.  Phase Cycling for the Bax-Summers[82] Proton Detected Long Range Heteronuclear Multiple Quantum Experiment Shown in Fig. 3-61.

| Acquisition | $\phi_1$ | $\phi_2$ | $\phi_R$ |
|---|---|---|---|
| 1 | 0 | 0 | 0 |
| 2 | 0 | 2 | 2 |
| 3 | 2 | 0 | 0 |
| 4 | 2 | 2 | 2 |
| 5 | 0 | 1 | 1 |
| 6 | 0 | 3 | 3 |
| 7 | 2 | 1 | 1 |
| 8 | 2 | 3 | 3 |

representing an acceptable compromise when a variety of types of carbon are present in the molecular structure. The duration of the $\Delta_2$ interval may be optimized as a function of $1/2(^{LR}J_{CH})$. Bax and Summers[82] suggest a duration of about 60 msec, which corresponds to an 8.3 Hz optimization. For polynuclear aromatics, we have found 50 msec to work nicely. There is some latitude in the choice of the long range optimization; values ranging from 20 to 100 msec work well. In general, short values of the $\Delta_2$ interval tend to give relatively more $t_1$ noise while longer values tend to give large intensity differences for protons that are singlets relative to those having complex multiplet structures.[191]

Several other considerations enter into the experimental set up beyond the selection of actual experimental parameters. For instance, the power output of the decoupler used to produce $^{13}C$ pulses may be a concern. It may be necessary to split the $^{13}C$ chemical shift range by performing two experiments if sufficient X($^{13}C$)-band decoupler power is unavailable. Second, the digitization of the $^{13}C$ chemical shift range ($F_1$) is dependent upon the number of increments of $t_1$ collected. We have found it useful to record data from a minimum of 200-300 increments of $t_1$ followed by zero filling to 512 points prior to the second Fourier transformation. Finally, we have found that the spectra generally contain less $t_1$ noise if the data is recorded with the sample not spinning.

**Data processing.** We have found it advisable to process the first free induction decay using sinusoidal multiplication or a similar function to match the shape of the weighting function to the echo-type FID. After the first Fourier transformation and transposition, it is best to locate an interferogram and to examine several alternative types of weighting prior to the second Fourier transformation.

As an alternative to absolute value presentation of the data, Bax and Marion[192] have recently proposed a method of computing spectra that are phase sensitive in one dimension through the use of a hypercomplex Fourier transformation. The examples shown in their work were clearly superior to the presentation afforded when the data was processed conventionally.

### Application to benzo[*b*]triphenyleno[1,2-*d*]thiophene.

The compound benzo[*b*]triphenyleno[1,2-*d*]thiophene (**3-15**) is interesting in that its proton spectrum contains three overlapping four spin systems in addition to the two spin system.[193] The assignment of the proton and carbon NMR spectra was rendered more difficult by the relatively low solubility of the material (10 mg/0.4 ml deuterochloroform) coupled with a very small total amount of material available. While it was possible to obtain a heteronucleus detected proton-carbon chemical shift correlation spectrum using the broadband homonuclear proton decoupling technique of Bax[52] shown in Fig. 3-25, in contrast, attempts to record a heteronucleus detected long range correlation spectrum using the pulse sequence shown in Fig. 3-51 were disappointingly unsuccessful. The spectrum recorded using the sequence shown in Fig. 3-51 after 70 hours of data acquisition is shown in Fig. 3-62. The sensitivity advantage inherent in the use of the proton detected pulse sequence shown in Fig. 3-61 is obvious from even a cursory comparison of the spectrum shown in Fig. 3-63 with that shown in Fig. 3-62. Based on the proton detected long range heteronuclear multiple quantum correlation spectrum it was possible to totally assign the spectra which would have been impossible by conventional heteronucleus detected means.

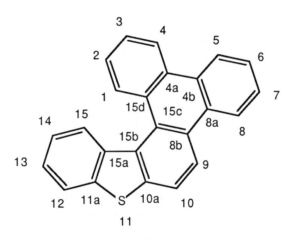

**3-15**

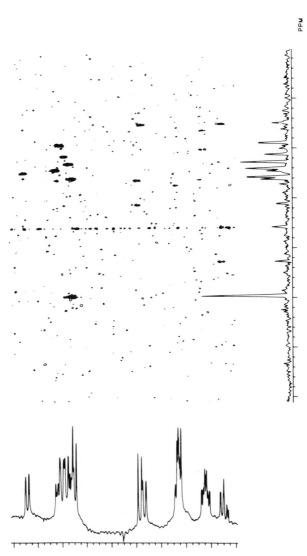

*Fig. 3-62*    Long range correlation spectrum of benzo[*b*]triphenyleno[1,2-*d*]thiophene (**3-15**) recorded using the heteronucleus detected pulse sequence shown in Fig. 3-51 modified to decouple on bond modulations. The data was acquired as 180 x 1K complex points with 1760 acquisitions taken per block. Total measuring time was approximately 70 hours.

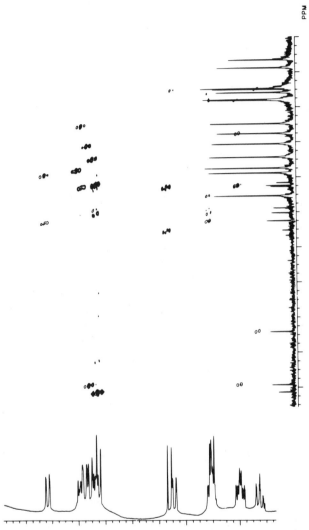

*Fig. 3-63.*　Proton detected long range heteronuclear multiple quantum chemical shift correlation spectrum of benzo[*b*]triphenyleno[1,2-*d*]thiophene (**3-15**) recorded using the pulse sequence of Bax and Summers[82] shown in Fig. 3-61. The low pass J-filter ($\Delta_1$) was optimized for an assumed average coupling of 165 Hz (3 msec); the $\Delta_2$ delay was optimized for 50 msec. The data was accumulated as 334 x 1K complex points with 278 acquisitions taken per $t_1$ increment. The sample was not spun during data accumulation. The data was processed using a 2.5 Hz exponential broadening prior to the first Fourier transform and was zero filled and processed without apodization prior to the second Fourier transform to give the 512 x 512 point matrix shown.

## A Brief Survey of Applications of the Proton Detected Long Range Heteronuclear Multiple Quantum Correlation Experiment

The applications of long range heteronuclear multiple quantum chemical shift correlation in the literature are thus far very few in number. Bax and coworkers have published three papers dealing with vitamin B-12 that employ the experiment.[82,83,85] In an unrelated study, Bax and a different group of coworkers have applied the experiment shown in Fig. 3-61 to the complex antibiotic desertomycin.[84] Kessler and coworkers[86] have reported the application of a simple variant of the experiment shown in Fig. 3-61 which did not employ the low pass J filter[67] in the assignment of peptide spectra. We have utilized the Bax and Summers[82] experiment in the total assignment of the proton and carbon NMR spectra of benzo[*b*]triphenyleno[1,2-*d*]thiophene (**3-15**).[193] Englert, Glinz and Liaaen-Jensen[194] have used the technique in assigning the carbon spectra of a series of caroten-20-als. Kusumi and coworkers[182] have used proton detected long range heteronuclear multiple quantum coherence in conjunction with the COLOC experiment in determining the structure of a cyclic peptide toxin from a blue-green alga. Alam and coworkers[195,196] have used the Bax-Summers experiment in conjunction with proton zero quantum coherence to determine the structures of a series of novel diterpenes. Kohn and coworkers[197,198] have exploited the sensitivity of the Bax-Summers experiment to assign the spectrum of bicyclomycin and a novel thiol adduct. Seto's group[199] has used the long range heteronuclear multiple quantum experiment to determine the structure of a new polyether antibiotic, portmicin. Kobayashi and coworkers[200] have employed the experiment to deduce the structure of a novel 25-membered macrocyclic lactone with antineoplastic activity. Finally, Cardellina, Martin and colleagues[201,202] have employed the technique in the determination of the structures of a series of brominated sesquiterpenes and a novel diterpene.

## REFERENCES

1.      B. Birdsall, N. J. M. Birdsall and J. Feeney, *J. Chem. Soc., Chem. Commun.*, 316 (1972).

2.      E.W. Hagaman, *Org. Magn. Reson.*, **8**, 389 (1976).

3.      R. T. Gampe, Jr., G. E. Martin, A. C. Pinton and R. A. Hollins, *J. Heterocyclic Chem.*, **18**, 155 (1981).

4.      R. Freeman, G. A. Morris and M .J. T. Robinson, *J. Chem. Soc., Chem. Commun.*, 754 (1976).

5.      G. Bodenhausen, R. Freeman and G. A. Morris, *J. Magn. Reson.*, **23**, 171 (1976).

6.      G. E. Martin, J. A. Matson, J .C. Turley and A. J. Weinheimer, *J. Am. Chem. Soc.*, **101**, 1888 (1979).

7.    F.W. Wehrli, "Organic Structure Assignments Using $^{13}$C Spin-Relaxation Data," in *Topics in Carbon-13 NMR Spectroscopy*, vol. 2, G.C. Levy, Ed., Wiley-Interscience, New York, 1976, pp. 343-391.

8.    R. S. Norton and A. Allerhand, *J. Am. Chem. Soc.*, **98**, 1007 (1976).

9.    M. Jay and G. E. Martin, *J. Heterocyclic Chem.*, **20**, 527 (1983).

10.    G. Commenges and R. C. Rao, *J. Magn. Reson.*, **58**, 496 (1984).

11.    K. G. R. Pachler, and P. L. Wessels, *J. Magn. Reson.*, **12**, 337 (1973).

12.    S. Sørensen, R. S. Hansen and H. J. Jakobsen, *J. Magn. Reson.*, **14**, 243 (1974).

13.    D. W. Nagel, K. G. R. Pachler, P. S. Steyn, R. Vleggaar and P. L. Wessels, *Tetrahedron*, **32**, 2625 (1976).

14.    K. G. R. Pachler, P. S. Steyn, R. Vleggaar, P. L. Wessels and D. B. Scott, *J. Chem. Soc., Perkin Trans. I*, 1182 (1976).

15.    G. A. Morris and R. Freeman, *J. Am. Chem. Soc.*, **101**, 760 (1979).

16.    R. Freeman and G. A. Morris, *Bull. Magn. Reson.*, **1**, 5 (1979).

17.    G. A. Morris, *J. Am. Chem. Soc.*, **102**, 428 (1980).

18.    D. M. Doddrell and D. T. Pegg, *J. Am. Chem. Soc.*, **102**, 6388 (1980).

19.    D. M. Doddrell, D. T. Pegg and M. R. Bendall, *J. Magn. Reson.*, **48**, 323 (1982).

20.    D. M. Doddrell, D. T. Pegg and M. R. Bendall, *J. Chem. Phys.*, **77**, 2745 (1982).

21.    R. Benn and H. Günther, *Angew. Chem., Int. Ed., Engl.*, **22**, 350 (1983).

22.    G.A. Morris, "Pulsed Methods for Polarization Transfer in $^{13}$C-NMR," in *Topics in Carbon-13 NMR Spectroscopy*, vol. 4, G.C. Levy, Ed., Wiley-Interscience, New York, 1984, pp. 179-196.

23.    E. Breitmaier and W. Voelter, *Carbon-13 NMR Spectroscopy*, 3rd ed., VCH, New York, 1987, pp. 78-84.

24.    A. A. Maudsley and R. R. Ernst, *Chem. Phys. Lett.*, **50**, 368 (1977).

25.    A. A. Maudsley, L. Müller and R. R. Ernst, *J. Magn. Reson.*, **28**, 463 (1977).

26.    G. Bodenhausen and R. Freeman, *J. Magn. Reson.*, **28**, 471 (1977).

27.    A. Bax, *Two-Dimensional Nuclear Magnetic Resonance in Liquids*, D. Reidel, Boston, 1982, pp. 50-63.

28.    G. Bodenhausen and R. Freeman, *J. Am. Chem. Soc.*, **100**, 320 (1978).

29.    R. Freeman and G.A. Morris, *J. Chem. Soc., Chem Commun.*, 684 (1978).

30.    A. Bax and G.A. Morris, *J. Magn. Reson.*, **42**, 501 (1981).

31.    A. Bax, *Two-Dimensional Nuclear Magnetic Resonance in Liquids*, D. Reidel, Boston, 1982, pp. 59-61.

32.    D.I. Hoult and R.E. Richards, *Proc. Roy. Soc. London*, **A344**, 311 (1975).

33.    J.S. Waugh, *J. Mol. Spectrosc.*, **35**, 298 (1970).

34.    G. A. Morris and K. I. Smith, *J. Magn. Reson.*, **65**, 506 (1985).

35.    G. A. Morris and K. I. Smith, *Mol. Phys.*, in press (1987).

36.    M. J. Musmar, A. J. Weinheimer, G. E. Martin and R. E. Hurd, *J. Org. Chem.*, **48**, 3580 (1983).

37.    M. J. Musmar, G. E. Martin, R. T. Gampe, Jr., V. M. Lynch, S. H. Simonsen, M. L. Lee, M. L. Tedjamulia and R. N. Castle, *J. Heterocyclic Chem.*, **22**, 545 (1985).

38.    P. H. Bolton, *J. Magn. Reson.*, **51**, 134 (1983).

39.    D. T. Pegg, M. R. Bendall and D. M. Doddrell, *J. Magn. Reson.*, **44**, 238 (1981).

40.    M. R. Bendall, D. T. Pegg, D. M. Doddrell and D. M. Thomas, *J. Magn. Reson.*, **46**, 43 (1982).

41.    M. R. Bendall and D. T. Pegg, *J. Magn. Reson.*, **53**, 144 (1983).

42.    M. H. Levitt, O. W. Sørensen and R. R. Ernst, *Chem. Phys. Lett.*, **94**, 540 (1983).

43.    M. R. Bendall and D. T. Pegg, *J. Magn. Reson.*, **53**, 272 (1983).

44.   T. C. Wong and V. Rutar, *J. Magn. Reson.*, **63**, 524 (1985).

45.   D. T. Pegg and M. R. Bendall, *J. Magn. Reson.*, **55**, 114 (1983).

46.   J. R. Garbow, D. P. Weitekamp and A. Pines, *Chem. Phys. Lett.*, **93**, 504 (1982).

47.   T. C. Wong, V. Rutar and J-S. Wang, *J. Am. Chem. Soc.*, **106**, 7046 (1984).

48.   A. Bax, *J. Magn. Reson.*, **53**, 517 (1983).

49.   T .C. Wong, *J. Magn. Reson.*, **63**, 179 (1985).

50.   G. Batta and A. Liptak, *J. Chem. Soc. Chem. Commun.*, 368 (1985).

51.   G. Batta and K. E. Kover, *Magn. Reson. Chem.*, **25**, 125 (1987).

52.   A. Bax, *J. Magn. Reson.*, **53**, 517 (1983).

53.   M. J. Quast, E. L. Ezell, G. E. Martin, M. L. Lee, M. L. Tedjamulia, J. G. Stuart and R. N. Castle, *J. Heterocyclic Chem.*, **22**, 1453 (1985).

54.   J. C. Carter, G. W. Luther, III and T. C. Long, *J. Magn. Reson.*, **15**, 122 (1974).

55.   W. J. Chazin, L. D. Colebrok and J. T. Edward, *Can. J. Chem.*, **61**, 1749 (1983).

56.   T. C. Wong, V. Rutar and J.-S. Wang, *J. Am. Chem. Soc.*, **106**, 7046 (1984).

57.   T. C. Wong and V. Rutar, *J. Magn. Reson.*, **63**, 524 (1985).

58.   T. C. Wong and G. R. Clark, *J. Chem. Soc., Chem. Commun.*, 1518 (1984).

59.   W. F. Reynolds, D. W. Hughes, M. Perpick-Dumont and R. G. Enriquez, *J. Magn. Reson.*, **64**, 304 (1985).

60.   H. Kessler, C. Griesinger, J. Zarbock and H. R. Loosli, *J. Magn. Reson.*, **57**, 331 (1984).

61.   W. F. Reynolds, D. W. Hughes, M. Perpick-Dumont and R. G. Enriquez, *J. Magn. Reson.*, **63**, 413 (1985).

62.   V. V. Krishnamurthy and J. E. Casida, *Magn. Reson. Chem.*, **25**, 837 (1987).

63.   E. Bartholdi and R. R. Ernst, *J. Magn. Reson.*, **11**, 9 (1973).

64.   T. C. Farrar and E. D. Becker, *Pulse and Fourier Transform NMR*, Academic Press, New York, 1971.

65.   L. Müller, *J. Am. Chem. Soc.*, **101**, 4481 (1979).

66.   G. Bodenhausen and D. J. Ruben, *Chem. Phys. Lett.*, **69**, 185 (1979).

67.   A. Bax, R. G. Griffey and B. L. Hawkins, *J. Magn. Reson.*, **55**, 301 (1983).

68.   A. G. Redfield, *Chem. Phys. Lett.*, **96**, 537 (1983).

69.   M. R. Bendall, D. T. Pegg and D. M. Doddrell, *J. Magn. Reson.*, **52**, 81 (1983).

70.   A. Bax, R. H. Griffey and B. L. Hawkins, *J. Am. Chem. Soc.*, **105**, 7188 (1983).

71.   D. H. Live, D. G. Davis, W. C. Agosta and D. Cowburn, *J. Am. Chem. Soc.*, **106**, 6104 (1984).

72.   D. H. Live, I. M. Armitage, D. C. Dalgarno and D. Cowburn, *J. Am. Chem. Soc.*, **107**, 1775 (1985).

73.   M. Frey, G. Wagner, M. Vasak, O. W. Sørensen, D. Neuhaus, E. Wörgötter, J. H. R. Kägi, R. R. Ernst and K. Wüthrich, *J. Am. Chem. Soc.*, **107**, 1775 (1985).

74.   J. D. Otvos, H. R. Engseth and S. Wehrli, *J. Magn. Reson.*, **61**, 579 (1985).

75.   L. Mueller, R. A. Schiksnis and S. J. Opella, *J. Magn. Reson.*, **66**, 379 (1986).

76.   D. Brühwiler and G. Wagner, *J. Magn. Reson.*, **69**, 546 (1986).

77.   A. Bax and S. Subramanian, *J. Magn. Reson.*, **67**, 565 (1986).

78.   J. A. Wilde, P. H. Bolton, N. J. Stolowich and J. A. Geralt, *J. Magn. Reson.*, **68**, 168 (1986).

79.   V. Sklenar and A. Bax, *J. Magn. Reson.*, **71**, 379 (1987).

80.   W. Leupin, G. Wagner, W. A. Denny and K. Wüthrich, *Nucleic Acids Res.*, **15**, 267 (1987).

81.   D. Neuhaus, J. Keeler and R. Freeman, *J. Magn. Reson.*, **61**, 553 (1985).

82.   A. Bax and M. F. Summers, *J. Am. Chem. Soc.*, **108**, 2093 (1986).

83.    M. F. Summers, L. G. Marzilli and A. Bax, *J. Am. Chem. Soc.*, **108**, 4285 (1986).

84.    A. Bax, A. Aszalos, Z. Dinya and K. Sudo, *J. Am. Chem. Soc.*, **108**, 8056 (1986).

85.    A. Bax, L. G. Marzilli and M. F. Summers, *J. Am. Chem. Soc.*, **109**, 566 (1987).

86.    W. Bermel, C. Griesinger, H. Kessler and K. Wagner, *Magn. Reson. Chem.*, **25**, 325 (1987).

87.    R. H. Griffey, A. G. Redfield, R. E. Loomis and F. W. Dahlquist, *Biochemistry*, **24**, 817 (1985).

88.    M. A. Weiss, A. G. Redfield and R. H. Griffey, *Proc. Nat. Acad. Sci., USA*, **83**, 1325 (1986).

89.    R. H. Griffey, C.D. Poulter, A. Bax, B. L. Hawkins, Z. Yamazuimi and S. Nihimura, *Proc. Nat. Acad. Sci., USA*, **80**, 5895 (1983).

90.    U. Hahn, R. Desaihah and H. Ruterjans, *Eur. J. Biochem.*, **146**, 705 (1985).

91.    K. Hallenga and G. Van Binst, *Bull. Magn. Reson.*, **2**, 343 (1980).

92.    P. E. Hansen, *Org. Magn. Reson.*, **12**, 109 (1979).

93.    P. E. Hansen, *Prog. NMR Spectrosc.*, **14**, 175 (1981).

94.    J. D. Memory and N. K. Wilson, *NMR of Aromatic Compounds*, Wiley-Interscience, New York, 1982.

95.    E. Breitmaier and W. Voelter, *Carbon-13 NMR Spectroscopy*, 3rd ed., VCH, New York, 1987.

96.    J. L. Marshall, *Carbon-Carbon and Carbon-Proton NMR Coupings: Applications to Organic Stereochemistry and Conformational Analysis*, "Methods in Stereochemical Analysis," vol. 2, A. P. Marshall, Ed., VCH, New York, 1983.

97.    W. F. Reynolds, R. G. Enriquez, L. I. Escobar and X. Lozoya, *Can. J. Chem.*, **62**, 2421 (1984).

98.    J. Wernly and J. Lauterwein, *J. Chem. Soc., Chem. Commun.*, 1221 (1985).

99.    C. Bauer, R. Freeman and S. Wimperis, *J. Magn. Reson.*, **58**, 526 (1984).

100.   M. J. Quast, A. S. Zektzer, G. E. Martin and R. N. Castle, *J. Magn. Reson.*, **71**, 554 (1987)

101.   A. S. Zektzer, B. K. John, R. N. Castle and G. E. Martin, *J. Magn. Reson.*, **72**, 556 (1987).

102.   A. S. Zektzer, B. K. John and G. E. Martin, *Magn. Reson. Chem.*, **25**, 752 (1987).

103.   M. Salazar, A. S. Zektzer and G. E. Martin, *Magn. Reson. Chem.*, **26**, 24 (1987).

104.   M. Salazar, A. S. Zektzer and G. E. Martin, *Magn. Reson. Chem.*, **26**, 28 (1987).

105.   M. Perpick-Dumont, R. G. Enriquez, S. McLean, F. V. Puzzuoli and W. F. Reynolds, *J. Magn. Reson.*, **75**, 416 (1987).

106.   G. Batta and K. E. Köver, *Magn. Reson. Chem.*, **25**, 125 (1987).

107.   C. Wynants, K. Hallenga, G. Van Binst, A. Michel and J. Zanen, *J. Magn. Reson.*, **57**, 93 (1984).

108.   J. C. Beloeil, M. A. Delsuc, J. Y. Lallemand, G. Dauphin and G. Jeminet, *J. Org. Chem.*, **49**, 1797 (1984).

109.   J. D. Connolly, C. O. Fakunle and D. S. Rycroft, *J. Chem. Res. (S)*, 368 (1984).

110.   J. N. Shoolery, *J. Nat. Prod.*, **47**, 226 (1984).

111.   J. P. Kintzinger, P. Maltese, M. Bourdonneau and C. Brevard, *Tetrahedron Lett.*, 6007 (1984).

112.   K. V. Schenker and W. von Philipsborn, *J. Magn. Reson.*, **61**, 294 (1985).

113.   T. Nishida, C. R. Enzell and G. A. Morris, *Magn. Reson. Chem.*, **24**, 179 (1986).

114. E. Giralt and M. Feliz, *Magn. Reson. Chem.*, **24**, 123 (1986).
115. R. Benn and R. Mynott, *Angew. Chem. Int. Ed., Engl.*, **24**, 333 (1985); *Angew. Chem.*, **97**, 330 (1985).
116. D. Lontsi, J. F. Ayafor, B. L. Sondengam, J. D. Connolly and D. S. Rycroft, *Tetrahedron Lett.*, 4249 (1985).
117. Y. Konda, Y. Harigaya and M. Onda, *J. Heterocyclic Chem.*, **23**, 877 (1986).
118. G. Cirrincione, G. Dattolo, A. M. Almerico, E. Aillo, G. Cusmano, G. Bacluso, M. Ruccia and W. Hinz, *J. Heterocyclic Chem.*, **23**, 1273 (1986).
119. B. H. Han, M. H. Park and S. T. Wah, *Tetrahedron Lett.*, 3957 (1987)
120. M. J. Quast, E. L. Ezell, G. E. Martin, M. L. Lee, M. L. Tedjamulia, J. G. Stuart and R. N. Castle, *J. Heterocyclic Chem.*, **22**, 1453 (1985).
121. A. W. H. Jans, C. Tintel, J. Cornelisse and J. Lugtenburg, *Magn. Reson. Chem.*, **24**, 101 (1986).
122. J. G. Stuart, M. J. Quast, G. E. Martin, V. M. Lynch, S. H. Simonsen, M. L. Lee, R. N. Castle, J. L. Dallas, B. K. John and L. F. Johnson, *J. Heterocyclic Chem.*, **23**, 1215 (1986).
123. A. S. Zektzer, J. G. Stuart, G. E. Martin and R. N. Castle, *J. Heterocyclic Chem.*, **23**, 1587 (1986).
124. A. L. Ternay, Jr. and J. S. Harwood, *J. Heterocyclic Chem.*, **23**, 1879 (1986).
125. H. Hopf, C. Mlynek, S. El-Tamany and L. Ernst, *J. Am. Chem. Soc.*, **107**, 6620 (1985).
126. S. Carmely and Y. Kashman, *Tetrahedron Lett.*, 511 (1985).
127. L. Mayol, V. Piccialli and D. Sica, *Tetrahedron Lett.*, 1253 (1985).
128. G. Cimino, G. Sodano, A. Spinella and E. Trivellone, *Tetrahedron Lett.*, 3389 (1985).
129. H. Nakamura, H. Wu, J. Kobayashi, Y. Nakamura, Y. Ohizumi and Y. Hirata, *Tetrahedron Lett.*, 4517 (1985).
130. C. Francisco, B. Banaigs, L. Codomier and A. Cave, *Tetrahedron Lett.*, 4919 (1985).
131. S. Carmely and Y. Kashman, *J. Org. Chem.*, **51**, 784 (1986).
132. S. Carmely and Y. Kashman, *Magn. Reson. Chem.*, **24**, 332 (1986).
133. S. Carmely and Y. Kashman, *Magn. Reson. Chem.*, **24**, 343 (1986).
134. H. Jacobs, F. Ramdayal, W. F. Reynolds and S. McLean, *Tetrahedron Lett.*, 1453 (1986).
135. K. K. Purushothaman, M. Venkatanarasimhan, A. Sarada, J. D. Connolly and D. S. Rycroft, *Can. J. Chem.*, **65**, 35 (1987).
136. B. F. Bowden, J. C. Coll, A. Heaton, G. König, M. A. Bruck, R. E. Cramer, D. M. Klein and P. J. Scheuer, *J. Nat. Prod.*, **50**, 650 (1987).
137. G. Cimino, G. Sodano and A. Spinella, *J. Org. Chem.*, **52**, 5326 (1987).
138. H. Kizu, Y. Imoto, T. Tomimori, K. Tsubono, S. Kadota and T. Kikuchi, *Chem. Pharm. Bull.*, **35**, 1656 (1987).
139. T. Kikuchi, K. Tsubono, S. Kadota, H. Kizu, Y. Imoto and T. Tomimori, *Chem. Lett.*, 987 (1987).
140. L. Somogyi, P. Herczegh and G. Batta, *Heterocycles*, **24**, 2735 (1986).
141. A. L. Waterhouse, I. Holden and J. E. Casida, *J. Chem. Soc., Perkin Trans. II*, 1011 (1985).
142. H. Bleich, S. Gould, P. Pitner and J. Wilde, *J. Magn. Reson.*, **56**, 515 (1984).
143. A. Bain, *J. Magn. Reson.*, **77**, 125 (1988).
144. Y. Sato, M. Geckle and S. J. Gould, *Tetrahedron Lett.*, 4019 (1985).
145. Y. Sato, R. Kohnert and S. J. Gould, *Tetrahedron Lett.*, 143 (1986).
146. T. C. Somers, J. D. White, J. J. Lee, P. J. Keller, C.-j. Chang and H. G. Floss, *J. Org. Chem.*, **51**, 464 (1986).

147.    D. Blasberger, D. Green, S. Carmely, I. Spector and Y. Kashman, *Tetrahedron Lett.*, 459 (1987).

148.    S. Carmely and Y. Kashman, *Tetrahedron Lett.*, 3003 (1987).

149.    Y.-I. Xu, H.-D. Suz, D.-Z. Wang, M. Kozuka, K. Naya and I. Kubo, *Tetrahedron Lett.*, 499 (1987).

150.    M. Kawasaki, T. Hayashi, M. Arisawa, M. Shimizu, S. Horie, H. Ueno, H. Syogawa, S. Suzuki, M. Yoshizaki, N. Morita, Y. Tezuka, T. Kikuchi, L. H. Berganza, E. Ferro and I. Basualdo, *Chem. Pharm. Bull .*, **35**, 3963 (1987).

151.    J. E. Hochlowski, S. J. Swanson, L. M. Ranfranz, D. N. Whittern, A. M. Buko and J. B. McAlpine, *J. Antibiotics*, **40**, 575 (1987).

152.    R. L. Halterman, N. H. Nguyen and K. P. C. Vollhardt, *J. Am. Chem. Soc.* , **107**, 1379 (1985).

153.    A. L. Waterhouse, *Magn. Reson. Chem.*, **26**, in press (1988).

154.    J. Wernly and J. Lauterwein, *Magn. Reson. Chem.*, **23**, 170 (1985).

155.    S. Wimperis and R. Freeman, *J. Magn. Reson.*, **58**, 348 (1984).

156.    H. Kogler, O. W. Sørensen, G. Bodenhausen and R. R. Ernst, *J. Magn. Reson.*, **55**, 157 (1983).

157.    P. H. Bolton, *J. Magn. Reson.*, **63**, 225 (1985).

158.    A. S. Zektzer, M. J. Quast, G. S. Linz, G. E. Martin, J. D. McKenney, M. D. Johnston, Jr. and R. N. Castle, *Magn. Reson. Chem.*, **24**, 1083 (1986).

159.    H. Kessler, W. Bermel and C. Griesinger, *J. Am. Chem. Soc.*, **107**, 1083 (1985).

160.    M. Salazar, M. S. in Pharmacy Thesis, University of Houston, 1987.

161.    L. R. Brown and J. Bremer, *J. Magn. Reson.*, **68**, 217 (1986).

162.    H. Kessler, C. Griesinger and J. Lautz, *Angew. Chem. Int. Ed., Engl.*, **23**, 444 (1984); *Angew. Chem.*, **96**, 434 (1984).

163.    V. V. Krishnamurthy and J. E. Casida, *Magn. Reson. Chem.*, **25**, 837 (1987).

164.    A. Bax and M. F. Summers, *J. Am. Chem. Soc.*, **108**, 2093 (1986).

165.    T. Nishida, G. A. Morris, I. Forsblom, I. Wahlberg and C. R. Enzell, *J. Chem. Soc., Chem. Commun.*, 998 (1986).

166.    H. Kessler and M. Bernd, *Liebigs Ann. Chem.*, 1145 (1985).

167.    H. Kessler and A. Müller, *Liebigs Ann. Chem.*, 1687 (1986).

168.    H. Kessler, W. Bermel, C. Griesinger and C. Kolar, *Angew. Chem., Int. Ed. Engl.*, **25**, 342 (1986); *Angew. Chem.*, **96**, 352 (1986).

169.    H. Kessler, W. Bermel, A. Müller and K.-H. Pook, "Modern Nuclear Magnetic Resonance Spectroscopy of Peptides," in *The Peptides*, vol. 7, V. J. Hruby, Ed., Academic Press, New York, 1985, pp. 437-473.

170.    H. Kessler and W. Bermel, "Conformational Analysis of Peptides by Two-Dimensional NMRF Spectroscopy," in *Applications of NMR Spectroscopy to Problems in Stereochemistry and Conformational Analysis*, vol. 6, Y. Takeuchi and A. P. Marchand, Eds., VCH, New York, 1986, pp 179-205.

171.    H. Kessler, C. Griesinger and K. Wagner, *J. Am. Chem. Soc.*, **109**, 6927 (1987).

172.    N. Müller, *Magn. Reson. Chem.*, **23**, 688 (1985).

173.    J. Wernly and J. Lauterwein, *J. Magn. Reson.*, **66**, 355 (1986).

174.    A. W. H. Jans, C. Tintel, J. Cornelisse and J. Lugtenburg, *J. Magn. Reson.*, **24**, 101 (1986).

175.    L. Müller and P. W. Jeffs, *J. Magn. Reson.*, **73**, 405 (1987).

176.    I. Uchida, S. Takase, H. Kayakiri, S. Kiyoto, M. Hashimoto, T. Tada, S. Koda and Y. Morimoto, *J. Am. Chem. Soc.*, **109**, 4108 (1987).

177.    C. Tringali, M. Piattelli, C. Geraci, G. Nicolosi and C. Rocco, *Can. J. Chem.*, **65**, 2369 (1987).

178.    A. Aiello E. Fattorusso, S. Magno and L. Mayol, *Tetrahedron Lett.*, 5929 (1987).

179. M. Murata, M. Kumagai, J. S. Lee and T. Yasumoto, *Tetrahedron Lett.*, 5869 (1987).
180. M. Kumagai, M. Murata, J. S. Lee, T. Yasumoto, *Tenne Yuki Kagobutsu Toronkai Koen Yoshishu*, **29**, 600 (1987); CA 108:70214v.
181. M. R. Kernan and D. J. Faulkner, *Tetrahedron Lett.*, 2809 (1987).
182. T. Kusumi, T. Ooi, M. M. Watanabe, H. Takahashi and H. Kakisawa, *Tetrahedron Lett.*, 4695 (1987).
183 D. M. Roll, J. E. Bikupiak, C. L. Mayne and C. M. Ireland, *J. Am. Chem. Soc.*, **108**, 6680 (1986).
184. W. B. Smith, *Magn. Reson. Chem.*, **25**, 981 (1987).
185. H. Jacobs, F. Ramadayal, S. McLean, M. Perpick-Dumont, F. Puzzuoli and W. F. Reynolds, *J. Nat. Prod.*, **50**, 507 (1987).
186. H. Jacobs, F. Ramadayal, W. F. Reynolds and S. McLean, *Tetrahedron Lett.*, 1453 (1986).
187. W. F. Reynolds, S. McLean, J. Poplawski, R. G. Enriquez, L. I. Escobar and I. Leon, *Tetrahedron*, **42**, 3419 (1986).
188. S. McLean, W. F. Reynolds and X. Zhu, *Can. J. Chem.*, **65**, 200 (1987).
189. A. Numata, G. R. Petit, M. Nabae, K. Yamamoto, E. Yamamoto, E. Matsumura and K. Tatsu, *Agric. Biol. Chem.*, **51**, 1199 (1987).
190. J. Cavanaugh and J. Keeler, *J. Magn. Reson.*, **77**, 356 (1988)
191. A. Bax, personal communication.
192. A. Bax and D. Marion, *J. Magn. Reson.*, submitted (1988).
193. A. S. Zektzer, L. D. Sims, R. N. Castle and G. E. Martin, *Magn. Reson. Chem.*, **26**, 287 (1988).
194. G. Englert, E. Glinz, S. Liaaen-Jensen, *Magn. Reson. Chem.*, **26**, 55 (1988).
195. M. Alam and P. Sharma, *J. Chem. Soc., Perkin Trans I*, in press (1988).
196. M. Alam, P. Sharma, A. S. Zektzer and G. E. Martin, *Magn. Reson. Chem.*, **26**, in press (1988).
197. S. Abuzar, H. Kohn, J. D. Korp, A. S. Zektzer and G. E. Martin, *J. Heterocyclic Chem.*, **25**, in press (1988).
198. S. Abuzar and H. Kohn, *J. Am. Chem. Soc.*, **110**, 3661 (1988).
199. H. Seto, K. Furihata, K. Saeki, N. Otake, Y. Kusakabe, C. Xu and J. Clardy, *Tetrahedron Lett.*, 3357 (1987).
200. U. Kobayashi, M. Ishibashi, M. R. Wälchli, H. Nakamura, Y. Hirata, T. Sasaki and Y. Ohizumi, *J. Am. Chem. Soc.*, **110**, 490 (1988).
201. D. E. Barnekow, J. H. Cardellina, II, A. S. Zektzer and G. E. Martin, *J. Am. Chem. Soc.*, submitted (1988).
202. R. Hendrickson, J. H. Cardellina, II, A. S. Zektzer and G. E. Martin, *J. Am. Chem. Soc.*, submitted (1988).

# CHAPTER 4

## RELAYED COHERENCE TRANSFER
## AND RELATED 2D-NMR EXPERIMENTS

### INTRODUCTION

The COSY and homonuclear zero and double quantum experiments described in Chapter 2 have been devised with the specific intent of establishing vicinal and/or geminal connectivities between coupled nuclear spins. Vicinal connectivity is inarguably useful for establishing molecular structure segments. Situations will still arise, however, when a more "global" view of the environment of a given atom would be useful or desirable. Hence, it becomes of interest to transfer or "relay" magnetization beyond a pair of directly coupled spins to the next nearest neighbor of each. Put another way, the relay process is capable of extending the connectivity network further than its predecessor experiments. It is important to note, however, that a COSY or similar experiment is necessary before the information contained in a relayed coherence transfer experiment may be digested. This makes the relay experiment, in terms of practical consideration, a second generation two-dimensional NMR experiment requiring another two-dimensional NMR experiment to facilitate its interpretation.

Credit for the origination of the concept of relayed coherence transfer (RCT) must be given to Eich, Bodenhausen and Ernst.[1] We will consider the homonuclear relayed coherence transfer experiment described by Eich, Bodenhausen and Ernst and derivatives of it before delving into the realm of heteronuclear relay. Within the manifold of homonuclear relay experiments, probably the most common at this point are the COSY-type experiments. Within this group, experiments have been reported to perform single or double relays of magnetization. Experiments in this first category used a sequence of pulses to establish the coherence between successively more distant spins. It is also possible, however, to establish coherences to more distant spins by using methods such as spin-locking fields. Techniques in this latter group include experiments such as TOCSY and HOHAHA and will be discussed later in this chapter. Although less commonly utilized at this point, relayed coherence transfer analogs of the proton zero and double quantum experiments have also been described.

Several years after the first homonuclear relayed coherence transfer experiments were described, the first heteronuclear relayed coherence transfer experiments were reported. Here there has been considerably less variety to consider. The basic heteronuclear relay experiment has spawned few analogs. Analogs that have been reported consist of only proton double quantum heteronuclear relayed coherence transfer, low pass, J-filtered heteronuclear relay and proton detected heteronuclear relay experiments. Each of these will be discussed in turn following the presentation of the basic heteronuclear relayed coherence transfer experiment below.

## HOMONUCLEAR RELAYED COHERENCE
## TRANSFER VIA SCALAR COUPING PATHWAYS

### Why Is Relayed Coherence
### Transfer Useful or Desirable?

An excellent illustration of the need for the information provided in an RCT experiment is presented by Wagner.[2] Consider the schematic representation of a COSY spectrum of a dipeptide segment shown in Fig. 4-1 in which the $\alpha_1$ and $\alpha_2$ proton chemical shifts are degenerate. Connectivities between the $NH_1$ and $NH_2$ protons will establish the correlation to the $\alpha$ protons. However, since the $\alpha$ protons have degenerate chemical shifts, there will be ambiguity in further connectivities from the $\alpha$ protons to the $\beta_1$ and $\beta_2$ protons. In contrast, consider the connectivity data provided from the RCT experiment, which is also presented schematically in Fig. 4-1 for the same dipeptide segment. Relayed coherence transfer in this case establishes the next nearest neighbor connectivities between $NH_1$ and $\beta_1$ and between $NH_2$ and $\beta_2$. In this fashion, any ambiguity engendered by the degenerate $\alpha$ proton resonance chemical shifts is successfully circumvented.

Another useful example of the benefits of relayed coherence transfer arises when we consider a system of three coupled spins, eg, an AMX spin system in which $J_{AX} = 0$. Here, quite obviously, the COSY spectrum will establish correlations between the A and M spins via $J_{AM}$ and between M and X via $J_{MX}$. There will be no correlation established between the A and X spins, however, since $J_{AX} = 0$. The correlation between the A and X spins of our AMX spin system is, however, established through the coupling of both to the intermediate M spin when the relayed coherence transfer experiment is utilized.

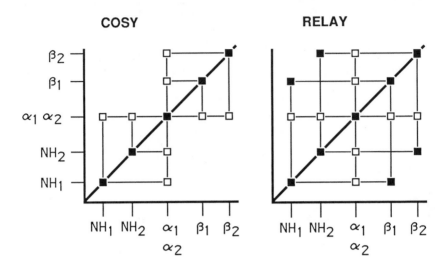

*Fig. 4-1.*    Schematic representation of the connectivities of a dipeptide linking amide NH resonances to α proton resonances with degenerate chemical shifts which are in turn linked to β proton resonances with discrete chemical shifts. In the COSY spectrum, connectivities of NH$_1$ to its corresponding β$_1$ proton and that of NH$_2$ to its corresponding β$_2$ proton are ambiguous because of the degeneracy of the α proton resonance shifts. In the relayed coherence transfer (RELAY) spectrum, coherence is relayed from the NH resonances to their respective β proton resonances, thereby circumventing the degeneracy of the α proton resonances and avoiding any ambiguity in the assignment.

## The Single Relay Experiment: RCOSY

Pulse sequencing for the single relayed coherence transfer experiment, sometimes designated by the acronyms RCOSY or RELAY (the former will be used in this chapter), builds on the pulse sequence used for the COSY experiment, which was shown in Fig. 2-1. Following the establishment of coherence between vicinally coupled protons via the first two 90° pulses of the sequence, which is shown in Fig. 4-2, magnetization is then relayed to the next nearest neighbor protons via a mixing period as defined by Eqn. [4-1].

$$- \tau_m - 180° - \tau_m - 90° - \qquad [4\text{-}1]$$

The mixing period described by Eqn. [4-1] is dependent on $J_{MX}$ in the latter case presented above and serves to optimally align components of M magnetization in antiphase prior to subsequent transfer to the X spin by the final 90° pulse shown in Eqn. [4-1]. The 180° refocusing pulse located midway through the mixing period (after $\tau_m$) serves to eliminate any further chemical shift evolution that would also transpire during the mixing period in the absence of the refocusing pulse.

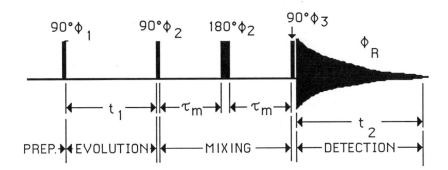

*Fig. 4-2.*     Relayed coherence transfer (RCOSY or RELAY) pulse sequence. The duration of the fixed intervals in the mixing period, $\tau_m$, are generally optimized for the couplings through which "relay" will occur. As an approximation, the function $1/4(J_{HH})$ may be used.

## Phase Cycling

The phase cycling scheme of the RCOSY experiment shown in Fig. 4-2 is somewhat more complicated than that employed with the COSY experiment (see Table 2-1). Although it is possible to employ a minimal sixteen step phase cycle to perform the RCOSY experiment, this is not advisable as it affords an intense antidiagonal passing through the spectrum. The most commonly utilized phase cycling scheme for the RCOSY experiment is given in Table 4-1 and consists of thirty-two steps. The phase cycle provides quadrature detection in both frequency domains and effective cancellation of axial responses.

Table 4-1.     Phase cycling scheme for the homonuclear relayed coherence transfer experiment (RCOSY) presented in Fig. 4-2.

| $\phi_1$ | 00000000 | 11111111 | 22222222 | 33333333 |
| --- | --- | --- | --- | --- |
| $\phi_2$ | 00112233 | 11223300 | 22330011 | 33001122 |
| $\phi_3$ | 02132031 | 13203102 | 20310213 | 21021320 |
| $\phi_R$ | 00220022 | 11331133 | 22002200 | 33113311 |

## Parameter Selection

**Basic parameters**. The selection of parameters for the RCOSY experiment differs from the normal COSY experiment only minimally. Parameters that are basically the same include evolution times in $t_1$, interpulse delays and levels of digitization in the data matrix.

With regard to the shape of the data matrix, it is still desirable to increment the duration of the evolution time, $t_1$, as a function of the dwell time in $t_2$ to ultimately provide a square data matrix that is amenable to symmetrization.[3] Next, relaxation considerations for the RCOSY experiment are identical to those of the COSY experiment. Typically, a minimum interpulse delay of 1.5 x $T_1$ is useful. In the absence of measured relaxation times, an interpulse delay of 1 sec plus the acquisition time provides a useful starting point. Finally, we must consider digitization in $t_1$. If the data matrix is to be symmetrized, a constraint is automatically imposed, necessitating a data matrix consisting of, for example, 256 x 512 or perhaps 512 x 1K points prior to processing. In the face of time constraints, it is possible to take fewer points and to zero-fill in $t_1$ to an appropriate level prior to the second Fourier transformation. If symmetrization is not a concern, under-digitization in $t_1$ will have the same consequences on response shape in $F_1$ after processing as it did in the example of the underdigitized plumericin COSY spectrum shown in Fig. 2-4.

**Optimization of the relay step**. A parameter of the RCOSY that is new is the optimization of the fixed duration delay, $\tau_m$ , in the mixing step of the pulse sequence shown in Fig. 4-2. Bax and Drobny have discussed the optimization of the RCOSY experiment in considerable detail.[4] Ideally, three goals should be considered in optimizing the experiment. First, it is, quite obviously, desirable to maximize the intensity of the relayed magnetization, since this is the new spectral information being sought in the experiment. Second, it is useful to minimize the intensity of the diagonal and COSY responses, since they provide no useful information. Finally, the optimization of the resolution is also desirable.

Let us consider as an example, an AMX spin system where we will relay magnetization from A to X. The intensity, I, of the relayed magnetization may be described by the expression:

$$I \sim \sin (\pi J_{AM} \tau) \sin (\pi J_{MX} \tau) \exp(-\tau /T_2) \qquad [4\text{-}2]$$

where $T_2$ is the spin-spin or transverse relaxation time of the M spin. It should also be noted that Bax and Drobny define the total duration of the mixing period as $\tau$. In the case where $J_{AM}$ and $J_{MX}$ are comparable in size and where $J_{AM}$, $J_{MX} \gg T_2^{-1}$, Bax and Drobny have shown that the optimal value of the delay, $\tau$, is near the condition shown in Eqn. [4-3]:

$$\tau = (J_{AM} + J_{MX})^{-1} \qquad\qquad [4\text{-}3]$$

When $J_{AM}$, $J_{MX} \ll T_2^{-1}$, the optimal value of the delay, $\tau$, is near:

$$\tau = 2T_2 \qquad\qquad [4\text{-}4]$$

Finally, in the intermediate condition where J is comparable to $T_2^{-1}$, a value for $\tau$ 10-30% shorter than that given in Eqn. [4-3] will optimize the amount of magnetization relayed between the A and X spins.

Assuming $J_{AM} = J_{MX} = 4.5$ Hz, Eqn. [4-3] leads to the expectation of a total mixing time of ~110 msec. The duration of $\tau_m$ would therefore be half of this value or approximately 55.5 msec. As a general rule of thumb, the duration of $\tau_m$ may be optimized as a function of $1/4(J_{HH})$ where $J_{HH}$ corresponds to the average couplings involved in a given transfer. Here, optimizing $\tau_m$ as a function of $1/4(J_{HH})$ gives a duration of $\tau_m = 58$ msec.

A number of other, more complicated, spin systems are considered in the comprehensive paper of Bax and Drobny.[4] Readers interested in optimization with respect to other spin systems are referred to that source for further details.

## Data Processing and Presentation

**Processing: Selection of weighting factors**. Having considered the acquisition parameters, it is next appropriate to examine the processing of RCOSY data. According to the work of Bax and Drobny,[4] near optimal results will be obtained if the data is processed using a non-phase-shifted sinusoidal multiplication of the data. Undesired diagonal and COSY responses will also tend to be minimized by this treatment of the data, and resolution in the absolute value presentation mode will also be near optimal.

The overall processing of the RCOSY data is essentially identical to the processing of the COSY experiment data. The first Fourier transformation is performed using a constant phase and scaling corrections. As noted above, best results will be obtained when sinusoidal multiplication is employed prior to the first Fourier transformation. Next, the RCOSY data is transposed to ready it for the second Fourier transformation. Once again, sinusoidal multiplication provides optimal weighting (see above).

**Symmetrization of RCOSY spectra**. RCOSY spectra, since they have the same symmetry relationships as a COSY spectrum, may be symmetrized.[3] As with COSY spectra, symmetrization requires a square matrix, for example defined by 256 x 256 points, or one of corresponding size in $F_1$,$F_2$.

**Presentation of RCOSY spectra**. RCOSY spectra are most easily interpreted as contour plots. Stack plots will have minimal utility, although there may occasionally be the need to extract individual slices of the RCOSY data matrix to examine spectral features.

## Interpretation of RCOSY Spectra

RCOSY spectra are what may be considered "second generation" two-dimensional NMR spectra. Since they are a derivative of a COSY experiment, it is generally expedient to have available a COSY spectrum to which the RCOSY spectrum may be compared so that relayed responses are not mistaken for the direct responses. As recently pointed out in a chapter by Hull,[5] RCOSY data acquired using phase sensitive methods, eg, TPPI (see Chapter 2), can also be used to distinguish relayed coherence transfer from direct coherence responses in an RCOSY spectrum. Hull does go on to suggest, however, that it is "...wise to perform a standard COSY experiment under identical conditions to ensure unambiguous assignments."

For the sake of simplicity, we will uniformly illustrate the utilization of RCOSY data beginning from a standard COSY spectrum, which will be used to identify direct coherence transfer responses. We will begin first with the cembranoid diterpene jeunicin (**4-1**) whose COSY spectrum was considered in Chapter 2 (see Fig. 2-8). We will subsequently consider the alkaloid strychnine (**4-4**) in the context of both single and double relayed coherence transfer experiments. In the latter example, we shall also go on to later consider proton double quantum relay and homonuclear Hartmann-Hahn (HOHAHA) experiments as well.

**Application of the RCOSY experiment to the cembranoid diterpene jeunicin**. The structure of jeunicin (**4-1**) in the region of the oxepin ring, the $\gamma$-lactone and the $\Delta^7$ double bond gives the most useful structural information in the COSY spectrum shown in Fig. 2-8.[6] To facilitate easy comparison of the connectivity information contained in the COSY and RCOSY spectra of jeunicin (**4-1**), a composite has been prepared by sandwiching half of each spectrum to the other along the diagonal in the presentation contained in Fig. 4-3. The COSY spectrum of jeunicin, which was presented in Fig. 2-8, is thus located below the diagonal, while the RCOSY spectrum acquired using the pulse sequence shown in Fig. 4-2 is presented above the diagonal.

Using the COSY portion of the spectrum and the H1 resonance located as 3.36 ppm as a starting point, we may track vicinal connectivities to the H14 resonance at 4.42 ppm and to the H2 anisochronous geminal methylene protons, which resonate at 2.36 and 1.89 ppm. The H14 proton is in turn coupled to H13, which resonates at 3.18 ppm just upfield of H1. Likewise, the anisochronous H2 resonances are coupled to the H3 resonance at 3.67 ppm just downfield of H1. In the COSY spectrum, we do not observe any responses that correlate H1 with either H3 or H13.

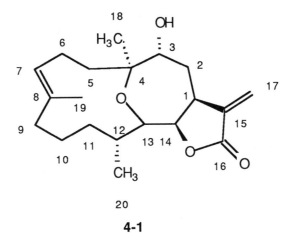

**4-1**

Having identified the vicinal connectivity network just described, we may construct the structural fragment shown by **4-2**. The RCOSY spectrum shown above the diagonal in Fig. 4-3 contains two responses associated with H1, which are circled. These responses correspond to the relayed coherence transfer between H1 and H3, the latter resonating at 3.67 ppm and between H1 and H13, the latter in this case resonating at 3.18 ppm. These connectivities are denoted by the arrows from H1 shown in **4-2**. Intervening protons are omitted for clarity.

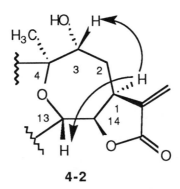

**4-2**

Vicinal proton-proton connectivities in the oxepin moiety of jeunicin are easily traced in the 300 MHz COSY spectrum, and the RCOSY data would not be necessary to locate either H3 or H13. Nonetheless, the illustration just presented is a useful one for introducing the technique. A more practical utilization of the RCOSY data comes into play when we consider the allylic and homoallylic H6 and H5 anisochronous geminal methylene protons. Connectivities in the COSY spectrum for the H7 proton, which resonates at 5.53 ppm, allow the location of the anisochronous H6 geminal methylene protons, which resonate at 2.27 and 2.08 ppm. No connectivities are observed for the homoallylic H5 methylene protons. Since the region containing the allylic and homoallylic proton resonances will frequently be quite congested, the COSY spectrum may not be amenable to analysis. The RCOSY spectrum, however, provides a convenient means for locating the homoallylic H5 methylene protons irrespective of the degree of

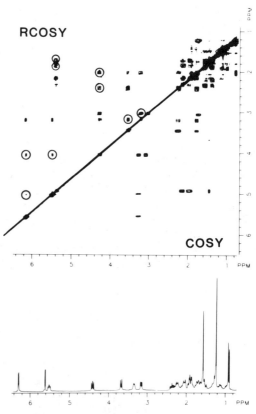

*Fig. 4-3.*     Composite COSY/RCOSY spectrum of the cembranoid diterpene jeunicin (**4-1**) recorded in deuterochloroform at 300 MHz.[6] Both spectra were taken on the same sample and consisted of final data matrices comprised of 512 x 512 points. Both spectra were symmetrized prior to plotting. The COSY spectrum shown below the diagonal was recorded using the 16 step phase cycle shown in Table 2-1, with 16 acquisitions/block; the RCOSY spectrum was recorded using the 32 step phase cycle shown in Table 4-1 with 32 acquisitions/block. Relayed coherence transfer responses between H1 and the H3 and H13 resonances and between H7 and the H5 geminal methylene protons are circled in the RCOSY spectrum.

congestion in the upfield region of the proton NMR spectrum. In the present case, the H5 anisochronous methylene protons, which resonate at 1.71 and 1.50 ppm, are denoted in the RCOSY portion of the spectrum shown in Fig. 4-3.

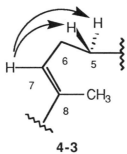

**4-3**

## A Brief Survey of
## RCOSY Applications

To date, the majority of RCOSY applications have been in the area of biomacromolecules. Illustrations of the application of RCOSY in the study of large biopolymers are found in studies of the Lac repressor headpiece[7-9] and the λ Cro repressor,[10,11] cytochrome c[12] and DNA oligomers.[13,14] Several studies applying the RCOSY experiment to oligosaccharides have also appeared.[15,16]

With regard to optimization, in addition to the work of Bax and Drobny[4] dealing with the optimization of the RCOSY experiment, Chazin and Wüthrich[17] have more recently discussed the optimization of the RCOSY experiment for protein studies in $H_2O$.

Small molecule, eg, natural products, applications of the RCOSY experiment are relatively few in number. Studies appearing to date that have utilized the RCOSY technique include the jeunicin example presented above[6]; a study reporting the complete proton and carbon NMR spectral assignment of oligomycin A[18] and finally, the elucidation of the structure of a novel cembranoid diterpene.[19]

## Long Range (Four Bond)
## Relayed Coherence Transfer

Recent applications of the RCOSY experiment to peptides have led to an interesting and unusual observation. Specifically, although experiments have been optimized in several cases for a single transfer of magnetization, for example from an α proton to its γ neighbor via the intervening β proton, longer range transfer to the proton has been observed. The first such observation was reported by Otter and Kotovych.[20] Their study uncovered α → δ correlations in both proline and arginine in an RCOSY spectrum recorded with an 80 msec mixing time. These responses were not visible, however, when a 40 msec mixing time was employed. More recently, Chary and Hosur[21] have reported similar long range (four bond) correlations in RCOSY spectra of peptides for leucine, tyrosine and tryptophan when the mixing time was optimized at 60 msec.

The utility of long range transfers, in the long term, remains to be seen. In the specific case of tyrosine and tryptophan, the long range correlation responses provide the means of specifically associating the α proton of these amino acids with the H2 aromatic and indolyl protons (δ), respectively. It should also be noted, however, that additional care in the acquisition and interpretation of RCOSY spectra should be exercised, since there may also be similar long range couplings in structural moities other than amino acid residues.

## Multiple Quantum Filtered Relay: MQF-RELAY

Quite recently, Weber and Müller[22] have described a pulse sequence for multiple quantum filtering of RCOSY spectra (MQF-RCOSY). While the RCOSY experiment is being supplanted in many applications by isotropic mixing experiments, there are some instances in which the use of the RCOSY experiment is still desirable, eg, selective optimization for various amino acids.[4,23-25]

One of the problems with applying RCOSY spectra to large molecules, obviously, is engendered by the diagonal responses. A substantial improvement in the utility of the data may be obtained by multiple quantum filtering. The pulse sequence described by Weber and Müller[22] is shown in Fig. 4-4, the attendant phase cycling given in Table 4-2. The experiment incorporates two multiple quantum filtering steps, the first immediately following the "COSY portion" of the experiment, the second following the mixing period. Typically the filtering inveral will be very short, usually just 5-10 μsec to allow for phase shifting.

The eventual utility of the MQF-RCOSY experiment remains to be determined. In all probability, the experiment will have considerable opportunities for usage in the study of peptides, oligosaccharides and other molecules that contain numerous protons with closely similar chemical shifts. The utility of this experiment with small molecules would seem to be substantially less, although this remains to be seen.

## Phase Cycling

The phase cycling for the multiple quantum filtered relayed coherence transfer experiment is given in Table 4-2 and consists of a sixteen step phase cycle. Quadrature phase detection in the $F_1$ frequency domain is obtained by collecting a second set of data with the phase of the first 90° pulse phase shifted by 90° according to the method of States, Haberkorn and Ruben.[26] The data matrix is then processed to yield a phase sensitive spectrum (see Chapter 2 for further discussion). It should be noted at this point that although two separate data sets are acquired for processing, the total number of acquisitions is 32, the same as in the conventional RCOSY experiment.

Table 4-2.      Phase cycling scheme for multiple quantum filtered relayed coherence transfer (MQF-RCOSY) presented in Fig. 4-4.

| $\phi_1$ | 0123 | 0123 | 0123 | 0123 |
|---|---|---|---|---|
| $\phi_2$ | 0000 | 1111 | 2222 | 3333 |
| $\phi_R$ | 0202 | 3131 | 2020 | 1313 |

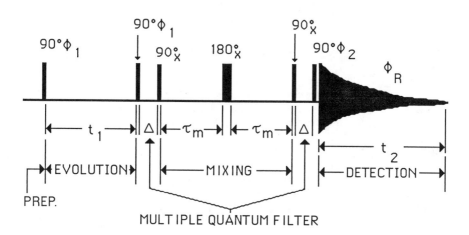

Fig. 4-4.      Multiple quantum filtered relayed coherence transfer (MQF-RCOSY) pulse sequence developed by Weber and Müller.[22]

## The Double Relay
## Experiment: 2RCOSY

We noted above in the section that deals with long range responses in single relay or RCOSY experiments that there are instances in which the intentional transfer of magnetization across four bonds can be useful.  One example already cited is in specifically linking the $\alpha$ proton of tyrosine with its H2/H6 aromatic protons. Another example is provided by the correlation of the amide protons with the methyl protons of valine. Rather that limiting our study of such responses to chance observations, it is possible to extend the RCOSY experiment to routinely provide such information by the inclusion of a second mixing period to give the double relay or 2RCOSY experiment.

Eich, Bodenhausen and Ernst[1] noted in their seminal communication that multiple relay of coherence was possible.  The experiment is based on the RCOSY experiment shown in Fig. 4-2 to which a second mixing period is added as shown in Fig. 4-5, with a commensurate increase in the complexity of the phase cycling to 64 steps as shown in Table 4-3 below.

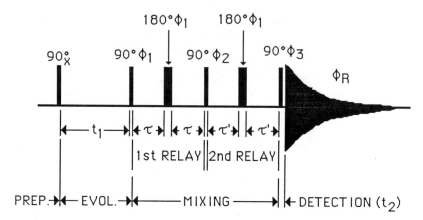

*Fig. 4-5.*    Pulse sequence for proton double relayed coherence transfer (2RCOSY) experiment. The duration of the two mixing times may be optimized separately as a function of $1/4(J_{HH})$.

## Phase Cycling

The phase cycling for the 2RCOSY experiment shown in Fig. 4-5 is given in Table 4-3. The first sixteen steps of the phase cycle provide the means to minimize unwanted COSY and single relay responses. The elimination of quadrature artifacts is obtained by the 90° increments of phase, which occur during the subsequent 48 steps in the phase cycle. Thus, the phase of the first pulse, labeled $90°_x$ in Fig. 4-5, would be incremented in 90° steps every sixteen acquisitions (ie, x for steps 1-16, y for steps 17-32, etc.).

## Parameter Selection

The selection of parameters for the 2RCOSY experiment is fundamentally no different from the RCOSY experiment discussed above. The optimization of the 2RCOSY experiment is considered in the work of Bax and Drobny.[4] The sole difference is that it becomes necessary to decide whether the same optimization will be employed for both mixing periods.

## Data Processing and Presentation

The data processing of 2RCOSY is identical to the RCOSY experiment. Typically it will be useful to employ sinusoidal or an equivalent treatment of the data prior to both Fourier transformations.

Likewise, presentation of the 2RCOSY data has the same implicit considerations as the RCOSY experiment discussed above.

Table 4-3.    Phase cycling scheme for the proton double relay (2RCOSY) experiment shown in Fig. 4-5.

| $\phi_1$ | 00001111 | 22223333 | 11112222 | 33330000 |
|---|---|---|---|---|
| | 22223333 | 00001111 | 33330000 | 11112222 |
| $\phi_2$ | 00221133 | 22003311 | 11332200 | 33110022 |
| | 22003311 | 00221133 | 33110022 | 11332200 |
| $\phi_3$ | 02021313 | 20203131 | 13132020 | 31310202 |
| | 20203131 | 02021313 | 31310202 | 13132020 |
| $\phi_R$ | 00002222 | 00002222 | 11113333 | 11113333 |
| | 22220000 | 22220000 | 33331111 | 33331111 |

## A Comparison of the COSY, RCOSY and 2RCOSY Spectra of Strychnine

In Chapter 2 we extensively utilized strychnine (**4-4**) as a model system to illustrate the capabilities of the various two-dimensional proton NMR experiments with a complex natural product. We shall continue that approach in treating the COSY/RCOSY/2RCOSY experiments.

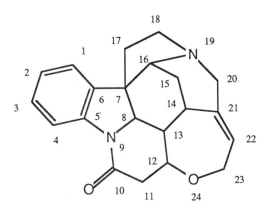

**4-4**

The structure of strychnine (**4-4**) is shown above, and we shall direct our consideration of the COSY, RCOSY and 2RCOSY experiments solely to the nonaromatic portion of the molecule. We shall begin our discussion with the COSY spectrum shown in Fig. 4-6 (the full spectrum including the aromatic region is presented in Fig. 2-11).

Responses in the COSY spectrum that provide additional useful information in the RCOSY spectrum principally involve H13 and H15. Hence, connectivity pathways for H13 and H15 are shown in the spectrum presented in Fig. 4-6 above and below the diagonal, respectively.

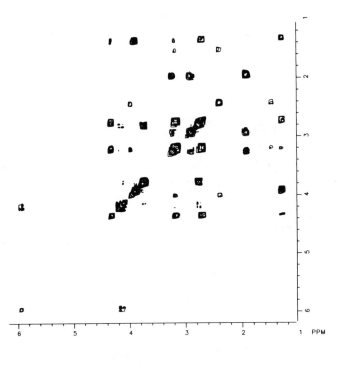

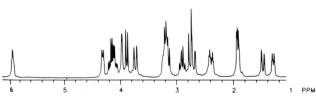

Fig. 4-6.     COSY spectrum of the aliphatic portion of the proton NMR spectrum of strychnine (4-4) recorded in deuterochloroform. Data was taken as 256 x 1K complex points with 16 acquisitions/block and was zero-filled to give a final matrix consisting of 512 x 512 points, which was symmetrized[3] prior to plotting.

Vicinal responses are observed between H13 and its neighbors H8 and H14. In a similar sense, responses that correlate with H15a/b include the geminal response and responses correlating H15a and H15b singly with its vicinal neighbors, H14 and H16.

Next we may compare the vicinal response networks in the COSY spectrum for H13 and H15a/b with the results obtained from the RCOSY spectrum shown in Fig. 4-7. The experiment was performed using the pulse sequence shown in Fig. 4-2 and the phase cycling given in Table 4-1. Data were collected as 256 x 1K points with 32 acquisitions/block and were processed using sinusoidal multiplication prior to the first Fourier transformation and sinusoidal multiplication and zero filling prior to the second transformation to afford a final matrix consisting of 512 x 512 points, which was symmetrized[3] prior to plotting. Perhaps most importantly, the mixing time in the relay step of the experiment was optimized for 4.5 Hz (55.5 msec).

Connectivities that, in general, involve H13 are shown above the diagonal, while those for H15a/b and other resonances are presented below the diagonal in Fig. 4-7. In addition to the vicinal correlations observed in the COSY spectrum, H13 also now exhibits responses corresponding to relay of magnetization to H11b and a weak response arising from relay to H15a (labeled 11b/13R and 15a/13R, respectively, in the figure) arising through relay via H12 and H14, respectively. Further interesting responses correlates H8 with H14 and H12 to H8, both through a relay via  H13.

Less relayed information arises from H15a/b. One response that is observed, however, is a relay from H16 to H14 through H15a/b. Also shown beneath the diagonal are responses arising from H22.  In addition to the normal responses (vicinal H22/H23a, H22/H23b and a long range between H22/H14, H22/H20a and H22/H20b) there are also relayed responses that correlate H23a and H23b with H14 via H22. Finally, a very weak response also correlates H20a with H14, again through H22.

It should be noted at this point that several of the responses observed in the RCOSY spectrum shown in Fig. 4-8 correspond to four bond relays of the type discussed in a section regarding the RCOSY experiment above. In particular, correlations passing through the H22-H14 and H22-H20a/b long range couplings would be grouped in this category.

Overall, somewhat less relayed coherence transfer information is gleaned from the RCOSY spectrum shown in Fig. 4-7 than from the proton double quantum relay spectrum of strychnine (**4-4**) acquired using the same 4.5 Hz optimization (compare Fig. 4-8 with Fig. 4-10). The proton double quantum relay experiment (Fig. 4-9) is discussed below.

Finally, the 2RCOSY spectrum of strychnine (**4-4**) is shown in Fig. 4-8.  The experiment was performed using the pulse sequence shown in Fig. 4-5 and the phase cycling specified in Table 4-3.  The data were collected as 180 x 1K points with 64 acquisitions/block. Optimization of both relay steps in the experiment was set for 4.5 Hz

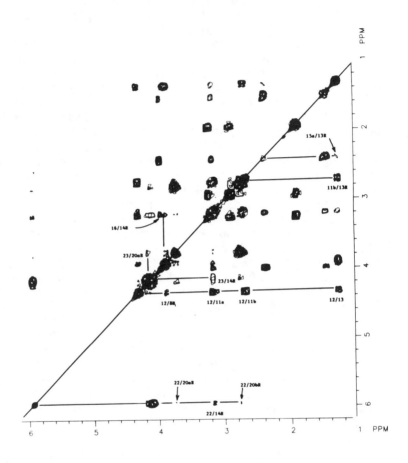

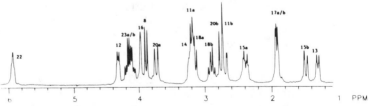

Fig. 4-7.        RCOSY spectrum of strychnine (4-4) recorded using the pulse
sequence shown in Fig. 4-2 and the phase cycling specified in Table 4-
1.    The data were collected as 256 x 1K points with 32
acquisitions/block.    The mixing period was optimized for a 4.5 Hz
coupling (55.5 msec).    The data were processed using sinusoidal
multiplication prior to the first transformation and sinusoidal
multiplication followed by zero filling prior to the second
transformation to afford a final data matrix consisting of 512 x 512
points, which was symmetrized[3] prior to plotting.

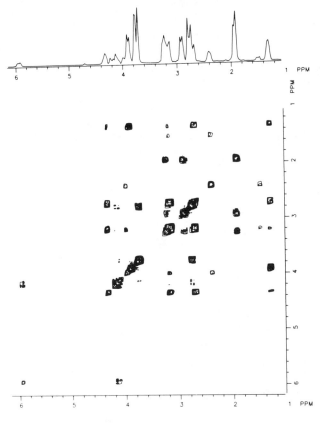

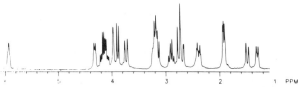

*Fig. 4-8*          2RCOSY spectrum of strychnine (**4-4**) recorded using the pulse sequence shown in Fig. 4-5 and the phase cycling specified in Table 4-3. The data was collected as 180 x 1K points with 64 acquisitions/block. The mixing periods were both optimized for a 4.5 Hz coupling (55.5 msec). The data were processed using sinusoidal multiplication prior to the first transformation and sinusoidal multiplication followed by zero filling prior to the second transformation to afford a final data matrix consisting of 512 x 512 points, which was symmetrized[3] prior to plotting.

(55.5 msec). Processing was identical to that used with the RCOSY experiment described above. Results obtained from this experiment were somewhat disappointing as no new information beyond that contained in the RCOSY spectrum was obtained. It is probable, however, that the lack of additional responses was a function of the choice for the second optimization of the second relay step in the experiment, which may have been unfortuitous.

## MULTIPLE QUANTUM RELAYED COHERENCE TRANSFER

Just as there are multiple quantum alternatives to the COSY experiment, so are there alternatives to the RCOSY experiment. Specifically, proton double and zero quantum relayed coherence transfer experiments have also been devised. We shall treat the proton double quantum relay experiment first, followed by the zero quantum relay experiment to maintain a treatment parallel to that contained in Chapter 2.

### Proton Double Quantum Relayed Coherence Transfer

Presently, two alternative proton double quantum relay experiments (DQCRCT) exist. Macura, Kumar and Brown[27] reported the

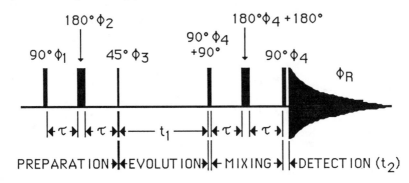

Fig. 4-9.     Proton double quantum homonuclear relayed coherence transfer pulse sequence devised by Müller and Pardi.[28] The fixed delays during the preparation and mixing periods ($\tau$) are normally optimized as a function of $1/4(J_{HH})$ and may be optimized independently of one another.

first such experiment and applied it to the cyclic peptide antibiotic polymyxin B. More recently, Müller and Pardi[28] reported an closely related pulse sequence, which has the advantage of being able to provide both proton double and zero quantum relay spectra if appropriate data storage steps are taken. The Müller-Pardi sequence is shown in Fig. 4-9 and its phase cycling is contained in Table 4-4 and discussed below.

The sequence proposed by Macura, Kumar and Brown[27] differs only in that the 45° pulse at the end of double quantum excitation in Fig. 4-9 is replaced by a 90° pulse. The 180° pulses contained in the sequence shown in Fig. 4-9 have also been replaced by composite 180° pulses in a study reported by Craig and Martin.[29]

## Phase Cycling

The phase cycling regimen for the proton double quantum homonuclear relay (DQCRCT) is given in Table 4-4. The minimally required phase cycle is contained in the first sixteen steps. Generally, however, better results are obtained when the entire sixty-four step cycle is used, time permitting.

It is important to note that the phase cycling specified in Table 4-4 does not provide quadrature detection in the $F_1$ frequency domain for the DQCRCT experiment. Thus, it is necessary to set the transmitter downfield of the lowest field resonance when the data is acquired. As a consequence of this type of data acquisition, the resultant spectrum will be symmetric about the $F_1 = 0$ Hz axis.

One further comment regarding the data storage for the phase cycling routine shown in Table 4-4 is warranted. Müller and Pardi[28], using the earlier work of Wokaun and Ernst,[30] have shown that both proton zero and double quantum homonuclear relay spectra may be obtained from the same data set. By storing even-and-odd numbered FIDs in different memory locations, it is possible, when the receiver phase cycling for zero quantum coherence is used, to obtain the zero quantum homonuclear relay spectrum from the data set by adding the two blocks of memory. Alternatively, the proton double quantum homonuclear relay spectrum may be obtained by subtraction of the two blocks of data in memory.

## Parameter Selection

**The $\tau$ delay used in exciting double quantum coherence**. The considerations in setting up the proton double quantum homonuclear relayed coherence transfer experiment differ somewhat from the RCOSY group of experiments described above. First, it is necessary to consider an average coupling constant for the excitation of double quantum coherence. Considerations in this aspect of setting up the experiment are considered in some detail in Chapter 2. For survey conditions, we have, however, found it useful to utilize an assumed average vicinal coupling constant of 7 Hz. Alternately, pseudo-uniform excitation of double quantum coherence can be achieved by using a 1.75 Hz optimization of the fixed delays during the excitation period.

Table 4-4.    Phase cycling for the proton double/zero quantum homonuclear relayed coherence transfer (DQCRCT) pulse sequence in Fig. 4-9.

| $\phi_1$ | 00000000 | 00000000 | 11111111 | 11111111 |
| | 22222222 | 22222222 | 33333333 | 33333333 |
| | | | | |
| $\phi_2$ | 00002222 | 00002222 | 11113333 | 11113333 |
| | 22220000 | 22220000 | 33331111 | 33331111 |
| | | | | |
| $\phi_3$ | 00000000 | 22222222 | 11111111 | 33333333 |
| | 22222222 | 00000000 | 33333333 | 11111111 |
| | | | | |
| $\phi_4$ | 01230123 | 01230123 | 12301230 | 12301230 |
| | 23012301 | 23012301 | 30123012 | 30123012 |

Receiver Phase Cycle

For Zero Quantum Coherence

| $\phi_R$ | 01230123 | 23012301 | 12301230 | 30123012 |
| | 23012301 | 01230123 | 30123012 | 12301230 |

For Double Quantum Coherence

| $\phi_R$ | 03210321 | 21032103 | 10321032 | 32103210 |
| | 21032103 | 03210321 | 32103210 | 10321032 |

**Incrementation of the evolution period.** Unlike RCOSY, the proton double quantum relayed coherence transfer experiments must be set up to provide a much wider frequency range in $F_1$. Indeed, typically the $F_1$ frequency range will be twice that encountered in $F_2$, necessitating that the evolution time be incremented by a factor of the dwell time x 0.5. If coupling of protons at the extremes of the $F_2$ frequency range is unlikely, slightly narrower $F_1$ frequency ranges may be utilized, eg, by incrementing as a function of dwell time x 0.75.

**Optimization of the mixing delay.** Cosiderations implicit in the optimization of the mixing period of the RCOSY experiment carry over to the DQCRCT experiment. Unfortunately, there has not been a study of relay response intensity for the double quantum homonuclear relay experiment comparable to the study reported by Bax and Drobny for the RCOSY experiment.[4]

## Data Processing and Presentation

Processing of the proton double quantum homonuclear relayed coherence transfer data (DQCRCT) makes the same demands as the proton double quantum experiment, which was discussed extensively in Chapter 2.

Presentation of the data in the DQCRCT experiment differs slightly from the other relay experiments that have been discussed thus far. Specifically, the data generated in this experiment is symmetrical about the $F_1 = 0$ Hz axis in the second dimension. An illustration of the symmetrical distribution of the data in $F_1$ is shown in Müller and Pardi's original work.[28] Because of symmetry, it is necessary to plot only half of the DQCRCT data matrix to obtain all of the information contained in the spectrum.

## Application of the Proton Double Quantum Relayed Coherence Transfer Experiment to Strychnine

To illustrate the distribution of data in the double quantum homonuclear relay experiment, we will once again consider the spectrum of strychnine (**4-4**). The conventional proton double quantum spectrum of strychnine was shown in Figs. 2-35 and 2-37 with the excitation delays optimized for 7 and 1.75 Hz, respectively. For comparison, the proton double quantum homonuclear relay spectrum of strychnine recorded with both the double quantum excitation and mixing delays optimized for 4.5 Hz is shown in Fig. 4-10. Responses in the spectrum shown in Fig. 4-10 are labeled according to the numbering scheme shown with the structure. It is useful to note that the 4.5 Hz optimized spectrum gave basically the same correlations as the 1.75 Hz experiment shown in Fig. 2-37. Importantly, the H22-H14 and H22-20a connectivities are observed. Some of the other correlations contained in the spectrum are also interesting and worthy of further comment.

We will concentrate only on the responses associated with the H22 vinyl proton. As in the normal proton double quantum spectrum optimized for 1.75 Hz, correlations were observed for the following pairs of connectivities: H22-H23a, H22-H23b, H22-H20a, H22-H14. Treating these connectivities individually, at the H22-H23a double quantum frequency, we observe relay responses to H20a/b and H14. At the H22-H23b double quantum frequency we observe relay responses only to H20a and H14. Most interesting is the H22-H14 connectivity, which exhibits relay responses to H23a/b, H20a/b and H15b. Hence in some cases it is possible to establish the relationship of protons up to five bonds removed from a given proton.

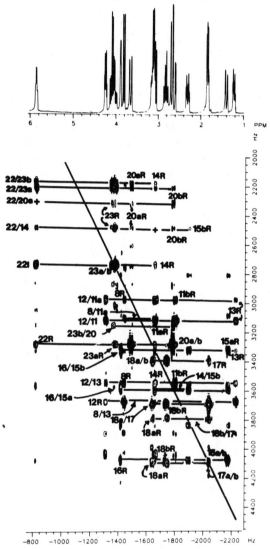

*Fig. 4-10.*     Proton double quantum homonuclear relay spectrum of strychnine (**4-4**) recorded using the pulse sequence shown in Fig. 4-9 and the phase cycling specified in Table 4-4. The double quantum creation and relay step in the experiment were both optimized for an assumed 4.5 Hz coupling. The data was collected as 512 x 2K points with 64 acquisitions/block.     The data were processed using sinusoidal multiplication prior to both Fourier transformationss and zero-filling prior to the second. Responses are labeled with the identities of the coupled protons using the format downfield/upfield. Relay responses are identified by the proton to which information is relayed followed by an upper case "R", eg, 14R.

## HOMONUCLEAR RELAYED COHERENCE: TRANSFER VIA SCALAR COUPLING PATHWAYS OR BY ISOTROPIC MIXING?

Relayed coherence transfer experiments have been widely utilized in the NMR investigation of peptides, oligosaccharides, nucleotides and similar compounds. The advent of experiments that employ isotropic mixing to effect coherence transfer initially, however, appeared likely to supplant the RCOSY type experiments. Recently, however, Drobny and coworkers[25] have pointed out that there are still good reasons for performing both relay and isotropic mixing experiments. For instance, several studies have shown that the relay experiment may be optimized for specific amino acids.[4,23,24] Conversely, the isotropic mixing experiments seem to be particularly well suited for identifying long amino acid side chain systems in proteins, eg, lysine, leucine, arginine, etc.

Overall, the field of two-dimensional homonuclear relay, whether by scalar coupling pathways or by isotropic mixing, is presently in a state of considerable flux. Unfortunately, as a result, some opinions in the following sections on experiments based on isotropic mixing and related techniques may become quickly dated as opinions shift. For this reason, the experiments below are presented largely without attempting to state which may or may not be best for a particular purpose.

## HOMONUCLEAR RELAYED COHERENCE TRANSFER VIA ISOTROPIC MIXING

The use of cross polarization (CP) was first recognized as a viable method for transferrence of coherence between coupled spins by Hartmann and Hahn in 1962.[31] In the intervening years, cross polarization has been extensively utilized in the area of solids NMR spectroscopy. Müller and Ernst[32] did, however, report a detailed analysis of the theory of coherence transfer in liquids as applied to heteronuclear cross-correlation spectroscopy. That work was followed in 1983 by the first example of the homonuclear application of isotropic mixing.[33] The prototypical experiment of Braunschweiler and Ernst has been given the acronym TOCSY corresponding to TOtal Correlation SpectroscopY and was the forerunner of what should become an increasingly important class of experiments.

### Total Correlation Spectroscopy: TOCSY

The TOCSY experiment is fundamentally quite simlar to the COSY experiment (see Fig. 2-1). An initial 90° pulse, which is phase cycled to suppress axial responses and quadrature phase images, is followed by an evolution time, $t_1$, incremented in the usual fashion to digitize the second frequency domain. The experiment differs in that the second 90° pulse of the COSY sequence is replaced by a mixing interval during which isotropic mixing and hence coherence transfer take place. The pulse sequence is shown schematically in Fig. 4-11. A

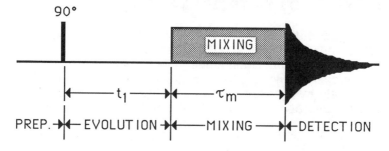

*Fig. 4-11.*    Pulse sequence for the TOCSY experiment of Braunschweiler and Ernst.[33] Possibilities presented for use during the mixing interval are shown in Fig. 4-12.

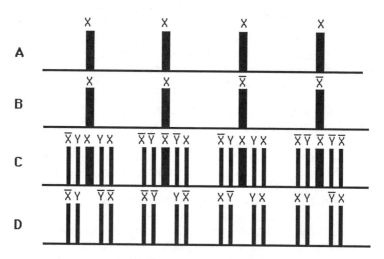

*Fig. 4-12.*    Isotropic mixing sequences proposed in the work of Braunschweiler and Ernst.[33] The sequences consist of (A) a repetitive sequences of 180° pulses with constant phase x; (B) a repetitive sequence of phase altered 180° pulses, which compensates for radiofrequency inhomogeneity; (C) cycle consisting of the minimum number of pulses (10) for isotropic mixing comprised of eight 90° pulses and two 180° pulses: the interval between successive 90° pulses is τ while that between the two identical sequences is 2τ; (D) sixteen step cycle consisting of 90° pulses distributed equally among the four possible phases.

variety of isotropic mixing schemes may be employed during the mixing interval, which is labeled generically in Fig. 4-11. Braunschweiler and

Ernst[33] suggested four possibilities.  More recently, Rance[34] has suggested a further variant that utilizes z-filters, which will be discussed further below.

The objective of the mixing period is to provide the efficient suppression of chemical shift terms with energy matching to allow spin exchange.  We may apply any of a variety of sequences of pulses that satisfy these requirements. The mixing period may, for example, consist of a single pulse, a series of pulses of constant phase to provide a spin lock or one of the sequences of phase shifted pulses such as MLEV-16, MLEV-17 or WALTZ-16.  In any case, as the duration of the mixing interval increases, we enhance the extent to which transfer between coupled spins can occur. When long mixing intervals are employed, the results are essentially analogous to multiple-step RELAY experiments (for example, see Fig. 4-9).  When mixing intervals reach a sufficient length, we will observe responses for all members of a spin system even without direct coupling.  This final case is the basis for the acronym given the experiment.

The TOCSY experiment of Braunschweiler and Ernst[33] was initially based on the utilization of one of the four isotropic mixing sequences shown in Fig. 4-12.  More recently, in a number of papers Bax and coworkers have described an experiment they have given the acronym HOHAHA, which corresponds to homonuclear Hartmann-Hahn spectroscopy.  The work by Bax and co-authors differs only in that it is based on more recent isotropic mixing sequences such as MLEV-16/17. However, since the terminology is now contained in the literature, we shall treat these experiments in a separate, appropriately headed section in an effort to preserve the sanity of the reader in dealing with acronym alphabet soup.

## Homonuclear Hartmann-Hahn Spectroscopy - HOHAHA

In work utilizing spin-locking fields for homonuclear nOe experiments, Bax and Davis[35] first discussed the problems associated with Hartmann-Hahn transfer between scalar coupled protons.  As a result of further studies of the undesired artifacts in the nOe spectra, Bax and Davis[36,37] later reported the coherence transfer via homonuclear Hartmann-Hahn effects where the isotropic mixing was obtained by using an MLEV-16 sequence,[38] which was initially devised for more efficient heteronuclear decoupling applications.  As with other experiments based on isotropic mixing, the essential feature is the removal of Zeeman contributions[39] during the evolution of the coherence.  One advantage of pulse sequences that rely on cross-polarization irradiation schemes during the mixing time is that the magnetization decay constant can be prolonged up to a factor of 2 as compared to $T_{1\rho}$.[36]  As described by Bax and Davis[35,36] once the Zeeman contribution is removed, the spin system will evolve under the influence of scalar coupling.[40] Exactly as in the TOCSY experiment described above, the duration of the mixing period controls the extent of propagation.

The pulse sequence schematic for HOHAHA is that shown in Fig. 4-13. Early work was done using an MLEV16 sequence, while more recent work by Bax and Davis[36] has been interpreted to suggest that an MLEV17 sequence, which incorporates trim pulses, may be preferable. Quite simply, the MLEV16 sequence consists of sixteen composite 180° pulses. Two types of composite 180° rotations are employed:

$$A = \quad 90°_{-y} - 180°_x - 90°_{-y} \qquad\qquad [4\text{-}5]$$

and

$$B = \quad 90°_y - 180°_{-x} - 90°_y \qquad\qquad [4\text{-}6]$$

where it will be quickly noted that B represents the phase inverted form of A. The original MLEV16 sequence consists of an integer number of these composite 180° pulses as shown in Eqn. [4-7].

$$\text{ABBA} \qquad \text{BBAA} \qquad \text{BAAB} \qquad \text{AABB} \qquad [4\text{-}7]$$

The modification to yield the MLEV17 sequence quite simply adds an uncompensated $180°_x$ pulse after the last composite 180° pulse in Eqn. [4-7]. Bax and Davis[36] argue that this final, uncompensated 180° pulse is quite efficient at removing pulse imperfections during the preceding MLEV16 cycle.

Ideally, the MLEV16 cycle will return the "tortured" spin components to the x-axis at its completion. The rationale presented by Bax and Davis[36] for resorting to the MLEV17 sequence was that imperfect phase shifts or imbalance in the x, y, -x and -y channels will result in a phase error between the vector and the x-axis at the completion of the MLEV16 cycle. Since the MLEV16 cycle will typically be repeated some integer number of times during the mixing period to afford the degree of propagation desired, the net phase error will be cumulative. The purpose of the uncompensated 180° pulse is to invert the phase error on successive passes through the MLEV16 cycle. Hence, if an even number of cycles through the MLEV16 sequence, which is recommended, are employed, the magnetization should ideally reside on the x-axis, as desired, at the termination of the mixing period. Drobny and colleagues[41] argue, in contrast, errors of the type that Bax and Davis have addressed do not dominate isotropic mixing spectra when the applied radiofrequency field is sufficiently strong. Hence, they conclude that the purge pulses are not required for obtaining good quality spectra. The final restitution of this argument remains to be determined.

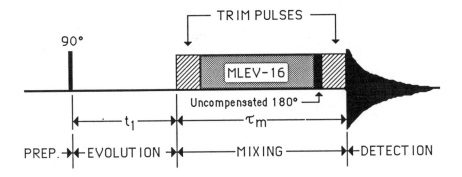

*Fig. 4-13.*    HOHAHA pulse sequence employing an MLEV-17 sequence flanked by trim pulses. The uncompensated 180° pulse which follows the normal MLEV16 portion of the mixing period refocuses errors due to phase shifter inaccuracies after an even number of cycles. The trim pulses are normally set to several msec and serve to defocus magnetization components not coaligned with the ±x axis.

## Phase Cycling

The phase cycling scheme utilized for the HOHAHA pulse sequence shown in Fig.4-13 requires only a sixteen step cycle when the data is to be processed using an absolute value computation following the second Fourier transformation. The first 90° pulse is cycled in a four step cycle ($\phi_1$) synchronously with the receiver ($\phi_R$) while the MLEV17 sequence iscycled in a sixteen step cyclops cycle ($\phi_2$) superimposed over the four step phase cycle for the receiver and the first 90° pulse to afford quadrature in the second frequency domain. The trim pulses are cycled in a sixteen step cycle phase shifted relative to the MLEV sequence by 90° ($\phi_3$). A z-filtered HOHAHA or TOCSY experiment, as the reader may prefer (see Fig. 4-20), can be accomodated using the phase cycling presented in Table 4-5. The 90° pulses used in the z-filter are cycled synchronously with the phase of the MLEV sequence.

Table 4-5.    Phase cycling scheme for TOCSY or HOHAHA experiments (see Figs. 4-11 and 4-13, respectively).

| $\phi_1$ | 0123 | 0123 | 0123 | 0123 |
|---|---|---|---|---|
| $\phi_2$ | 0000 | 1111 | 2222 | 3333 |
| $\phi_3$ | 1111 | 2222 | 3333 | 0000 |
| $\phi_R$ | 0123 | 0123 | 0123 | 0123 |

## Parameter Selection

Parameter selection in the HOHAHA experiment is similar to the now familiar COSY experiment. Thus, the proton dwell time should be utilized to increment the duration of the evolution period to provide a final data matrix that will be square.

Digitization in the second frequency domain may differ somewhat from the COSY experiment. Generally in the latter it is desirable to digitize to afford a matrix with equal digitization in both frequency domains. Indeed, this is a necessity of symmetrization is to be employed. In contrast, with the HOHAHA experiment collected in the phase sensitive mode, it may not be advantageous to symmetrize the spectrum. Indeed, it may even be disadvantageous where very weak responses are involved. Thus, levels of digitization in the second or $t_1$ frequency domain of the HOHAHA experiment may be selected with a greater degree of freedom than in the case of the COSY experiment when symmetrization is to be employed.

A further departure from the COSY experiment is found in the establishment of the duration of the isotropic mixing interval. In large part, the choice of mixing times becomes experiential for the class of molecule being examined. For small molecules, we have generally found it advantageous to utilize mixing times in the range of 20-50 msec to establish connectivities between a proton, its vicinal neighbors and the protons one step further removed. Longer mixing intervals in the range of 50-90 msec will extend the correlations further by one or two additional protons. Mixing periods ranging from 100 to 150 msec have generally been sufficient to provide all of the information we could easily utilize. Indeed, with longer intervals, it is generally necessary to have at one's disposal at least a COSY spectrum and occasionally a HOHAHA spectrum recorded with a shorter mixing period to establish the identities of shorter range connectivities.

## Data Processing

Processing HOHAHA data, whether it is acquired conventionally or in a phase sensitive fashion, is identical to the processing utilized with the corresponding COSY data set.

## Applications of HOHAHA

The spectra that we will present in the following sections were recorded using the MLEV17 sequence as suggested by Bax and Davis.[36] In particular, we shall illustrate the application of the HOHAHA experiment to strychnine (**4-4**), once again to allow direct comparison of the results obtained using the HOHAHA experiment with other experiments described here and in Chapter 2. Finally, we will also discuss HOHAHA spectra of the cyclic peptide didemnim-B (**4-5**).

## Identification of large connectivity networks in the proton NMR spectrum of strychnine using HOHAHA.

We have employed strychnine (**4-4**) extensively as a model compound because the complex spin-coupling network found in the aliphatic portion of its proton NMR spectrum. The aliphatic region of the HOHAHA spectrum of strychnine recorded at 500 MHz using an MLEV17 sequence for isotropic mixing flanked by 2.5 msec spin locking intervals is shown in Fig. 4-14. The data was collected in the phase sensitive mode and the duration of the mixing interval, corresponding to the sum of all of the applied pulses, was 19 msec in this first case.

Coherence is established between H22 and its vicinal neighbor protons H23a/b. It is more interesting to note that coherence is also established between H22 and its long range coupled neighbors H14 and H20a/b, which is comparable to the results obtained in the 1.75 Hz optimized proton double quantum experiment (see Fig. 2-37) and the 4.5 Hz optimized RCOSY experiment (see Fig. 4-7).

Slices taken through H22 and several other resonances shown in the contour plot in Fig. 4-14 are presented in Fig. 4-15. In addition to the responses already discussed, it will be noted that even with a relatively short 19 msec mixing time, coherence has been transferred as far as the H15a resonance, although this response is a barely perceptible distortion in the baseline denoted by the arrow. Undoubtedly, this response would be missed in lower signal-to-noise ratio spectra. In the experiments we have considered thus far, only the proton double quantum relayed coherence transfer experiment (see Fig. 4-10) has been able to establish this connectivity.

Doubling the duration of the isotropic mixing period to 38 msec does not add substantially to the information that we already have. Slices taken from the spectrum recorded using a 38 msec mixing time are presented in Fig. 4-16. Substantially more information is obtained, however, by increasing the duration of the mixing time to 116 msec. As wil be noted from the traces shown in Fig. 4-17, H22 shows now easily discernible responses for transfer to H15a/b and weak responses for transfer to protons further removed which include H13 and even H8 and H12! Regrettably, no connectivities are noted between any of the protons and the ethylene bridge protons H17a/b and H18a/b

## Application of the HOHAHA experiment to the cyclic depsipeptide didemnim-B.

The complex proton NMR spectrum of the small, cyclic depsipeptide didemnim-B (**4-5**) provides the grounds for a more powerful demonstration of the capabilities of the HOHAHA experiment.

Peptides, if interresidue couplings are ignored, can be considered as nothing more than an assemblage of potentially overlapped spin systems. Herein lies one of the difficulties in assigning their spectra: the overlap of noncoupled resonances can be troublesome, particularly with COSY spectra. To an extent, these problems are ameliorated with proton zero and double quantum coherence techniques. A further useful alternative is provided by the

HOHAHA experiment, which, in principle, can be utilized to generate subspectra of individual amino acid residues.

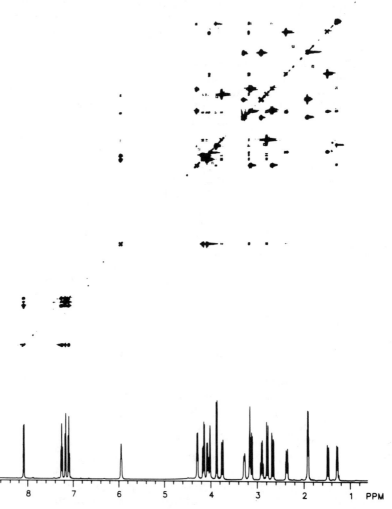

*Fig. 4-14.*    Phase sensitive HOHAHA spectrum of strychnine in deuterochloroform recorded at 500 MHz. The data was acquired using the pulse sequence shown in Fig. 4-13 in which an MLEV17 sequence was used for isotropic mixing flanked by 2.5 msec trim pulses or spin-locking intervals.    The data was taken using 8 acquisitions/block with a mixing time of 19 msec requiring six passes through the MLEV17 cycle.    A total of 256 x 2K complex points was taken, giving an acquisition time of 1.5 hours.

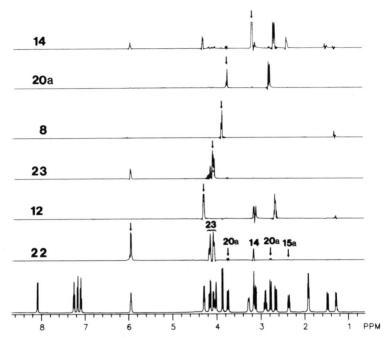

*Fig. 4-15.*     Slices taken from the 19 msec HOHAHA spectrum shown as a contour plot in Fig. 4-14. Identities of the proton at whose chemical shift the slice was taken are labeled on each trace.

The HOHAHA spectrum of didemnim-B (**4-5**) recorded in deuterochloroform is shown in Fig. 4-18. The three NH protons that appear as doublets downfield are attributable to leucine, threonine and isostatine, respectively. We shall consider the identification of the proton resonances of each in turn.

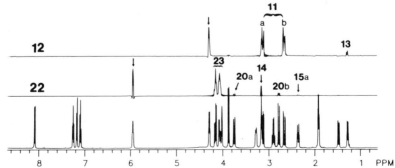

*Fig. 4-16.*     Slices taken from the phase sensitive HOHAHA spectrum of strychnine recorded using a 38 msec mixing interval. Proton identities are labeled on each slice and are identical to those shown in Fig. 4-15 to facilitate comparison.

**4-5**

Beginning from the NH resonating furthest downfield at 7.81 ppm, the $\alpha$ proton of leucine is easily identified as the proton resonating at 4.78 ppm. On the basis of response intensity, the $\beta_1$ and $\beta_2$ protons may be ascribed to the signals observed at 1.59 and 1.20 ppm, the $\gamma$ proton at 1.50 and finally, the $\delta$ methyl protons furthest upfield at approximately 0.9 ppm. Similar connectivities are observed when the corresponding slice is taken at the $\alpha$ proton chemical shift. This slice is shown in Fig. 4-19.

Proceeding upfield, the threonine NH proton is encountered next, resonating at 7.65 ppm. A slice taken at the chemical shfit of the NH proton reveals the identity of its associated $\alpha$ proton, which resonates at 4.52 ppm and gives a weak response for the hydroxyl bearing $\beta$ proton resonating at 5.41 ppm. There is, however, no trace of a connectivity to the threonine methyl observed in the subspectrum generated from the NH proton. In contrast, slices taken at the chemical shift of both the $\alpha$ and $\beta$ protons give subspectra that identify all components of the threonine spin system.

The third NH bearing amino acid, isostatine (**4-6**) is one of the so-called "rare" amino acids. Beginning from the NH resonating at 7.20 ppm, we observe responses in the slice at the chemical shfit of the $\gamma$ proton, which resonates at 4.12 ppm, and other protons resonating at 4.08 (shoulder of the 4.12 ppm resonance),3.27, 2.64, 1.83 and 0.95 ppm. Beyond the $\gamma$ proton, none of the others may be readily assigned on the basis of response intensity. A slice taken at 4.08 ppm, the chemical shift of the hydroxyl bearing $\beta$ proton, readily identifies the vicinal $\alpha_1$ and $\alpha_2$ protons and gives slightly better response intensity at

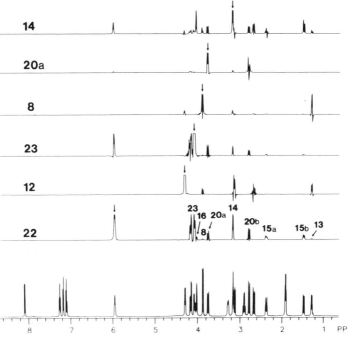

Fig. 4-17.    Slices taken from the phase sensitive HOHAHA spectrum of strychnine
recorded using a 116 msec mixing interval.   Proton identities are
labeled on each slice and are identical to those shown in Fig. 4-15 to
facilitate comparison.

the responses noted above as well as further responses at 1.43, 1.40
and 1.23 ppm, the first and last very weak.   Although they have not been
unequivocally assigned, all of the constituent resonances of the
isostatine spin system are identified at this point.   Tentatively, we may
attribute the δ proton to the resonance noted at 1.83 ppm.   It is likely that
one of the ε methylene protons may be assigned to the resonance at
1.43 and the other at 1.23 ppm.   Finally, only the methyl protons remain
to be assigned.   Since the resonance at 0.95 ppm appears in the slice
taken at the chemical shift of the NH proton, it is quite probable that this
resonance may be assigned as the ε methyl group leaving the ζ methyl
assigned as the doublet resonating at 1.40 ppm.

Other amino acid residues in the proton NMR spectrum of
didemnim-B (4-5) may be assigned in a manner similar to that just
described.   In a particularly "long" amino acid, it may be necessary to
use more than one point of entry to identify all of its constituent spins as
in the case of isostatine (4-6) just described.   In any case, the HOHAHA
experiment has obvious utility in disentangling the complex proton NMR
spectra of molecules such as peptides and small proteins.

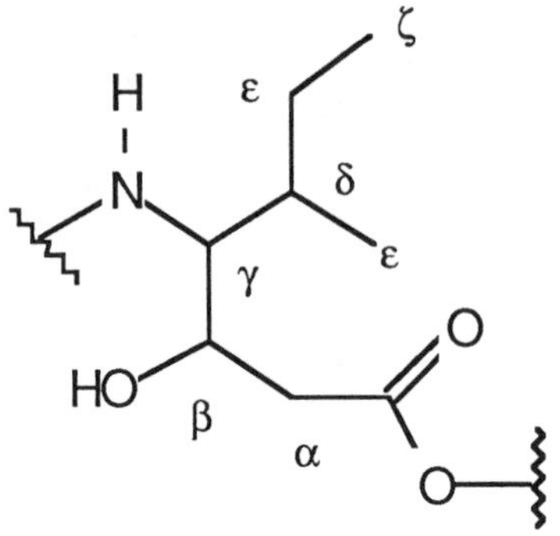

**4-6**

## A Brief Survey of Applications
## of the HOHAHA Experiment

Although the HOHAHA (TOCSY) experiment is relatively new, there are already a substantial number of applications contained in the literature. In addition to the application to strychnine (**4-4**) described above, Edwards and Bax[42] have reported the use of HOHAHA in the determination of the alkaloid gephrytoxin. Other natural products applications include a study of vitamin B-12[43] and the elucidation of the structure of the macrocyclic antibiotic desertomycin.[44] Small peptide applications include angiotensin II,[36] the cyclic peptide toxin cyanoviridin RR[45] and didemnim-B (**4-5**) presented above. Small proteins are also a fertile area for the application of the HOHAHA experiment. Applications that have appeared thus far include human interleukin-1β,[46] human growth hormone releasing factor[47] and aponeocarzinostatin.[41] The entire sugar spin systems of nucleic acids will undoubtedly also prove to be an area in which the HOHAHA experiment finds useful application.[48]

### Isotropic Mixing Using
### Other Sequences

The heteronuclear composite pulse decoupling sequence MLEV16 and the derivative sequence MLEV17 are by no means the only sequences that can be used effectively for isotropic mixing. Several others proposed in the early work of Braunschweiler and Ernst[33] are shown above in Fig. 4-12. Another is the WALTZ-16 sequence of Freeman and coworkers,[49] which was compared in one study to MLEV16 by Waugh.[50] More recently, Drobny[41] and coworkers have made a similar comparison, concluding that MLEV16 gave better results. Expanding on the theory of phase alternated broadband decoupling presented by Waugh,[51,52] Feng[53] has proposed a number of

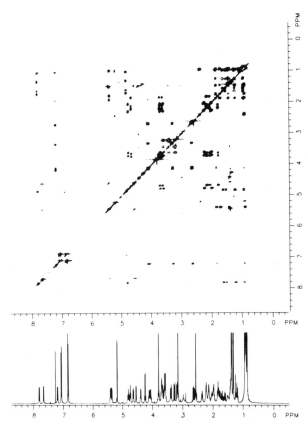

*Fig. 4-18.*   Phase sensitive HOHAHA spectrum of the cyclic depsipeptide didemnim-B **(4-5)** recorded in deuterochloroform at 500 MHz.

additional phase alternated sequences, some of which might presumably find application as isotropic mixing sequences.

### TOCSY With z-Filters

One of the problems associated with the acquisition of phase sensitive TOCSY consists of phase anomolies which can significantly degrade spectral quality. Recently, Rance[34] proposed a technique for the elimination of these problems based on the utilization of z-filtration.[54,55] Rance's modified pulse sequence is shown in Fig. 4-20. The mixing interval in the z-filtered TOCSY experiment may utilize MLEV-17 or WALTZ-16. The former was generally noted to be superior, the latter yielding unacceptably phase twisted line shapes.

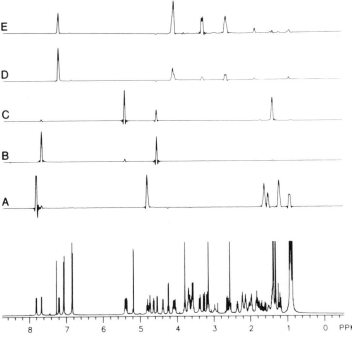

Fig. 4-19.    Slices taken through the HOHAHA spectrum of didemnim-B (**4-5**) shown in Fig. 4-17. (A) Slice taken at 7.81 ppm corresponding to the chemical shift of the leucine NH proton. (B and C) Slices taken at the chemical shift of the threonine NH and β protons resonating at 7.65 and 5.41 ppm, respectively. (D and E) Slices taken at the chemical shift of the isostatine (**4-6**) NH and β protons resonating at 7.20 and 4.08 ppm, respectively.

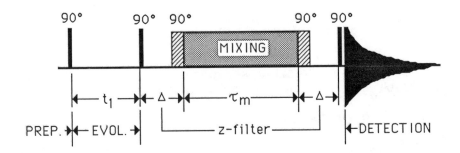

Fig. 4-20.    z-Filtered TOCSY pulse sequence proposed by Rance.[34] The z-filter delay, Δ, may be set to the minimal necessary for phase shifting or may be lengthened as desired. The 90° pulses flanking the mixing interval may be deleted for some purposes. The phase of these pulses will be dependent upon the type of experiment being performed and the mixing sequence being utilized. One possible set of phase cycles is given in Table 4-5. The interested readers are also referred to the literature for additional details of phase cycling possibilities.

## HETERONUCLEAR RELAYED COHERENCE TRANSFER

Given the ability to establish heteronuclear chemical shift correlations (Chapter 3), it is frequently useful to have information available about protons that are vicinally coupled to a proton attached to a given carbon atom. Experiments of this type are categorized as heteronuclear relayed coherence transfer experiments. These experiments were introduced in 1982 in a pair of papers by Bolton.[56,57] A modification was reported shortly thereafter by Bax,[58] as were the first applications.[59,60] Since these initial communications, there have been a reasonable number of applications of the experiment. There are, however, some shortcomings, which it is appropriate to discuss while also considering the operation of the experiment.

Since the development of the heteronuclear relayed coherence transfer experiment, a wide variety of other variants have also been developed. For example, Bolton[61] has reported a proton double quantum heteronuclear relay experiment. More recently, Turner[62] has reported the zero quantum equivalent of the Bolton double quantum experiment. Sequences that provide for proton detection[63-65] as well as coherence transfer via spin-locking sequences have also been described.[66] Finally, low pass J filters have been introduced to eliminate direct responses in the heteronuclear relay spectra.[67]

## Conventional Heteronuclear Relayed Coherence Transfer

A nice treatment of the workings of the heteronuclear relayed coherence transfer experiment is found in the early work of Bax.[58] The pulse sequence for the heteronuclear relayed coherence transfer experiment (HC-RELAY) is found in Fig. 4-21, and it is useful to consider the function of the various pulses contained in the sequence.

Quite obviously, the first 90° proton pulse rotates magnetization into the transverse plane, at which point precession occurs during the evolution period. The 180° $^{13}$C pulse midway through the evolution period serves to eliminate the effects of heteronuclear scalar coupling at the end of the evolution period just before the second 90° pulse. In this sense, the 180° $^{13}$C pulse serves the same function as in the conventional heteronuclear chemical shift correlation experiment (see also Fig. 3-10). The second 90° pulse, in Bax's[58] description of the experiment, may be considered as a cascade of two semiselective pulses,[68] one applied to nucleus A, the other to nucleus M. Here perhaps some additional explication is warranted. Consider first the effects of a hypothetical, selective pulse applied to A. This pulse will change the longitudinal magnetization of the A transitions and, as a consequence, the magnetization of the M transitions as well. The hypothetic 90° M pulse that follows will rotate the longitudinal M magnetization, which originates from spin A, to the y-axis. It should further be noted that immediately following transfer, the magnetization

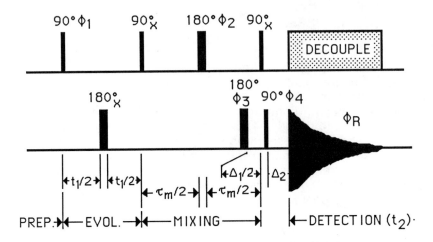

Fig. 4-21.    Heteronuclear relayed coherence transfer (HC-RELAY) pulse sequence devised by Bax.[58] This experiment differs from its predecessor reported by Bolton[56,57] by the inclusion of the 180° carbon pulse (compensated refocusing) midway through the $\Delta_1$ interval.

components will be antiphase and colinear with the ±y-axis. This series of operations has carried us through the evolution period of the experiment.

Next, following the first half of the mixing period, $\tau_m/2$, which corresponds to $1/2J_{AM}$, the transferred M magnetization vector components will be parallel once again. There will also be a net M magnetization which originates from A in the transverse plane. The 180° proton pulse inserted at the midpoint of the mixing period serves to suppress average precessions due to proton chemical shift terms. It is also important to note that precession due to scalar (J) coupling is unaffected. Hence, proton M magnetization vectors will be colinear with the ±x axis at the end of the mixing period, $\tau$, as in the experiments described by Bolton.[56,57]

The sequence described by Bax[58] (Fig. 4-21) applies a 180° pulse midway through the $\Delta_1$ interval ($1/4(J_{MX})$ before the end of the mixing period). By flipping the spin state of the $^{13}C$ nucleus, heteronuclear couplings are collapsed at the end of the mixing period, providing a gain in sensitivity relative to earlier versions of the experiment.[56,57]

Finally, as in the conventional heteronuclear chemical shift correlation experiment (Fig. 3-10) described in Chapter 3, the pair of 90° pulses applied to proton and carbon effectively transfer magnetization from the M nucleus to X ($^{13}C$).

The net result of the manipulations just described is to produce a heteronuclear correlation spectrum in which responses appear at the frequency of the directly attached proton (H_d) and at the chemical shift of those protons vicinally coupled ($H_v$) to $H_d$. Schematically, these results are shown in Fig. 4-22, below, where they are discussed in conjunction with the conventional heteronuclear chemical shift correlation spectrum, which must be obtained prior to the interpretation of the heteronuclear relayed cohrence transfer data.

## Phase Cycling

The phase cycling for the heteronuclear relayed coherence transfer experiment proposed by Bax[58] and shown in Fig. 4-21 was originally described as a four step phase cycle. Somewhat better results may be obtained, however, using the slightly more complex eight step phase cycle shown in Table 4-6.

## Parameter Selection

The selection of parameters for the HC-RELAY experiment is somewhat more complex than for the conventional heteronuclear chemical shift correlation experiment (for comparison, see discussion pertaining to Fig. 3-10). There are, however some similarities so we shall consider the points that are similar first.

Table 4-6.    Phase cycling for the heteronuclear relayed coherence transfer experiment pulse sequence (HC-RELAY) shown in Fig. 4-21.

| Acquisition | $\phi_1$ | $\phi_2$ | $\phi_3$ | $\phi_4$ | $\phi_R$ |
|---|---|---|---|---|---|
| 1 | 0 | 0 | 0 | 0 | 0 |
| 2 | 2 | 0 | 0 | 0 | 3 |
| 3 | 1 | 0 | 2 | 1 | 2 |
| 4 | 3 | 0 | 2 | 1 | 1 |
| 5 | 0 | 1 | 0 | 0 | 2 |
| 6 | 2 | 1 | 0 | 0 | 1 |
| 7 | 1 | 1 | 2 | 1 | 0 |
| 8 | 3 | 1 | 2 | 1 | 3 |

Evolution, $t_1$, for the proton frequency domain ($F_1$) in the HC-RELAY experiment is set in a fashion identical to the normal chemical shift correlation experiment. Specifically, the incremented delay for the evolution time is set as a function of half the proton dwell time. Likewise, the refocusing delay, $\Delta_2$, is also set in a manner identical to the chemical shift correlation experiment. For compounds with varied carbon multiplicity, a compromise setting for this delay corresponding to $1/3(^1J_{CH})$ must be employed. When the sample contains only methine carbons as in an aromatic system, the $\Delta_2$ delay may be optimized as a function of $1/2(^1J_{CH})$.

Delays for the mixing period are new relative to the conventional heteronuclear chemical shift correlation experiment and warrant further discussion. Implementation of this experiment on the authors Nicolet

NT-300 spectrometer utilizes two delay intervals to establish the duration of the various periods in the mixing interval. The mixing portion of the experiment is defined by the delays shown in Eqn. [4-8].

$$- 90°(^1H) - D5 - D4 - 180°(^1H) - D4 - 180°(^{13}C) - D5 - \qquad [4-8]$$

Typical durations of these intervals are set as follows. The period D5 corresponds to one half of the normal $\Delta_1$ interval. Hence, D5 should be set as a function of $1/4(^1J_{CH})$. The interval D4 is optimized as a function of $J_{AM}$ referring to the discussion above, which we normally have found it useful to optimize at $1/10(^3J_{HH})$, which is consistent with the suggestion of Kessler and coworkers.[59]

## Efficiency of the Relay Step

One factor that must be considered is the transfer efficiency of the relay step in the HC-RELAY experiment. The efficiency of the transfer process is discussed in some detail by Kessler and coworkers.[59] If we consider the isolated structural fragment $C_AH_A$-$C_BH_B$ then we must examine the transfer pathway $H_A \rightarrow H_B \rightarrow C_B$. As discussed above, $H_B$ magnetization is initially antiphase with respect to $J_{AB}$ and comes into phase at the end of the mixing period as a function of:

$$\sin(\pi J_{AB}\tau_m) \qquad [4-9]$$

The transfer process is most efficient when the duration of the mixing interval is chosen to optimize the transfer function shown in Eqn. [4-10]:

$$I^{Relay} = \sin(2\pi J_{AB}\tau_m)\cos^{m-1}(2\pi J_{AB}\tau_m)\cos^k x$$

$$(2\pi J_{BC}\tau_m)\exp(-2\tau_m/T_2) \qquad [4-10]$$

Where $m = k = 2$, $J_{AB} = 7$ Hz and $T_2 = 0.16$ sec, $I^{Relay} = 0.26$, which implies that the HC-RELAY transfer efficiency is essentially one quarter that of the conventional heteronuclear chemical shift correlation experiment. It is also important to note that in extended spin networks it may be necessary to factor in additional cosine terms for other non-participating protons

Practically, sensitivity reduces to a function of $\sin(\pi J_{AM}\tau_m)$ as noted by Kessler et al.[59] This is to say that relay response intensity will be best when the transfer function is properly optimized and correspondingly worse as the disparity between the ideal and the value optimized for increases. Put simply, where there is a range of couplings present in a molecule, such as would typically be encountered with a natural product, it may be difficult to select a single value on which to optimize the HC-RELAY experiment.

## Data Acquisition, Processing
## and Presentation

Acquistion of HC-RELAY data, as implied above, requires, if feasible, larger samples and/or longer acquisition times than the conventional heteronuclear relayed coherence transfer experiment. In any event, once the data is acquired it is necessary to consider processing. Since relay response intensity will generally be relatively low (see examples below), it is desirable to process the data using generous exponential multiplication in most cases to maximize signal-to-noise ratios in the spectrum. Data processing routines that tend to enhance resolution at the expense of signal-to-noise ratio should be avoided.

Presentation of the HC-RELAY data is comparable to the conventional heteronuclear chemical shift correlation spectrum. Contour plots provide the most convenient means of displaying data if sufficiently high signal-to-noise ratios are available. Ideally, especially when complex molecules are being studied, it may be helpful to have identically prepared contour plots of both the direct and relayed correlation experiments. Where less than ideal signal-to-noise ratios are obtained in the experiment, it may be preferable to prepare a stack plot of slices from the data matrix taken at the individual carbon chemical shifts so that there is no chance of missing a low intensity relay response.

## Interpretation of Heteronuclear
## Relayed Coherence Transfer Spectra

Heteronuclear relayed coherence transfer spectra are perhaps best described as "second generation" two-dimensional NMR experiments. The term "second generation" in this case refers to the notion that these experiments require a prior two-dimensional NMR experiment for their interpretation. In the case of HC-RELAY, a proton-carbon heteronuclear correlation spectrum is necessary for the former to be interpreted.

As an example, let us consider an eight spin heteronuclear system such as would be typically encountered in a 1,2-disubstituted aromatic ring. A typical structural fragment of this type is shown by **4-7**. The location of direct responses in the HC-RELAY spectrum of **4-7** are first identified from the conventional heteronuclear correlation spectrum. Once the directly attached proton resonances have been identified, other proton resonances at a given carbon chemical shift in the data matrix will arise from protons vicinally coupled to the directly attached proton. A schematic presentation of the HC-RELAY spectrum is shown in Fig. 4-22.

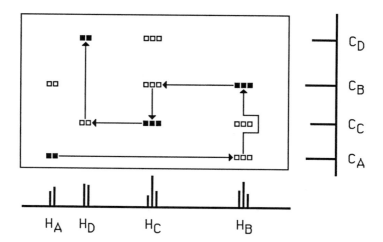

**4 - 7**

*Fig. 4-22.*     Schematic representation of the HC-RELAY spectrum of **4-7**. Direct responses are shown as solid squares, while relay responses are shown as open squares.

Beginning from the $H_A$-$C_A$ pairing, we note that at the chemical shift of $C_A$ in the matrix there is a response at $H_B$-$C_A$, which is as expected since $H_B$ is the vicinal neighbor of $H_A$. Following the arrow vertically, we next locate the direct response for the $H_B$-$C_B$ pair. Next, at the chemical shift of $C_B$ we note two vicinal responses, one corresponding to $H_A$ where we began, the other corresponding to the other vicinal neighbor, $H_C$. Once again we may locate the $H_C$-$C_C$ direct response. At the chemical shift of $C_C$ we observe responses for the two vicinal neighbors, $H_B$ and $H_D$. Finally, following the arrow vertically, we locate the $H_D$-$C_D$ direct response. The last point to be noted is the redundant vicinal response for $H_C$ at the chemical shift of $C_D$.

Using the relatively facile set of operations just described, we have mapped out an eight spin heteronuclear system. If we have available one assignment, eg, $H_D$, we may completely assign the balance of the aromatic ring to which it is attached given only the direct proton-carbon chemical shift correlation spectrum (as would be represented in Fig. 4-22 if only the responses shown as solid squares were present) to aid in the interpretation of the HC-RELAY spectrum shown.

## Applications of HC-RELAY

To illustrate the utility of the HC-RELAY experiment, let us consider two examples. We shall begin by considering our now familiar model alkaloid, norharmane (**4-8**). Next we will examine a complex polynuclear heteroaromatic compound where the vicinal couplings are quite uniform in character, allowing the HC-RELAY experiment to perform quite well. Finally, we will consider an alkaloid that has a range of vicinal couplings, which leads to breaks in the vicinal connectivity network.

### Application of HC-RELAY to a model alkaloid, norharmane. We have been using the simple alkaloid norharmane (**4-8**) to illustrate a variety of heteronuclear two-dimensional NMR experiments (see Chapter 3). The direct heteronuclear chemical shift correlation spectrum of norharmane is shown in Fig. 3-18 and must be obtained to establish the identity of the directly attached protons. The heteronuclear relay spectrum recorded using the pulse sequence shown in Fig. 4-21 is presented in Fig. 4-23. The data were acquired using a sample containing 200 mg of norharmane dissolved in 0.4 ml of $d_6$-DMSO in approximately 2 hours, the high concentration making this an ideal sample for first setting up the HC-RELAY experiment.

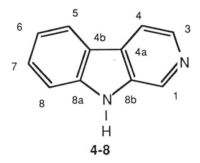

**4-8**

In considering the structure of norharmane, we would expect several features to appear in the HC-RELAY spectrum. First, the H1 resonance, since it does not have a vicinal coupling partner, will not have a relay response. The H3-H4 AX spin system will give relay responses at their respective carbon chemical shifts. Finally, the four spin system, which fundamentally resembles our hypothetical spin system shown schematically in Fig. 4-22, should be easily tracked and assigned with the HC-RELAY spectrum.

From even the most cursory examination of Fig. 4-23, we first note that the isolated H1 proton gives a response only at the chemical shift of the carbon to which it is directly attached. Next, H3 resonances gives a vicinal relay response to the H4 proton, which would identify this visible neighbor if its identity were not already known. Beginning with H5, which resonates furthest downfield in the four spin system, we may begin to map out this more complex spin network. There is a vicinal relay to the "triplet" furthest upfield, which corresponds to H6. Following the connectivity arrow to the direct response for H6, we next note that there are relay responses at the chemical shift of C6 that correspond to the H5 proton from which we began the assignment and to H7 the other vicinal neighbor of H6. Continuing in the connectivity network, the H7 proton gives vicinal neighbor relay responses for H6 and finally H8, which completes the mapping of the four proton spins while consequently assigning their respective carbon resonances. You will note by comparing the operations that we have just gone through that we have, save for differences in chemical shifts, traced a spin system distributed exactly the same as that shown schematically in Fig. 4-22.

Overall, when homonuclear vicinal couplings are quite regular, as in polynuclear aromatics, the HC-RELAY experiment provides a good means of tracing proton-proton connectivities. In particular, as we shall see from the following example, spin networks may be successfully traced despite very substantial overlap in the proton NMR spectrum.

**Phenanthro[4,3-_a_]dibenzothiophene**. Referring back to the COSY spectrum of phenanthro[4,3-_a_]dibenzothiophene (**4-9**), which was shown in Fig. 2-13, we note that there is a very high degree of congestion engendered by nine of the fourteen protons resonating within a region only 0.25 ppm wide. Hence the COSY spectrum is nearly useless, making it difficult to assign the spectra. Using the HC-

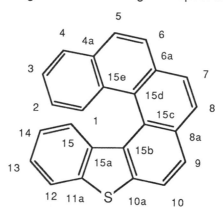

**4-9**

RELAY spectrum shown in Fig. 4-24, Martin and coworkers[69] were able to successfully subgroup all of the carbon resonances into the respective heteronuclear spin systems.

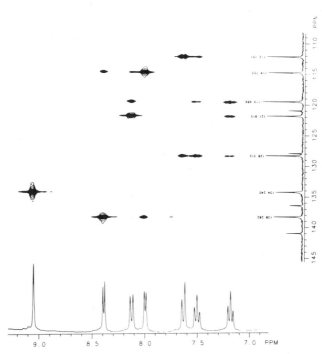

*Fig. 4-23.*        Heteronuclear relay spectrum of norharmane (**4-8**) acquired using a sample containing 200 mg/0.4 ml $d_6$-DMSO. The data was acquired as 96 x 1K data points and was zero-filled to 128 x 1K points during processing. Total acquisition time was 2.5 hours.

Beginning with the carbon resonating furthest upfield, which was ultimately assigned as the protonated carbon adjacent to the thiophene sulfur, C10, H10 is identified from the heteronuclear chemical shift correlation spectrum as the proton resonating furthest downfield. We may next locate the vicinal neighbor proton, H9, using the HC-RELAY spectrum shown in Fig. 4-24. In the process, we also have avaliable the means to assign C9 as the carbon resonating at 126.83 ppm once H9 is assigned.

Continuing to the carbon resonating at 122.13 ppm, we note from the spectrum shown in Fig. 4-24 that it has a vicinal relay response to one of the triplets resonating upfield and hence must be a member of one of the four spin systems. Ultimately, the directly correlated pair of resonances we began from in this case was assigned as C15/H15 on the basis of a carbon-carbon double quantum spectrum.[70] The vicinal neighbor, H14, correlates directly with the carbon resonance observed at 125.25 ppm, which is assigned as C14. Vicinal relay neighbor responses from H14 appear at the H15 proton where we began and at the triplet resonating furthest upfield, which is thus assigned as H13. The C13 carbon resonates at 122.41 ppm. Vicinal relay responses from H13 again appear for H14 and finally H12, which resonates upfield at 6.81 ppm, C12 resonating at 126.30 ppm.

The balance of the heteronuclear spin systems shown in the spectrum of **4-9** in Fig. 4-24 may be traced in a fashion analogous to that just described. It should also be noted that these vicinal proton-proton connectivities may be established despite a degree of overlap that renders the COSY spectrum intractable (see Fig. 2-13).

At this point, we should also note a shortcoming of the HC-RELAY experiment. Specifically, the HC-RELAY spectrum cannot bridge heteroatoms or quaternary carbon sites. The HC-RELAY spectrum does provide an effective means of identifying all the five heteronuclear spin systems of **4-9**, and of establishing the connectivity within each. Unfortunately, the HC-RELAY experiment cannot orient the spin systems it identifies, nor can it locate them within the molecular framework. For example, H15 could be assigned as H12 in the absence of chemical shift arguments or for that matter as either H1 or H4 as well. However, once the spin systems have been identified through the use of HC-RELAY, they may be oriented and linked together across quaternary carbons by either long range heteronuclear correlation (see Chapter 3) or via carbon-carbon double quantum coherence (see Chapter 5) and across heteroatoms by long range heteronuclear correlation.

**Isopteropodine**. In addition to problems encountered due to the presence of quaternary carbons and heteroatoms, problems with the HC-RELAY experiment may arise when there is substantial disparity between the actual vicinal coupling and the optimization of the mixing period. To illustrate how breaks in the connectivity network may occur due to misoptimization, we will finally consider the alkaloid isopteropodine (**4-10**).

**4-10**

There are numerous starting points from which an assignment or, if this were an unknown compound, structure elucidation could begin.

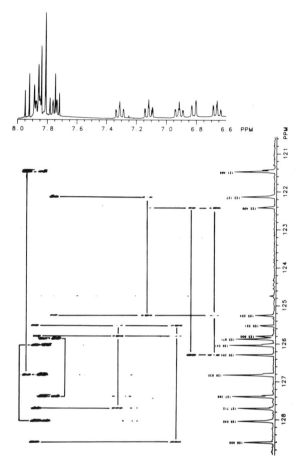

*Fig. 4-24.*     Heteronuclear relay spectrum of phenanthro[4,3-a]dibenzothiophene
(4-9) recorded using the pulse sequence shown in Fig. 4-21. The data
was acquired overnight using a sample containing 100 mg of the
compound dissolved in 0.4 ml of deuterochloroform as 200 x 1K
points. The data was zero-filled to 256 x 1K points during processing.

Perhaps the simplest and most informative starting point is the methyl
doublet H18, which is clearly visible in the high resolution proton
reference spectrum plotted beneath the stacked slice summary shown in
Fig. 4-25. We note first that the methyl doublet is vicinally correlated
with a proton resonating at 4.34 ppm, which corresponds to the H19
methine proton of the dihydropyran ring, C19 resonating at 72.12 ppm.
H19 is coupled, in turn, to a proton resonating in a complex multiplet
upfield at 1.57 ppm, which may be assigned as H20, C20 resonating at
37.88 ppm. Finally, H20 is vicinally correlated to the anisochronous
H21 methylene protons resonating at 2.43 and 3.48 ppm, C21
resonating at 53.3 ppm. The connectivity between H20 and H15 does
not give any response in the HC-RELAY.

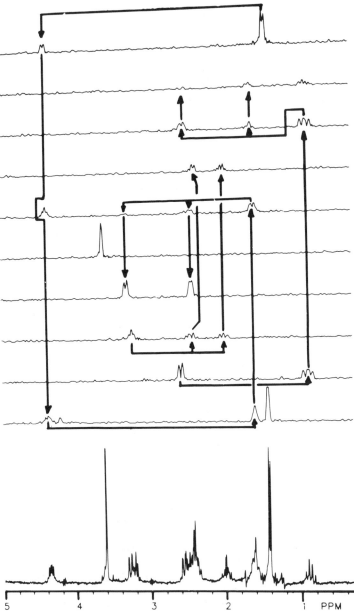

*Fig. 4-25.*        Heteronuclear relay spectrum of isopteropodine (**4-10**) presented as a slice summary. The data was acquired using the pulse sequence shown in Fig. 4-21 and a sample containing 80 mg of the alkaloid dissolved in deuterochloroform. The data was acquired overnight as 180 x 1K points and was zero-filled to 256 x 1K points during processing.

In similar fashion, beginning from H3, which resonates at 2.54 ppm, we establish a connectivity to the H14 proton resonating at 0.86 ppm and then to H15 but no further.

The HC-RELAY data shown in Fig. 4-25 was generated using a 7 Hz optimization of the mixing period. Although the actual coupling has not been measured, the vicinal coupling between the H15 and H20 protons, which are cis, can be expected to be significantly less than 7 Hz, leading, in this case, to a break in the connectivity pathway. This observation underscores a further weakness of the HC-RELAY pathway, a sensitivity to the optimization of the mixing period that may be responsible for breaks in the connectivity network being assembled.

## A Brief Summary of Applications
## of the HC-RELAY Experiment

The shortcomings of the HC-RELAY experiment, namely its lower sensitivity relative to conventional heteronuclear chemical shift correlation and the tendency for breaks in the connectivity network due to misoptimization of the mixing period, have perhaps contributed to the somewhat low utilization of this experiment. There have been some applications of the experiment in natural products spectral assignment and structure elucidation studies, which have included peptides,[59] alkaloids,[71,72] terpenoids,[73-75] and the antibiotic erythromycin.[76] The classes of molecules with which HC-RELAY may be employed most confidently are probably the oligosaccharides[60] and polynuclear aromatic compounds, the latter category having been extensively probed in this fashion by Martin and coworkers.[69,70,77-80]

## Techniques to Eliminate Direct
## Responses from Heteronuclear Relayed
## Coherence Transfer Spectra

Two methods to eliminate unnecessary direct responses from heteronuclear relayed coherence transfer spectra have been proposed. The earliest approach was that of Ernst and coworkers,[67] which utilized a low pass J-filter. More recently, Bolton[81] has proposed an alternative means of removing directly coupled proton responses using a BIRD pulse[82] to achieve semiselective refocusing of unwanted components of magnetization. Both of these experiments will be discussed separately in the following sections.

## Low Pass, J-Filtered Heteronuclear
## Relayed Coherence Transfer

Responses arising from directly attached protons can be undesirable. One such instance is the long range heteronuclear correlation experiment. Heteronuclear relay spectra represent another such category. The amplitudes of signals from directly attached protons may dominate the spectrum, partially masking  signals arising from vicinal neighbor (remote) protons, which are generally much weaker in intensity (see slices contained in Fig. 4-25). Hence, Ernst and coworkers[67] have proposed a modified version of the heteronuclear relay that utilizes a mechanism known as low pass J-filtering to remove responses arising from directly attached protons.

The low pass J-filter consists of a 90° carbon pulse that is phase alternated along the ±x-axis following a fixed delay, $\tau_p$, which is optimized as a function of $1/2(^1J_{CH})$ as shown in Fig. 4-26. The delay, $\tau_p$, serves to bring undesired components of magnetization into antiphase, after which they are converted to unobservable heteronuclear multiple quantum coherence by the first 90° carbon pulse. The single step J-filtered pulse sequence shown can be expected to result in attenuation of desired responses by < 7% and a maximum residual amplitude for the undesired direct responses of < 10%.

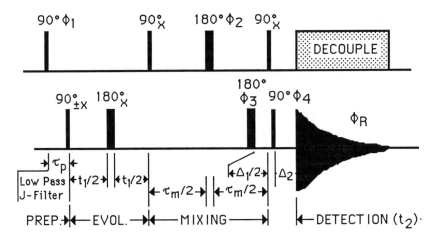

*Fig. 4-26.*    Pulse sequence for removing direct responses from heteronuclear relayed coherence transfer spectra via low pass J-filtration as proposed by Ernst and coworkers.[67]

**Phase cycling**. The phase cycling prescription for the low pass J-filtered heteronuclear relay spectrum is given in Table 4-6. The only difference is the phase of the 90° J-filtering pulse, which is alternated between the ±x-axis on successive scans.

**Parameter selection**. Parameters required for the pulse sequence shown above in Fig. 4-26 are essentially the same as for the conventional HC-RELAY experiment, which was discussed above. The sole difference is the optimization of the fixed duration delay, $\tau_p$. Normally, this delay will be optimized as a function of $1/2(^1J_{CH})$, in which case the interval may be set equal to the $\Delta_2$ interval. For multiple step J-filtration it may be advisable to use the ranges of durations given in the original paper by Ernst and coworkers.[67]

**Comparison of the low pass, J-filtered and conventional HC-RELAY spectra.** The spectrum obtained using the pulse sequence shown in Fig. 4-26 is presented in Fig. 4-27. As will be noted when the spectrum shown in Fig. 4-27 is compared with the conventional spectrum shown in Fig. 4-23, the low pass J-filter has effectively removed most of the direct responses. Only direct responses for H1/C1 and a very weak response for H3/C3 remain visible when the experiment is optimized for a 170 Hz average one bond heteronuclear coupling constant and a 7 Hz vicinal proton-proton coupling.

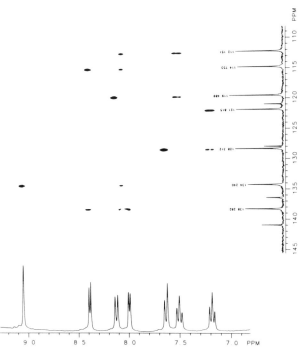

*Fig. 4-27.*     Low pass J-filtered heteronuclear relay spectrum of norharmane (**4 - 8**) recorded using the pulse sequence shown in Fig. 4-26. The experiment was optimized for an average vicinal proton-proton coupling of 7 Hz and a 170 Hz one bond heteronuclear coupling. The data was acquired using a sample consisting of 200 mg of norharmane dissolved in 0.4 ml of $d_6$-DMSO. The data was acquired as 96 x 1K complex points with 32 acquisitions taken per block. The projected proton spectrum is plotted beneath the contour plot while the normal, high resolution carbon spectrum is plotted vertically beside the contour plot.

## HC-RELAY with Semiselective Refocusing

Semiselective refocusing to eliminate direct responses in HC-RELAY spectra is achieved by the utilization of a BIRD pulse[82] midway through the evolution period, $t_1$.[81] The concept is analogous to that of the heteronuclear chemical shift correlation experiment devised by Bax,[83] which was shown in Fig. 3-25. The modified HC-RELAY pulse sequence is shown in Fig. 4-28.

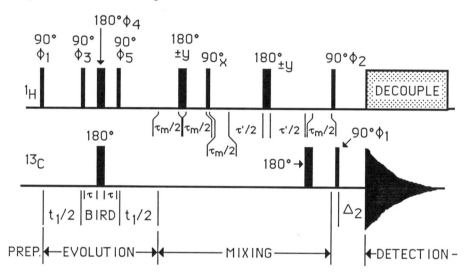

*Fig. 4-28*     Heteronuclear relayed coherence transfer with semiselective refocusing of magnetization terms for directly attached protons described by Bolton.[81]

The experiment operates by semiselective refocusing of the magnetization terms arising from the directly attached proton. Since the directly coupled protons are refocused, the precession of the vicinal neighbor protons will be effectively decoupled from the scalar coupling to the direct proton. Because of the homonuclear decoupling provided by the BIRD pulse, an additional delay is necesitated to allow the remote protons to become out of phase with respect to the direct-vicinal neighbor scalar coupling so that an efficient transfer of magnetization can be achieved.

In addition to eliminating responses due to directly attached protons, the experiment shown in Fig. 4-28 has one other attribute that is potentially even more beneficial. Specifically, since couplings between the vicinal neighbor and directly attached protons are eliminated by the semiselective refocusing, the relayed responses arising from the vicinal neighbor protons are semiselectively proton decoupled, which leads to a corresponding increase in sensitivity relative to the conventional HC-RELAY experiment discussed above.

With regard to multiplet structure in the spectral data, protons at the terminus of a spin system that are coupled only to the vicinal neighbor (directly attached in this case) will appear as singlets. Protons in the interior of a spin system that are coupled to protons in addition to the directly attached protons will appear as multiplets, the coupling in these cases arising only from coupling to other remote protons.

**Phase cycling**. Magnetization from the direct protons is of no interest in this modification of the HC-RELAY experiment and can be expediently eliminated by phase cycling the BIRD pulse so that all components arising from refocusing are canceled. Complete phase cycling for the pulse sequence shown in Fig. 4-28 is given in Table 4- 7.

Table 4-7.    Phase cycling for the HC-RELAY experiment with semiselective proton refocusing via a BIRD pulse.[81]

| $\phi_1$ | 01230123 | 01230123 | 01230123 | 01230123 |
|---|---|---|---|---|
|  | 01230123 | 01230123 | 01230123 | 01230123 |
| $\phi_2$ | 02020202 | 02020202 | 02020202 | 02020202 |
|  | 02020202 | 02020202 | 02020202 | 02020202 |
| $\phi_3$ | 01231230 | 23013012 | 01231230 | 23013012 |
|  | 01231230 | 23013012 | 01231230 | 23013012 |
| $\phi_4$ | 12302301 | 30120123 | 12302301 | 30120123 |
|  | 12302301 | 30120123 | 12302301 | 30120123 |
| $\phi_5$ | 23013012 | 01231230 | 23013012 | 01231230 |
|  | 23013012 | 01231230 | 23013012 | 01231230 |
| 1st $^1$H 180° | 13131313 | 13131313 | 31313131 | 31313131 |
|  | 13131313 | 13131313 | 31313131 | 31313131 |
| 2nd $^1$H 180° | 13131313 | 13131313 | 13131313 | 13131313 |
|  | 31313131 | 31313131 | 31313131 | 31313131 |

## Heteronuclear Relayed Hartmann-Hahn Spectroscopy

Conventional heteronuclear RELAY experiments suffer from sensitivity and optimization problems, especially in the case of saturated cyloalkyl systems such as alkaloids and steroids. Furthermore, there is a 90° phase difference between the direct and relayed response in such spectra, precluding the acquisition of phase sensitive data. To partially overcome these problems, Bax, Davis and Sarkar[66] have proposed the pulse sequence shown in Fig. 4-29, which uses net magnetization transfer among protons via homonuclear Hartmann-Hahn cross polarization.

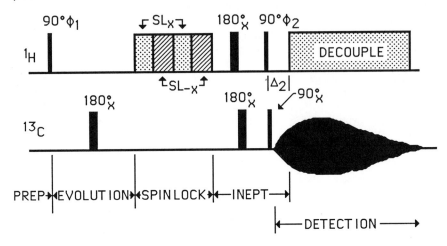

*Fig. 4-29*    Pulse sequence for heteronuclear relayed Hartmann-Hahn spectroscopy devised by Bax, Davis and Sarkar.[66] Note that decoupling is begun $\Delta_2$ into the acquisition period, where typically $\Delta_2 = 1/2(J_{CH})$.

Homonuclear Hartmann-Hahn (HOHAHA) transfer of magnetization among coupled protons is particularly suited for the mixing period of heteronuclear relay spectra. First, consider a decoupling field whose strength is defined by Eqn. [4-11]

$$\nu = \gamma H_2/2\pi \qquad [4\text{-}11]$$

which is continuously aligned along the +x-axis. Next, let us examine the case of two protons, A and B, with offset frequencies of $\Delta_A$ and $\Delta_B$, respectively, which experience effective fields of $\nu_A$ and $\nu_B$. Given that $\nu \gg \Delta_A$ and $\Delta_B$, then $\nu_A$ and $\nu_B$ are approximated by the following relations:

$$\nu_A = \nu + \Delta_A^2/(2\nu) \qquad [4\text{-}12]$$

$$\nu_B = \nu + \Delta_B^2/(2\nu) \qquad [4\text{-}13]$$

If $|\Delta_A| = |\Delta_B|$, then the effective field experienced by the two spins is equivalent and a perfect Hartmann-Hahn match occurs, allowing oscillatory magnetization exchange between spins A and B. In the case where $\nu_A \neq \nu_B$, the mismatch may be compensated by phase alternation of the spin-locking field. In general, the duration of the spin-lock interval, $\tau$, should be on the order of $1/(4J_{AB})$.

Following magnetization transfer, Bax and coworkers[66] utilized an INEPT type transfer (see also Fig. 3-4 and attendant discussion) to relay proton magnetization to the $^{13}C$ nucleus attached directly to $H_B$ (this is assuming that $H_A$ is the vicinal neighbor spin).

## Phase Cycling

The phase cycling scheme described by Bax and colleagues[66] for the pulse sequence shown in Fig. 4-29 is intended to provide a 2D spectrum with pure absorption mode responses according to the methods described by Müller and Ernst[84] and States and coworkers.[85] Phases are cycled according to Table 4-8 with odd- and even-numbered scans stored in separate locations for phase sensitive processing.

Table 4-8.    Phase Cycling Scheme for Heteronuclear Relayed Hartmann-Hahn Experiment Shown in Fig. 4-29.

| Acquisition | $\phi_1$ | $\phi_2$ | Receiver | |
|---|---|---|---|---|
| 1 | 0 | 0 | 0 | |
| 2 | 1 | 0 | | 0 |
| 3 | 2 | 0 | 2 | |
| 4 | 3 | 0 | | 2 |
| 5 | 0 | 2 | 0 | |
| 6 | 1 | 2 | | 0 |
| 7 | 2 | 2 | 2 | |
| 8 | 3 | 2 | | 2 |

## Applications of Heteronuclear Relayed Hartmann-Hahn Spectroscopy

To date, other than the application of this experiment to quinine which was presented in the communication by Bax, Davis and Sarkar[66], no applications of the heteronuclear relayed Hartmann-Hahn experiment have appeared in the literature. In a comparison with the conventional heteronuclear relay experiment (see Fig. 4-21), Bax and coworkers claim greater sensitivity for the pulse sequence shown in Fig. 4-29 with one quarter the acquisition time. Whether this improvement in sensitivity is generally realized remains to be confirmed.

## Proton Detected (Reverse) Heteronuclear Relayed Coherence Transfer

One obvious problem inherent with the HC-RELAY experiment is its sensitivity. In part, the sensitivity problem should be overcome by the semiselective refocusing technique proposed by Bolton,[81] which was described above. Greater gains in sensitivity may be realized, however, by resorting to proton detection of the signals for only those protons which are coupled to $^{13}C$. To date, there have been four different pulse sequences reported for this purpose. Other than the papers in which each has been described, there have been no applications reported for any of these experiments in the literature. Determining which of these experiments is the best for any given purpose is clearly beyond the scope of this chapter. Thus, the experiments will be presented briefly below without further comment as to utility or relative sensitivity.

## The Bolton Sequence for Proton Detected Heteronuclear Relay

The earliest pulse sequence for proton detected heteronuclear relay was proposed by Bolton[63] in 1985. The pulse sequence is shown in Fig. 4-30. The first portion of the experiment is essentially a heteronuclear zero quantum chemical shift correlation sequence.[86,87] This portion of the experiment is followed by a mixing time to allow proton magnetization to go out of phase relative to both homo- and heteronuclear couplings, which is necessary for efficient magnetization transfer to take place. Hence, signals will be obtained with $F_2$ frequencies of direct and vicinal neighbor protons at their respective $^{13}C$ $F_1$ frequencies. This situation is identical to that obtained with the conventional HC-RELAY experiment save the fact that the $F_1$ and $F_2$ labels have been exchanged with the change in the detection scheme.

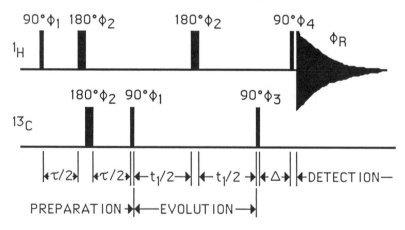

Fig. 4-30.    Proton detected heteronuclear relay pulse sequence proposed by Bolton.[63]

It should be noted that the pulse sequence shown in Fig. 4-30 does not provide for heteronuclear decoupling of the proton responses in $F_2$. Broadband heteronuclear decoupling can be provided, however, by inserting an additional delay and applying $^{13}C$ decoupling during acquisition.

**Phase cycling**. The phase cycling suggested by Bolton for the pulse sequence shown in Fig. 4-30 is quite extensive, constituting a 256 step cycle. In principle, however, it should be possible to limit the phase cycle to the first 64 steps without penalty. The reader interested in implementing this experiment is referred to the original work of Bolton[63] for further details regarding the cycling of the various phases in the experiment.

### The Bax Sequences for Proton Detected Heteronuclear Relay

Lerner and Bax[64] followed the initial work of Bolton[63] with two additional pulse sequences for proton detected heteronuclear relay. Both sequences begin with a BIRD pulse to invert proton magnetization arising from protons not directly coupled to $^{13}C$. The pulse sequences shown in Figs. 4-31 and 4-32 share similarities. The first, shown in Fig. 4-31, is quite similar to the conventional heteronuclear relay experiment (see Fig. 4-21), where magnetization is transferred between coupled protons by means of a 90° pulse. The second variant is similar to the heteronuclear relayed Hartmann-Hahn experiment described above (see Fig. 4-29) in that it uses a homonuclear Hartmann-Hahn (HOHAHA) mechanism to transfer magnetization between coupled protons.

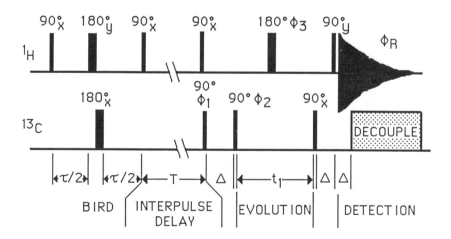

*Fig. 4-31.* Proton detected analog of the conventional heteronuclear relayed coherence transfer pulse sequence shown in Fig. 4-21. Phases are cycled according to Table 4-9. The transfer of magnetization between coupled protons is provided by a 90° pulse as in the conventional, heteronucleus detected experiment.

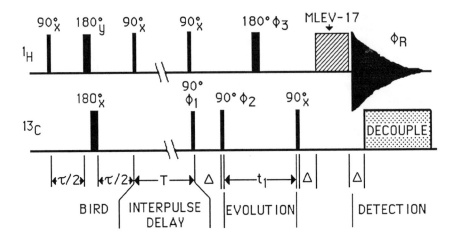

Fig. 4-32.     Proton detected analog of the heteronuclear relayed Hartmann-Hahn pulse sequence shown in Fig. 4-29. Phases are cycled according to Table 4-9. The transfer of magnetization between coupled protons is provided by a HOHAHA mechanism as in the heteronuclear relayed Hartmann-Hahn experiment shown in Fig. 4-29.

The BIRD pulse that begins the experiments shown in Figs. 4-31 and 4-32, is utilized to saturate protons not directly bound to $^{13}C$. Typically, the BIRD pulse should be applied about two-thirds through the interpulse delay, where it serves to invert z-magnetization of all protons not coupled to $^{13}C$ while leaving those which are coupled as essentially unperturbed z-magnetization. Subsequently, at the point of the 90° $^{13}C$ $\phi_2$ pulse that actually begins the experiment, protons attached to $^{13}C$ are at equilibrium, while those not attached to $^{13}C$ are close to saturation, which makes overcoming the large dynamic range problem much easier by precluding the phase cycling that would otherwise be necessary.

The active part of the heteronuclear relay begins with the $90°_x$ proton pulse. Magnetization created by the $90°_x$ proton pulse is converted to zero and double quantum heteronuclear coherences for proton $H_A$ directly coupled to $C_A$ by the 90° $\phi_2$ pulse applied following the fixed delay $\Delta = 1/2(^1J_{CH})$. Zero and double quantum coherences are interchanged by the 180° proton pulse as in the long range heteronuclear multiple quantum coherence experiment discussed in Chapter 3. The final $90°_x$ $^{13}C$ pulse converts heteronuclear multiple quantum coherence into observable $^1H$ magnetization, which has now been modulated during $t_1$ as a function of the $^{13}C$ chemical shift. Finally, again following a fixed delay, $\Delta$, the application of a proton $90°_y$ pulse transfers magnetization to the proton vicinal coupling partners whose magnetization is in turn also modulated by the $^{13}C$ chemical shift frequency.

As with conventional HC-RELAY experiments, the data at this point contains components of magnetization from both direct and remote or vicinal neighbor protons. Generally, the former will be unnecessary and more intense. Therefore, Lerner and Bax[64] have demonstrated that these responses may be suppressed by initiating $^{13}C$ decoupling after a third fixed delay, $\Delta$, at which point the direct components of magnetization are antiphase and hence canceled when decoupling is initiated. It should be noted that complete suppression of direct responses is possible only when a narrow range of $^1J_{CH}$ values is encountered in a sample, e.g. polynuclear aromatics and saccharides. As the size of the one bond coupling deviates from ideality, direct response intensity will increase proportionately.

**Phase cycling**. The phase cycling schemes for the pulse sequences shown in Figs. 4-31 and 4-32 are given in Table 4-9. Lerner and Bax[64] make the recommendation in their communication that phase cycling of the observe pulses should be minimal since changes in the phase of one proton pulse relative to another creates a new "steady-state" z-magnetization at the beginnning of the actual experiment (not at the BIRD pulse) thereby aggravating the problem of suppressing responses for protons not directly coupled to $^{13}C$.

The phase of the 180° proton pulse is generally held constant if 32 or fewer acquisitions are to be performed per $t_1$ increment. If 128 scans are to be performed, the phase of the 180° pulse should be stepped by 90° every 32 acquisitions. For scans 1-32 use $180°_x$, for scans 33-64 use $180°_y$, etc. In addition, each time the phase of the 180° pulse is incremented, the phase of the receiver should be inverted.

Table 4-9.    Phase cycling scheme for the pulse sequences for proton detected heteronuclear relay shown in Figs. 4-31 and 4-32.

| Acquisition | $\phi_1$ | $\phi_2$ | $\phi_3$[a] | $\phi_R$[b] |
|---|---|---|---|---|
| 1 | 0 | 0 | 0 | 0 |
| 2 | 0 | 1 | 0 | 0 |
| 3 | 0 | 2 | 0 | 2 |
| 4 | 0 | 3 | 0 | 2 |
| 5 | 2 | 0 | 0 | 0 |
| 6 | 2 | 1 | 0 | 0 |
| 7 | 2 | 2 | 0 | 2 |
| 8 | 2 | 3 | 0 | 2 |

[a]    The phase of the 180° pulse should be minimally cycled in 90° steps after every 32 acquisitions. For example, hold phase = 0 for steps 1-32, phase = 1 for 33-64, etc.

[b]    The phase of the receiver should be inverted whenever the phase of the 180° proton pulse ($\phi_3$) is incremented. For example, beginning with step 33, the phase will shift from 0 to 2.

## The Brühwiler-Wagner Proton
## Detected Heteronuclear
## Relay Experiment

The most recently reported variant of the proton detected heteronuclear relay experiment is that of Brühwiler and Wagner.[65] In addition, the application that they show with their report is the most sophistocated thus far, the small protein BPTI (MW = 6500 daltons) used as the example.

The Brühwiler-Wagner[65] sequence, as with those described above, begins with a BIRD pulse to aid in the suppression of responses arising from protons not directly coupled to $^{13}C$. The balance of the experiment operates in a fashion comparable to the description of the experiments originated by Lerner and Bax.[64] In principle, a refocusing period following the second 90° $^{13}C$ pulse in the sequence consisting of $\tau_o$ - 180° - $\tau_o$ followed by a 90°$_y$ pulse would provide relayed coherence transfer to a coupled proton. This operation, however, would counter-productively reorient all magnetization along the z-axis into the xy-plane, thus necessitating much more extensive phase cycling. To allow sufficient defocusing to $^1H$-antiphase magnetization prior to the last 90°$_y$ pulse, the $\tau_2$ interval may be longer than $\tau_o$, thereby precluding the rotation of unwanted components of magnetization into the xy-plane.

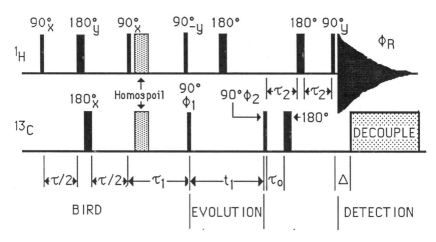

*Fig. 4-33.*    Pulse sequence for proton detected heteronuclear relay proposed by Brühwiler and Wagner.[65]

**Phase cycling.** Table 4-10 gives the phase cycling required in the implementation of the Brühwiler-Wagner[65] proton detected heteronuclear relay experiment shown in Fig. 4-33. Fundamentally it is no more complex than that required for the Lerner-Bax[64] experiments described above (see Table 4-9 for comparison).

Table 4-10.    Phase cycling scheme for the proton detected heteronuclear relay pulse sequence shown in Fig. 4-33.[a]

| Acquisition | $\phi_1$ | $\phi_2$ | $^1H$-$90°_y$ | $\phi_R$ |
|---|---|---|---|---|
| 1 | 0 | 0 | 1 | 0 |
| 2 | 2 | 0 | 1 | 2 |
| 3 | 0 | 2 | 1 | 2 |
| 4 | 2 | 2 | 1 | 0 |
| 5 | 0 | 0 | 3 | 0 |
| 6 | 2 | 0 | 3 | 2 |
| 7 | 0 | 2 | 3 | 2 |
| 8 | 2 | 2 | 3 | 0 |

[a]    In addition to the phase cycling shown, the phases of the 180° pulses in the sequence should be inverted independently of the basic eight step cycle shown.

## Comparative Evaluation of the Various Proton Detected Heteronuclear Relay Experiments

A comparison of the relative utility of the proton detected heteronuclear relay experiments now available remains to be conducted. When performed, such a comparison will undoubtedly prove both interesting and useful. Currently there are no guidelines to the selection of one of the experiments just described relative to the others available.

## Multiple Quantum Heteronuclear Relayed Coherence Transfer Spectroscopy

A final category of experiments worthy of some comment while treating the subject of heteronuclear relayed coherence transfer, consists of the multiple quantum based experiments proposed by Bolton[61] and Turner.[62] Here the fundamental idea is to provide a different means of distributing informaiton in one of the frequency domains, typically $F_1$. The basis for these experiments is the same that has prompted the development of proton zero and double quantum coherence experiments discussed in Chapter 2: avoiding undesirable signal overlap.

## Proton Double Quantum Heteronuclear Relayed Coherence Transfer

The first "alternative quantum" approach to heteronuclear relay spectroscopy was proposed by Bolton[61] in 1983. Information in the $F_1$ (proton) frequency domain, rather than being distributed as a function of the proton chemical shift, is instead distributed as a function of the proton double quantum frequency, which corresponds to the algebraic sum of the proton shifts relative to the transmitter. This approach, in principle, provides an alternative means of distributing information in $F_1$ that could be useful in the event of especially crowded HC-RELAY spectra.

The experiment is shown in Fig. 4-34 and first utilizes the standard 90°- $\tau$ - 180° - $\tau$ - 90° sequence to generate proton double quantum coherence. However, since the experiment will ultimately be heteronuclear, 180° $^{13}$C pulses must be inserted midway through both of the $\tau$ intervals to make the experiment independent of the heteronuclear spin coupling. Proton double quantum coherence generated in this fashion is then allowed to evolve through the $t_1$ interval, again with a $^{13}$C 180° pulse midway through the interval to maintain freedom from heteronuclear coupling terms. The evolved proton double quantum coherence is converted back into single quantum coherence at the end of the evolution period by the application of a 45° proton pulse, at which point the mixing period begins. During the mixing interval, proton single quantum coherences are brought into phase with respect to homonuclear coupling terms and are brought to an antiphase condition with respect to heteronuclear couplings as in the conventional HC-RELAY experiment described above. Transfer from proton to carbon is provided by the pair of 90° pulses applied to $^1$H and $^{13}$C prior to detection.

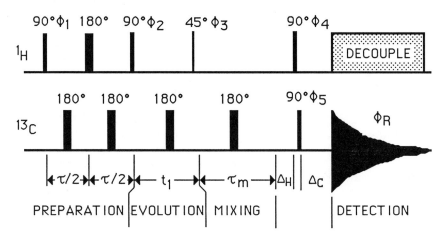

Fig. 4-34.    Proton double quantum heteronuclear relay experiment proposed by Bolton.[61] Proton double quantum coherence is generated with 180° carbon pulses midway through the $\tau/2$ delays to eliminate heteronuclear coupling. Double quantum coherence evolves, again with a 180° pulse at $t_1/2$, to suppress heteronuclear couplings. Magnetization is transferred to vicinal neighbors during the mixing interval and then to $^{13}$C in the usual fashion.

**Phase cycling**. The phase cycling regimen for the double quantum heteronuclear relay pulse sequence shown in Fig. 4-34 utilized a sixteen step phase cycle. In the experience of the authors, this phase cycling still left artifacts in the spectrum, which could be eliminated by inverting the phase of the 180° $^{13}$C pulse applied midway through the evolution period for scans 9-16. Readers interested in implementing the experiment are referred to the literature for the appropriate phase cycling. These readers should also note the comment on sensitivity in the following section.

**Sensitivity and applications**. One of the shortcomings of the HC-RELAY experiment is its low sensitivity relative to the conventional heteronuclear chemical shift correlation experiment (see Fig. 3-10). Sensitivity problems have prompted the development of proton detected analogs of the HC-RELAY experiment, which were discussed in the preceding segments of this chapter. Relative to the HC-RELAY experiment, the proton double quantum heteronuclear relay experiment, in the experience of the authors, has still lower sensitivity which is likely to severely restrict application of the experiment in its present form. Indeed, other than the initial communication describing the pulse sequence, there have been no reported applications.

## Proton Zero Quantum Heteronuclear Relayed Coherence Transfer

Following Bolton's proton double quantum heteronuclear relay experiment by several years was an experiment described by Turner[62] that exploits proton zero quantum rather than double quantum coherence. In this sense, information is distributed in the $F_1$ dimension as the algebraic difference of the offsets of the protons relative to the transmitter, rather than as the sum, as in the experiment described immediately above. Again, the alternative means of distributing spectral information in $F_1$ could conceivably be useful when dealing with crowded spectra. Furthermore, the inherent insensitivity of proton zero quantum coherence to magnetic field inhomogeneities could also provide an effective increase in resolution due to narrow linewidths.

The pulse sequence is shown in Fig. 4-35 and begins with the creation of proton zero quantum coherence via a 90° - $\tau$ - 180° - $\tau$ - 90° sequence with a 180° $^{13}$C pulse applied $1/2J_{CH}$ before the second 90° pulse. Here again, the intent is to eliminate the effects of heteronuclear scalar coupling, although the approach used here differs from that of Bolton.[61] Zero quantum coherence, following creation, is allowed to evolve. A 180° $^{13}$C pulse is applied $1/2J_{CH}$ before the midpoint of the evolution period, which will leave the zero quantum coherence in antiphase at the end of the evolution period. Single quantum coherence is restored to the protons by a second refocusing sequence identical to the first, followed by transfer to $^{13}$C with the final 90° pulse sandwich. Decoupling and acquisition are initiated after a fixed delay with a duration of $1/4J_{CH}$.

**Phase cycling**. Specific details of the phase cycling for the experiment shown in Fig. 4-35 were not given in Turner's original communication.[62]

**Sensitivity and applications**. To date, there have been no reported applications of the proton zero quantum heteronuclear relay experiment described by Turner,[62] nor is there any indication of the sensitivity of this experiment relative to either the proton double quantum or conventional HC-RELAY analogs.

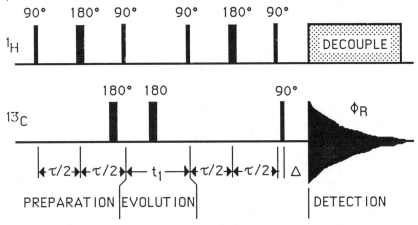

*Fig. 4-35.*     Proton zero quantum heteronuclear relay pulse sequence proposed by Turner.[62] The intervals in the preparation and final refocusing periods are normally set to $1/4J_{HH}$ as in other zero and double quantum coherence experiments. The 180° carbon pulses are applied at time $1/2J_{CH}$ prior to the end of the preparation period and prior to the middle of the evolution period, respectively.

# REFERENCES

1.     G. Eich, G. Bodenhausen and R. R. Ernst, *J. Am. Chem. Soc.*, **104**, 3731 (1982).
2.     G. Wagner, *J. Magn. Reson.*, **55**, 151 (1983).
3.     R. Baumann, G. Wider and R. R. Ernst, *J. Magn. Reson.*, **44**, 402 (1981).
4.     A. Bax and G. Drobny, *J. Magn. Reson.*, **61**, 306 (1985).
5.     W. E. Hull, "Experimental Aspects of Two-Dimensional NMR," in *Two-Dimensional NMR Spectroscopy - Applications for Chemists and Biochemists*, W. R. Croasmun and R. M. K. Carlson, Eds., VCH, New York, 1987, pp. 67-258.
6.     R. Sanduja, G. S. Linz, M. Alam, A. J. Weinheimer, G. E. Martin and E. L. Ezell, *J. Heterocyclic Chem.*, **23**, 529 (1986).

7.  E. R. P. Zuiderweg, R. Kaptein and K. Wüthrich, *Eur. J. Biochem.*, **137**, 279 (1983).
8.  E. R. P. Zuiderweg, M. Billeter, R. Boelens, R. M. Scheek, K. Wüthrich and R. Kaptein, *FEBS Lett.*, **174**, 243 (1984).
9.  R. Kaptein, E. R. P. Zuiderweg, R. M. Scheek, R. Boelens and W. F. van Gunsteren, *J. Mol. Biol.*, **182**, 179 (1985).
10. P. L. Weber, G. Drobny and B. R. Reid, *Biochemistry*, **24**, ###(paper before pg 4553) (1985).
11. P. L. Weber, D. E. Wemmer and B. R. Reid, *Biochemistry*, **24**, 4553 (1985).
12. A. J. Wand and S. W. Englander, *Biochemistry*, **24**, 5290 (1985).
13  D. W. Hughes, R. A. Bell, T. Neilson and A. D. Bain, *Can. J. Chem.*, **63**, 3133 (1985).
14. D. R. Hare and B. R. Reid, *Biochemistry*, **25**, 5341 (1986).
15. J. Dabrowski, A. Ejchart, M. Kordowicz and P. Hanfland, *Magn. Reson. Chem.*, **25**, 338 (1987).
16. S. W. Homans, R. A. Dwek, D. L. Fernandes and T. W. Rademacher, *Proc. Natl. Acad. Sci., U.S.A.*, **81**, 6286 (1984).
17. W. J. Chazin and K. Wüthrich, *J. Magn. Reson.*, **72**, 358 (1987).
18. G. A. Morris and M. S. Richards, *Magn. Reson. Chem.*, **23**, 676 (1985).
19. G. S. Linz, R. Sanduja, A. J. Weinheimer, M. Alam and G. E. Martin, *Tetrahedron Lett.*, 4833 (1986).
20. A. Otter and G. Kotovych, *J. Magn. Reson.*, **69**, 187 (1986).
21. K. V. R. Chary and R. V. Hosur, *J. Magn. Reson.*, **74**, 352 (1987).
22. P. L. Weber and L. Müller, *J. Magn. Reson.*, **73**, 184 (1987).
23. P. L. Weber, G. Drobny and B. R. Reid, *Biochemistry*, **24**, 4549 (1985).
24. T. A. Holak and J. H. Prestegard, *J. Magn. Reson.*, **73**, 530 (1987).
25. P. L. Weber, L. C. Sieker, T. S. Anantha Samy, B. R. Reid and G. P. Drobny, *J. Am. Chem. Soc.*, **109**, 5842 (1987).
26. P. J. States, R. A. Haberkorn and D. J. Ruben, *J. Magn. Reson.*, **48**, 286 (1982).
27. S. Macura, N. G. Kumar and L. R. Brown, *J. Magn. Reson.*, **60**, 99 (1984).
28. L. Müller and A. Pardi, *J. Am. Chem. Soc.*, **107**, 3484 (1985).
29. D. A. Craig and G. E. Martin, *J. Nat. Prod.*, **49**, 456 (1986).
30. A. Wokaun and R. R. Ernst, *Chem. Phys. Lett.*, **52**, 407 (1977).
31. S. R. Hartmann and E. L. Hahn, *Phys. Rev.*, **128**, 2042 (1962).
32. L. Müller and R. R. Ernst, *Mol. Phys.*, 38, 963 (1979).
33. L. Braunschweiler and R. R. Ernst, *J. Magn. Reson.*, **53**, 521 (1983).
34. M. Rance, *J. Magn. Reson.*, **74**, 557 (1987).
35. A. Bax and D. G. Davis, *J. Magn. Reson.*, **63**, 207 (1985).
36. A. Bax and D. G. Davis, *J. Magn. Reson.*, **65**, 355 (1985).
37. D. G. Davis and A. Bax, *J. Am. Chem. Soc.*, **107**, 7197 (1985).
38. M. Levitt, R. Freeman and T. A. Frenkiel, *J. Magn. Reson.*, **47**, 313 (1982).
39. D. P. Weitekamp, A. Bielecki, D. Zax, K. Zilm and A. Pines, *Phys. Rev. Lett.*, **50**, 1807 (1983).
40. J. S. Waugh, *J. Magn. Reson.*, **68**, 189 (1986).
41. P. L. Weber, L. C. Sieker, T. S. Anantha Samy, B. R. Reid and G. P. Drobny, *J. Am. Chem. Soc.*, **109**, 5842 (1987).
42. M. W. Edwards and A. Bax, *J. Am. Chem. Soc.*, **108**, 918 (1986).
43. M. F. Summers, L. G. Marzilli and A. Bax, *J. Am. Chem. Soc.*, **108**, 4285 (1986).
44. A. Bax, A. Aszalos, Z. Dinya and K. Sudo, *J. Am. Chem. Soc.*, **108**, 8056 (1986).

45.  T. Kasumi, T. Ooi, M. M. Watanabe, H. Takahashi and H. Kakisawa, *Tetrahedron Lett.*, 4695 (1987).
46.  A. M. Gronenborn, G. M. Clore, U. Schmeissner and P. Wingfeld, *Eur. J. Biochem.*, **161**, 37 (1986).
47.  G. M. Clore, S. R. Martin and A. M. Gronenborn, *J. Mol. Biol.*, **191**, 553 (1986).
48.  G. P. Drobny, manuscript in preparation cf. ref. 41.
49.  A. J. Shaka, J. Keller and R. Freeman, *J. Magn. Reson.*, **53**, 313 (1983).
50.  J. S. Waugh, *J. Magn. Reson.*, **68**, 189 (1986).
51.  J. S. Waugh, *J. Magn. Reson.*, **49**, 517 (1982).
52.  J. S. Waugh, *J. Magn. Reson.*, **50**, 30 (1982).
53.  B. M. Feng, *J. Magn. Reson.*, **59**, 275 (1984).
54.  S. Macura, Y. Huang, D. Suter and R. R. Ernst, *J. Magn. Reson.*, **43**, 259 (1981).
55.  O. W. Sørensen, M. Rance and R. R. Ernst, *J. Magn. Reson.*, **56**, 527 (1984).
56.  P. H. Bolton, *J. Magn. Reson.*, **48**, 336 (1982).
57.  P. H. Bolton and G. Bodenhausen, *Chem. Phys. Lett.*, **89**, 139 (1982).
58.  A. Bax, *J. Magn. Reson..*, **53**, 149 (1983).
59.  H. Kessler, M. Bernd, H. Kogler, O. W. Sørensen, G. Bodenhausen and R. R. Ernst, *J. Am. Chem. Soc.*, **105**, 6944 (1983).
60.  P. Bigler, W. Ammann and R. Richarz, *Org. Magn. Reson.*, **22**, 109 (1984).
61.  P.H. Bolton, *J. Magn. Reson..*, **54**, 333 (1983).
62.  D. L. Turner, *J. Magn. Reson.*, **65**, 169 (1985).
63.  P. H. Bolton, *J. Magn. Reson.*, **62**, 143 (1985).
64.  L. Lerner and A. Bax, *J. Magn. Reson.*, **69**, 375 (1986).
65.  D. Brühwiler and G. Wagner, *J. Magn. Reson.*, **69**, 546 (1986).
66.  A. Bax, D. G. Davis and S. K. Sarkar, *J. Magn. Reson.*, **63**, 230 (1985).
67.  H. Kogler, O. W. Sørensen, G. Bodenhausen and R. R. Ernst, *J. Magn. Reson.*, **55**, 157 (1983).
68.  G. Bodenhausen and R. Freeman, *J. Magn. Reson..*, **36**, 221 (1977).
69.  M. J. Musmar, G. E. Martin, M. L. Tedjamulia, H. Kudo, R. N. Castle and M. L. Lee, *J. Heterocyclic Chem.*, **21**, 929 (1984).
70.  M. J. Musmar, G. E. Martin, R. T. Gampe, Jr., M. L. Lee, R. E. Hurd, M. L. Tedjamulia, H. Kudo and R. N. Castle, *J. Heterocyclic Chem.*, **22**, 219 (1985).
71.  G. E. Martin, R. Sanduja and M. Alam, *J. Nat. Prod.*, **49**, 406 (1986).
72.  S. R. Johns, J. A. Lamberton, H. Suares and R. I. Willing, *Aust. J. Chem.*, **38**, 1091 (1985).
73.  C. J. Turner, *Magn. Reson. Chem.*, **22**, 531 (1984).
74.  A. San Feliciano, A. F. Barrero, M. Medarde, J. M. Miguel del Corral, A. Aramburu, A. Perales and J. Faykos, *Tetrahedron Lett.*, 2369 (1985).
75.  G. S. Linz, M. J. Musmar, A. J. Weinheimer, G. E. Martin and J. A. Matson, *Spectroscc. Lett.*, **19**, 545 (1986).
76.  J. R. Everett and J. W. Typler, *J. Chem. Soc., Perkin Trans. I*, 2599 (1985).
77.  M. J. Musmar, G. E. Martin, R. T. Gampe, Jr., V. M. Lynch, S. H. Simonsen, M. L. Lee, M. L. Tedjamulia and R. N. Castle, *J. Heterocyclic Chem.*, **22**, 545 (1985).
78.  J. G. Stuart, M. J. Quast, G. E. Martin, V. M. Lynch, S. H. Simonsen, M. L. Lee, R. N. Castle, J. L. Dallas, B. K. John and L. F. Johnson, *J. Heterocyclic Chem.*, **23**, 1215 (1986).
79.  A. S. Zekzter, J. G. Stuart, G. E. Martin and R. N. Castle, *J. Heterocyclic Chem.*, **23**, 1587 (1986).
80.  M. J. Musmar, A. S. Zektzer, R. T. Gampe, Jr., M. L. Lee, M. L. Tedjamulia, R. N. Castle and R. E. Hurd, *Magn. Reson. Chem.*, **24**, 1039 (1986).

81.    P. H. Bolton, *J. Magn. Reson.*, **63**, 225 (1985).
82.    J. R. Garbow, D. P. Weitekamp and A. Pines, *Chem. Phys. Lett.*, **93**, 504 (1982).
83.    A. Bax, *J. Magn. Reson.*, **53**, 517 (1983).
84.    L. Müller and R. R. Ernst, *Mol. Phys.*, **38**, 963 (1979).
85.    D. J. States, R. A. Haberkorn and D. J. Ruben, *J. Magn. Reson.*, **48**, 286 (1982).
86.    P. H. Bolton, *J. Magn. Reson.*, **57**, 427 (1984).
87.    A. Bax, R. H. Griffey and B. L. Hawkins, *J. Magn. Reson.*, **55**, 301 (1983).

# CHAPTER 5

## 13C-13C DOUBLE QUANTUM COHERENCE 2D-NMR: THE INADEQUATE EXPERIMENT

### INTRODUCTION

Information derived from $^{13}C$-$^{13}C$ spin-spin coupling can be used to establish the identity of adjacent carbon atoms in a molecule. Connectivity information about the carbon skeleton of a compound can be extremely valuable in the elucidation of complex structures. Additionally, long range $^{13}C$-$^{13}C$ spin-spin coupling can be used in the study of conformation. This information is normally lost in conventional $^{13}C$ NMR spectrum due to the weak intensity of the signals from $^{13}C$-$^{13}C$ couplings, which at natural abundance have a statistical probability of 1:10,000. Therefore, the signals due to $^{13}C$-$^{13}C$ couplings are 1/200th the intensity of the signal from a single $^{13}C$. These problems were addressed and overcome by a pulse sequence reported by Bax, Freeman and Kempsell in 1980.[1] This experiment was given the acronym INADEQUATE standing for Incredible Natural Abundance DoublE QUAntum Transfer Experiment.

We shall consider the carbon-carbon INADEQUATE experiment in the context of the proton double quantum experiment described late in Chapter 2. Useful comparisons between the proton and carbon variants of the experiment can be drawn. First, let us consider sensitivity. The proton double quantum experiment can hardly be described as one of "incredible natural abundance." Every proton is magnetically susceptible, and hence the proton double quantum experiment is quite easy to perform. Indeed, one of the problems inherent to the proton double quantum experiment is the appearance signals from AMX and more complex spin systems, which may not be desirable. Schemes have been devised to suppress these responses. In contrast, the carbon variant of the experiment is exclusively limited to AB through AX pairings, since the statistical probability of three $^{13}C$ nuclides adjacent to one another is 1:1,000,000 and consequently such an occurrence would be undetectable! A similarity between the proton and carbon double quantum experiments does, however, exist. It is also necessary to suppress undesired responses that in the carbon version of the experiment arise from molecules containing isolated ¹³C nuclides. This task is, in principle, accomplished by a four step phase cycle. In practice, phase cycles ranging from a minimum of 32 steps to 256 steps are utilized, with 128 steps perhaps the most common.

## THE ONE-DIMENSIONAL INADEQUATE EXPERIMENT

The basis for the ¹³C-¹³C double quantum experiment proposed by Bax, Freeman and Kempsell is the discrimination between isolated and coupled ¹³C nuclides on the basis of the presence or absence of the homonuclear spin-spin coupling, $J_{CC}$. Their pulse sequence, shown in Fig. 5-1, excites double quantum coherence[2-5] between two coupled ¹³C nuclei. Since the phase properties of single quantum and other multiple quantum coherences are different than those of double quantum coherence,[3,6] the unwanted signals are removed through phase cycling. The generation of multiple quantum coherence is accomplished through a relatively simple pulse train[7], shown in Eqn. [5-1], which is identical to that shown in Eqn. [2-7] regarding the proton double quantum experiment.

$$90°(x) - \tau - 180°(y) - \tau - 90°(x)　　　　　　[5-1]$$

Discrimination between single- and multiple-quantum coherence is accomplished by setting $\tau$ equal to $1/(4J_{CC})$.[7]

### Parameter Selection

The one-dimensional double quantum experiment requires the selection of a minimal number of parameters. The sole variable to be selected is the duration of the fixed delay, $\tau$, used during the excitation of double quantum coherence. Typically, this interval is set as a function of $1/4J_{CC}$. Generally employed values range from about 6 msec for aliphatic compounds to 5 msec for aromatic compounds, with 5.5 msec providing a usable compromise for aryl/alkyl containing compounds. The fixed delay, $\Delta$, in the one-dimensional INADEQUATE experiment should be long enough to accommodate accurate phase shifts. Normally a delay of about 10 μsec should be sufficient.

### Phase Cycling

The phase cycle used to remove the undesired components is rather simple, consisting of the four step cycle shown in Table 5-1. This simple phase cycle, of course, assumes perfect pulses over the entire carbon spectral width. In practice, however, the phase cycling required to remove the single quantum component is more complex than the

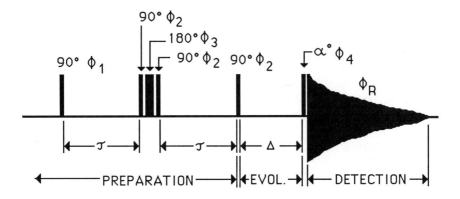

Fig. 5-1.      The pulse sequence used in the observation of carbon double quantum
               coherence. The phase cycling of the pulses and the receiver are shown
               below in Table 5-3. The delay $\Delta$ is set to ~10 $\mu$sec for one dimensional
               spectra. For two-dimensional spectra, the fixed delay, $\Delta$, is replaced
               by the incremented evolution time, $t_1$, which is employed to encode
               the double quantum frequency domain. The pulse sequence is shown
               with a composite 180° pulse, the use of which is recommended to
               eliminate offset effects possible because of the broad carbon spectral
               width and/or the effects of miscalibration of the 180° pulse.

Table 5-1.     Minimum phase cycling for the observation of double quantum
               coherence.

| $\phi_1{}^a$ | $\phi_R{}^b$ | $2Q^c$ |
|:---:|:---:|:---:|
| 0 | 0 | 0 |
| 1 | 3 | 3 |
| 2 | 2 | 2 |
| 3 | 1 | 1 |

a   $\phi_1$ is the phase of the final or "read" pulse.

b   $\phi_R$ is the receiver phase.

c   2Q is the phase of the double quantum coherence.

original four step scheme shown.    In their original communication Bax,
Freeman and Kempsell suggested that the phase of the 180° pulse
should be shifted by 180° on successive scans, giving an eight step
cycle. In addition, the phase of all pulses should be incremented 90°
after every eighth scan, giving a 32 step phase cycle.  This was later
increased to a 128 step phase cycle by using a four step phase cycle of
the 180° pulse that is incremented 90° after every 32nd scan. To insure

an accurate phase shift between the final two carbon pulses, a short (10 μsec) delay should be used.  The use of this type of delay to ensure an accurate phase shift was discussed in Chapter 2 for the double quantum filtered COSY experiment.

## DOUBLE QUANTUM COHERENCE
## AND THE DENSITRY MATRIX

Representation of magnetization components during the double quantum experiment by a vector model breaks down at the point of the creation of multiple quantum coherence by the third pulse of the excitation pulse train of Eqn. [5-1].  Therefore, it is necessary to use the more rigorous and intellectually intimidating density matrix to properly represent the evolution of the coupled spins.  An abbreviated treatment will be given here to provide verification of the minimum phase cycling necessary to detect double quantum coherence while filtering out the single quantum coherence. An excellent development of the density matrix for the double quantum coherence experiment  is to be found in a chapter by Mateescu and Valeriu[8] to which the interested reader is referred.

The pulse train presented in Eqn. [5-1] excites the desired double quantum coherence as well as an undesired residual single quantum components that may arise due to noncoupled (isolated) ¹³C signals. After the third pulse of the excitation sequence, shown in Eqn. [5-1], the density matrix contains elements representing single quantum coherence which are on the diagonal or in the $M_{11}$ and $M_{44}$ locations of the matrix while those corresponding to double quantum coherence are in the $M_{14}$ and $M_{41}$ locations of the matrix.  Treating the diagonal and nondiagonal elements separately, it can be demonstrated that the fourth pulse in Eqn. [5-1] produces an equalized population for the diagonal components.  In contrast, the matrix elements due to double quantum coherence, when acted on by this same 90° pulse, produce a detectable single quantum signal.  The equations that represent these matrix elements are shown in Eqn. [5-2] and Eqn. [5-3], respectively:

$$M_{T1} = m\,(1Q)\,[\exp(i\,\Omega_{12}\,t_d) + \exp(i\,\Omega_{13}\,t_d)$$
$$- \exp(i\,\Omega_{24}\,t_d) - \exp(i\,\Omega_{34}\,t_d)] \qquad [5\text{-}2]$$

where  $m\,(1Q) = (M_0/2)\,(\exp\text{-}i\phi)(-\cos\Omega_{14}\Delta)$, and

$$M_{T2} = m\,(2Q)(4\exp(i\,\Omega_{14}\,t_d)) \qquad [5\text{-}3]$$

where $m\,(2Q) = (-i\,\Omega M_0/2)\,\exp(i\phi)$.  Based on these expressions, the phase cycling that produces the desired coherence at the receiver is as given in Table 5-2.

Table 5-2.     Basic phase cycling for the observation of double quantum coherence as predicted by a density matrix treatment.

| $\phi_1{}^a$ | $\phi_R{}^b$ | $(1Q)^c$ | $(2Q)^d$ $(\exp(i\phi_1)\exp(-i\phi_R))$ | $(\exp(-i\phi_1)\exp(-i\phi_R))$ |
|---|---|---|---|---|
| +x | +x | 1 | 1 | |
| +y | -y | -1 | 1 | |

| sum of acquired signal intensity:$^e$ | 0 | 2 |
|---|---|---|

a)     $\phi_1$ is the phase of the final or "read" pulse.

b)     $\phi_R$ is the receiver phase.

c)     1Q is the acquired signal intensity of single quantum coherence using the phases of the first pulse and the receiver shown.

d)     2Q signifies the acquired signal intensity of the double quantum coherence.

e)     Therefore, after the second scan the net accumulated signal of double quantum is 2, while that of single quantum is zero.

Although a two step phase cycle is predicted by the density matrix treatment to be sufficient to discriminate between single and double quantum coherence, as was cited earlier, Bax, Freeman and Kempsell suggest a minimum 32 step phase cycle to remove the effects of imperfect phases. Most applications cited in this chapter use a 128 step cycle. The phase cycle for this extended cycle is shown in Table 5-3.

Table 5-3.     Extended phase cycling for the observation of double quantum coherence. The phase of the first 90° pulse ($\phi_1$) is held at 0° and thus, is not phase cycled. The phases of the other pulses in Fig. 5-1 are incremented by 90° every 32$^{nd}$ scan.

| $\phi_2$ | 00000000 | 00000000 | 00000000 | 00000000 |
|---|---|---|---|---|
| $\phi_3$ | 00000000 | 00000000 | 22222222 | 22222222 |
| $\phi_4$ | 00001111 | 22223333 | 00001111 | 22223333 |
| $\phi_R$ | 01233012 | 23011230 | 23011230 | 01233012 |

## THE TWO-DIMENSIONAL INADEQUATE EXPERIMENT

Following the initial report by Bax, Freeman and Kempsell of the one-dimensional $^{13}C$-$^{13}C$ double quantum coherence experiment,[1] Bax, Freeman, Frenkiel and Levitt reported a two-dimensional version of this sequence.[9] This two-dimensional version of the earlier one-dimensional experiment allows the identification of adjacent carbons through exploitation of the double quantum frequency ($F_1$) domain. Coherence between coupled carbon nuclei appears as responses in the double quantum dimension at the algbraic sum of the offset frequencies of the coupled ¹³C nuclei.  Therefore, the ambiguity of the one-dimensional experiment is removed through the distribution of the data in two discrete frequency domains. The pulse sequence that provides two-dimensional double quantum coherence is essentially the same sequence used to obtain a one-dimensional double quantum spectrum. To encode the double quantum frequency domain, the pulse sequence shown in Fig. 5-1 is used with the fixed delay, $\Delta$, replaced by a systematically incremented evolution time, $t_1$, between the third and fourth pulses.

There are, however, two disadvantages with this method. First since signals due to double quantum coherence appear in the double quantum domain at the algebraic sum of the offsets of the coupled resonances from the transmitter, there are thus locations in the matrix that cannot possibly contain a response. Second, the sequence proposed by Freeman and coworkers lacks quadrature detection in the double quantum frequency domain and hence this sequence cannot differentiate the sign of the double quantum frequency.[9]  The consequence of this second point is that in the absence of quadrature detection, responses will be symmetrically disposed about $F_1 = 0$ Hz if the transmitter is set in the middle of the spectrum.

Bax and co-workers suggested that the first disadvantage could be overcome by using one-half the frequency necessary for the complete calculated range of double quantum coherences possible.  If this suggestion is employed, some of the responses are "folded" into regions of the final matrix normally devoid of responses, making better use of the available digitization. This is possibly due to the fact that location of the responses in the double quantum frequency domain can be calculated from the offset in the $F_2$ dimension of the coupled signals. An example of this will be discussed later in this chapter.  The second problem, the lack of quadrature detection in the double quantum dimension, requires more effort to circumvent and will be discussed in the following section.

## Quadrature Detection in the F1 Dimension Using Composite z-Pulses

Bax et al. [10] reported a sequence for the acquisition of carbon-carbon double quantum two-dimensional NMR spectra that provided for quadrature detection in the double quantum frequency domain. Modification of the two-dimensional experiment reported earlier[9] and shown in Fig. 5-1 was accomplished by the addition of a composite 45° z-pulse on alternate scans.  This modification facilitates a reduction of the size of the final data matrix by a factor of 4.   The two pulse sequences are shown in Fig. 5-2. The sequence in Fig. 5-2a can be

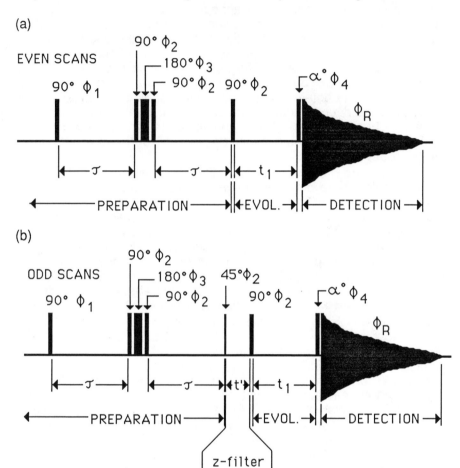

*Fig. 5-2.*    The pulse sequence used to obtain two-dimensional double quantum correlation spectra with quadrature detection in both frequency domains. The signal from both sequences are coadded for each $t_1$ value. The delay, t, is the same as in the preceding sequences, and t' delay is approximately 10 μsec delay to allow accurate phase shifting. The phase of the second 90° pulse is shifted 90° from the 45° pulse. Phase cycling for this sequence is the same as in Table 5-3. It should be noted that the phases are shifted only on alternate scans, while the order of the two sequences is arbitrary. The upper sequence is identical to the sequence shown in Fig. 5-1 except that $t_1$ replaces $\Delta$.

used alone to produce a spectrum without quadrature detection in the double quantum frequency domain, or can be used on alternate scans in conjunction with the pulse sequence in Fig. 5-2b to afford quadrature detection in the double quantum frequency domain. As will be discussed below, quadrature detection in the double quantum frequency domain can also be obtained in a different manner through a deceptively simple modification to the sequence shown in Fig. 5-2a.

## Autocorrelated Two-Dimensional INADEQUATE

After the work reported by Bax and coworkers concerning two-dimensional ¹³C-¹³C double quantum experiments, Turner reported two pulse sequences for two-dimensional ¹³C-¹³C double quantum correlation.[11] For these pulse sequences it was suggested that 45° phase shifts be performed to acheive quadrature detection in the double quantum frequency domain.

Turner reported a pulse sequence that "autocorrelates" the carbon-carbon coherence in two dimensions.[12] Turner's pulse sequence uses an additional $t_1$ delay after the final pulse, before acquisition begins. This effectively reduces the size of the double quantum domain by half and eliminates the regions where there could not possibly be any signal present. Unlike earlier INADEQUATE pulse sequences, 45° phase shifts are unnecessary to provide for quadrature detection in the double quantum dimension. Instead, a 45° rotation of the coherence is provided by the second portion of the evolution period ($t_1$). Turner's modified pulse sequence is shown in Fig. 5-3.

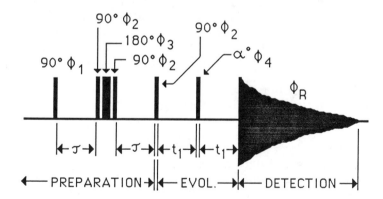

Fig. 5-3.      Pulse sequence proposed by Turner[12] to provide for an autocorrelated two dimensional double quantum experiment. In this experiment both of the $t_1$ incremented delays are equal in value and are set to digitize the double quantum domain with the same spectral width as the original spectrum. The $\tau$ delay is set as before. Use of the additional $t_1$ increment serves to provide for an autocorrelated format in the final data matrix.

Unfortunately, early claims touted the pulse sequence shown in Fig. 5-3 as having superior sensitivity to the conventional INADEQUATE based experiments, which employ but a single evolution stage. A more objective evaluation of the experiment, however, has concluded that rather than improve sensitivity, Turner's experiment instead suffers from somewhat diminished sensitivity due to the decay of magnetization occurring during the second portion of the evolution period after reconversion to single quantum coherence.[13]

## Obtaining Quadrature Detection in $F_1$ Using a 135° Read Pulse

A major improvement in the implementation of two-dimensional double quantum coherence experiments was suggested by Mareci and Freeman.[14] They proposed using a 135° pulse in place of the 90° read pulse (labeled $\alpha$ in Figs. 5-2a and 5-3) to provide quadrature detection in the double quantum frequency domain. Although deceptively simple, this modification eliminates the need for 45° phase shifts and/or co-addition of data collected with a 45° phase shift provided by a composite z-rotation such as that obtained by the concerted use of the two sequences shown in Fig. 5-2. In addition to the obviation of the need for 45° phase shifts, the use of the 135° pulse suggested by Mareci and Freeman[14] provides a modest improvement in sensitivity of about 20% relative to the use a conventional 90° reconversion pulse. Slightly higher sensitivity can be gained using a 120° read pulse, although this is at the expense of discrimination between the echo and antiecho components of magnetization which is generally an undesirable trade-off. Thus, the 135° pulse is preferred. Benefits of using a 135° read pulse accrue in both the conventional INADEQUATE experiment proposed by Bax and coworkers[9] (eg, Fig. 5-2a) and the autocorrelated version proposed by Turner[12] (Fig. 5-3).

## Excitation of Double Quantum Coherence Using "Double Selection"

Reports by Mareci and Freeman[14] concentrated on the effects produced by varying the flip angle of the read pulse. Lallemand and co-workers[15] suggested that modification of both the final pulse of the excitation sequence and the read pulse would provide greater double quantum signal intensity while at the same time suppressing quadrature images better than the excitation scheme shown in Fig. 5-1.

The sequence originally proposed by Müller to excite double quantum coherence took the form

$$90° - \tau - 180° - \tau - \alpha° \qquad\qquad [5\text{-}4]$$

where $\alpha = 90°$.[7] Lallemand and coworkers explored the technique of "double selection" of double quantum excitation and subsequent reconversion to observable single quantum coherence. In the double selection technique, the flip angle of $\alpha$ is set equal to the flip angle of the final read pulse. Work by Mareci and Freeman[14] demonstrated that the optimum value for sensitivity of the reconversion pulse was 120°. However, as noted above, the flip angle of the read pulse should be increased to 135° to ensure a higher selection ratio of the coherence transfer echo over the coherence transfer antiecho. Mareci and Freeman also noted that the price in sensitivity of using 135° versus 120° is less than 10%.

Using the double selection method of Lallemand,[15] both the read pulse (labeled $\alpha$ in Figs. 5-2a and 5-3) and the final pulse in the excitation sequence (also labeled $\alpha$ in Eqn. [5-4]) are set to 120°. With these flip angles the selection ratio of the coherence echo over the coherence antiecho is calculated to be 9:1 which is sufficient for most applications. If greater suppression of the coherence antiecho is necessary, the flip angles of these two pulses can be increased to 135°. A smaller decrease in sensitivity results with the double selection method than when only the read pulse flip angle is changed as proposed was by Mareci and Freeman.[14] As with the method of Mareci and Freeman, the double selection method of double quantum coherence provides quadrature detection in the $F_1$ dimension. However, the method of Mareci and Freeman still leads to the detection of some of the anticoherence transfer echo, and if sufficient levels of the data matrix are plotted these signals become apparent.

With the double selection method there are two benefits to reducing the chance of these signals interfering with the data in the matrix; the first is the greater intensity of the signal from the coherence transfer echo and the second is the more complete suppression of the responses from the anticoherence transfer echo, which would produce a quadrature image if not suppressed.

As before the modification of the final pulse of the excitation scheme is applicable to both the two-dimensional INADEQUATE version of Freeman and coworkers[9] and the autocorrelated version of Turner.[12] These sequences are shown in Figs. 5-2a and 5-3, respectively.

## Practical Aspects of Double Quantum NMR Spectroscopy

The practical aspects of the ¹³C-¹³C double quantum NMR experiment have been discussed by Bax and Mareci,[13] Turner[16] and Levitt and Ernst.[17] Of the two fundamental pulse sequences that should be compared, Figs. 5-2a and 5-3, there are potential advantages to each. Clearly, the sequence shown in Fig. 5-2a, which employs a single evolution period, is more sensitive than the autocorrelated variant shown in Fig. 5-3. In contrast, it might be argued that the response format of the data obtained using the sequence shown in Fig. 5-3 is more familiar, resembling that of the COSY experiment. Regardless of which experiment is employed, the comments of Levitt and Ernst[17] regarding the use of composite pulses should doubtless be heeded. Since carbon spectra generally involve wide spectral widths, it will be advantageous to compensate for offset effects from the transmitter through the use of a composite 180° pulse as shown in both Figs. 5-2a and 5-3.

## Sample Preparation and Solubility

Quite obviously, one of the problems associated with performing $^{13}C$-$^{13}C$ double quantum coherence NMR experiments is the inherently low sensitivity regardless of which variant of the experiment is utilized. Thus, for a model compound we would quite obviously like to have one with very generous solubility and quite favorable, relaxation characteristics. An excellent model compound is provided in the form of menthol (**5-1**). Solutions approaching 80% w/v can easily be obtained in $d_6$-benzene. Relaxation times are also quite favorable since we are dealing exclusively with protonated carbon resonances. Furthermore, relaxation behavior can be further enhanced by increasing the temperature to somewhere in the range of 35-50°C.

## Parameter Selection

Clearly, the first choice we must make is which variant of the experiment to use and how it is to be modified. Generally, we will be choosing between the sequences shown in Figs. 5-2a and 5-3. Beyond this, we must select a flip angle for the read pulse and also for the final pulse in the double quantum excitation train if double selection is to be employed. In part, we have discussed these considerations above and we shall hence leave further comment on the merits of the possible choices to the following sections, wherein the results obtained with the various permutations of the experiment are presented.

Beyond the choice of the experiment, there are obviously a number of parameters that must be chosen prior to the accumulation of $^{13}C$-$^{13}C$ double quantum coherence data. The first consideration is encountered in the optimization of the fixed duration delay, $\tau$, necessary to excite the double quantum coherence. While proton-proton homonuclear couplings are familiar and hence easily selected in the case of the proton analog of the double quantum experiment (see Chapter 2), their carbon homonuclear counterparts are considerably less familiar. Fortuitously, however, carbon-carbon coupling constants fall in a relatively narrow range and hence needn't be measured prior to performing the experiment. The duration of the fixed delay, $\tau$, should, ideally, be optimized as a function of $1/4(J_{CC})$. Typical carbon-carbon coupling constants range from about 40 Hz for a pair of coupled aliphatic carbons to about 70 Hz for coupled vinylic or aromatic carbons.

A very useful compilation of carbon-carbon homonuclear coupling constants is found in the monograph by Marshall.[18] In the absence of prejudices to optimize for a specific carbon-carbon homonuclear coupling constant, we find it useful to optimize the $\tau$ delay for 6 msec in the case of a purely aliphatic compound and 5 msec for an aromatic system. In the case of a compound that is aralkyl in nature, we will typically optimize the duration of $\tau$ at 5.5 msec as a compromise.

Given the means to optimize the delays necessary in the excitation of double quantum coherence, we must next decide on the spectral width that we desire in the $F_1$ frequency domain. Assuming that we have quadrature detection in $F_1$, the possible frequency range will be twice that in $F_2$. If we care to use this spectral width in $F_1$, we may do so by simply incrementing the duration of the evolution time, $t_1$, as a function of the dwell time in $F_2/2$. However, since the spectral width in $F_1$ will typically be quite large and hence severely underdigitized because of the relatively low number of increments of $t_1$ usually recorded (typically 64 or fewer), it may be quite advantageous to select for a substantially narrower $F_1$ spectral width. The ramifications of this choice will be made clear below in the case of example spectra of menthol recorded with various $F_1$ frequency widths.

Next, we must consider the interpulse delay to be used between acquisitions. Typically, two-dimensional NMR experiments may be cycled fairly rapidly. This does not apply, however, in the case of the INADEQUATE experiment. For the experiment to work properly, giving responses for all possible connectivities, we must allow 1.5 x the $T_1$ relaxation time of the most slowly relaxing carbon. In the experience of the authors with polynuclear aromatics, this can range to the 10's of seconds, clearly necessitating a compromise. Basically, we have available to us two choices: select a shorter interpulse delay and live with the potential loss of some of the responses; dope the sample with a relaxation reagent such as $Cr(acac)_3$. We have used both approaches and cannot necesssarily recommend one over the other on a general basis. Perhaps the most important determining factor is the relative value of the sample to the researcher if it proves difficult to separate it from the relaxation reagent. Clearly this is an individual choice.

As alluded to in the paragraph above, we must select how many increments of $t_1$ will actually be accumulated. Typically, because of the low sensitivity of the experiment, we will be restricted to an overnight or at most a weekend accumulation of data. Depending upon the quantity of material available and its solubility, we must then make a decision on how many increments of the evolution time we can afford while maintaining adequate signal-to-noise levels. All studies in the literature, to the best of our knowledge, have utilized 128 or fewer increments of the evolution period, with most typically employing 64 or fewer.

## Data Processing

Processing considerations in the case of carbon-carbon double quantum coherence experiments are somewhat different from those of other experiments. Prior to the first Fourier transformation, the free induction decay can be mathematically manipulated in whatever manner produces the best response intensity and resolution, the latter generally being the less important consideration. After transposition of the data matrix we have generally found it disadvantageous to employ any mathematical massage of the interferograms. It may be desirable, however, to zero fill to augment digitization in the second frequency domain. Care should also be taken not to employ a baseline correction routine as these do not operate effectively on very limited interograms.

## Double Quantum NMR
## Spectra of Menthol

Using the pulse sequences in Figs. 5-2a and 5-3, two-dimensional INADEQUATE NMR spectra of menthol (**5-1**) were obtained to compare the position of responses produced with the two different pulse sequences.

**5-1**

## The Conventional INADEQUATE
## Spectrum of Menthol

The spectrum of menthol shown in Fig. 5-4 was recorded using a modified form of the pulse sequence shown in Fig. 5-2a. Specifically, the sequence originally reported by Freeman and coworkers[10] was used with double selection[15] and 135° pulse widths.[14]

The spectrum presented in Fig. 5-4 was recorded with twice the spectral width in the $F_1$ dimension as in $F_2$, to preclude folding in the double quantum frequency domain. The evolution period, $t_1$, was thus incremented as a function of $F_2$ dwell/2. Chemical shifts in ppm downfield of TMS are plotted beneath the conventional broadband proton decoupled carbon spectrum. Frequency relative to the transmitter is shown in hertz beneath the contour plot. Double quantum frequencies relative to the transmitter are plotted in hertz along the vertical axis. From a cursory comparison of the two axes it should be noted that the frequency range in $F_1$, as expected, is twice that in $F_2$.

The version of Freeman and coworkers[10] allows one to calculate the position of a set of coupled responses in the $F_1$ dimension. As stated earlier, coupled carbons will have responses that are present at the sum of the offsets of the resonances from the transmitter in the $F_1$ dimension and at the chemical shift of each resonance in the $F_2$ dimension. Thus, the double quantum correlation axis runs from lower right to upper left with a slope of -2. This feature of the experiment has some utility when dealing with an unknown structure. If there is any question whether a given pair of resonances are coupled, the requisite $F_1$ frequency can be calculated and then the appropriate slice examined.

To illustrate the interpretation of the spectrum of menthol (**5-1**) shown in Fig. 5-4, let us establish several of the connectivities. As with the interpretation of any other two-dimensional NMR experiment, it is useful to locate a starting point. In the case of menthol, a convenient starting point is provided by the hydroxyl bearing methine carbon, C3. By simple inspection, we note in Fig. 5-4 that there are two responses located at the chemical shift of C3 in the data matrix. We also might wish to note that the C3 resonance +2180 Hz from the transmitter in $F_2$. Horizontally displaced from the responses at the $F_2$ frequency of C3, we note responses at $F_2$ frequencies of +580 and +220 Hz that correlate the C3 methine carbon resonance to resonances for a methine and a methylene carbon, respectively. Based on these $F_2$ frequency pairs, the $F_1$ responses should be observed at +2760 and +2300 Hz. In the $F_1$ frequency domain, responses are located at frequencies of -2760 and -2300 Hz according to the $F_1$ axis labels. At this point it should be noted that the orientation of the positive and negative frequency labels along the $F_1$ axis in the plot shown in Fig. 5-4 is reversed. The correct orientation could, however, be reversed by spectrum rotation about $F_1$ = 0 Hz if cosmetically necessary.

Through the set of operations just completed, which establish two connectivities, C3 is linked to the C2 methylene and C4 methine carbon resonances that flank it in the menthol structure. Continuing from the C4 chemical shift in $F_2$, we note that C4 correlates further to two additional carbons, a methylene carbon resonating at 26.4 ppm, which is identifiable as C5, and a methine carbon resonating at 28.7 ppm, which corresponds to C8 of the isopropyl substituent attached to C4. Connectivities from C5 and C8 next locate C6 and the C9 and C10 methyls, respectively. Next, the C6 to C1 connectivity is established with C1 further linked to the C7 methyl and the C2 methylene, the latter connectivity completing the assembly of the structure.

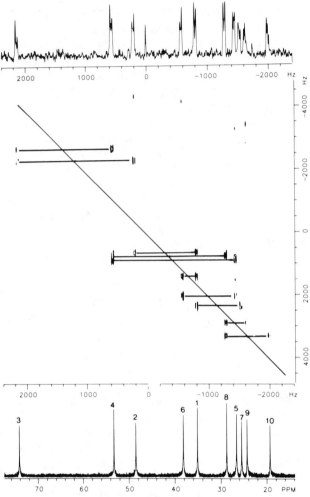

*Fig. 5-4.*          The two-dimensional INADEQUATE NMR spectrum of menthol **(5-1)** acquired using the pulse sequence shown in Fig. 5-2a modified for double selection as suggested by Lallemand and coworkers.[15] The broadband proton decoupled carbon spectrum is plotted beneath the contour plot and chemical shifts are reference in ppm downfield from TMS. The axis immediately beneath the contour plot labeled in hertz gives resonance locations relative to the transmitter frequency. Frequencies in $F_1$ are also labeled in hertz with positive and negative labels interchanged. Correct orientation of the frequency labeles could be established by rotation about $F_1 = 0$ Hz. The carbon coupled carbon spectrum is recovered by 0° projection through $F_2$ and is plotted above the contour plot.

## The Autocorrelated INADEQUATE
## Spectrum of Menthol Recorded
## Using the Turner Pulse Sequence

The spectrum of menthol obtained with the modified sequence of Turner[12] (shown in Fig. 5-3) is shown in Fig. 5-5. The data was generated with the duration of each evolution period, $t_1$, set to the $F_2$ dwell time. The consequence of using this set of parameters is to produce a spectrum in which the spectral width of $F_1 = F_2$. Additionally, the correlation axis, rather than having a slope of ±2 as would normally be expected for the skew diagonal, now has a slope of ±1. A second feature of the Turner experiment that may not be immediately apparent is the level of digitization in $F_1$. Since the frequency range in $F_1$ is halved relative to the conventional pulse sequence, effective digital resolution will be doubled if the same number of increments of $t_1$ are employed in both data collections, as in the present case. The most noticeable result of the improved digitization is the enhanced appearance of the responses in the spectrum shown in Fig. 5-5 relative to their counterparts in Fig. 5-4. Compare, for example, the responses correlating C3 with C2 and C4 in both spectra.

Beyond cosmetic differences in appearance, the distribution of responses in the spectrum shown in Fig. 5-5 precludes the calculation of $F_1$ response frequencies in the fashion described above. Furthermore, since responses are now distributed perpendicular to the diagonal axis, individual slices cannot be examined for the presence of weak correlations, which might not be observed in the contour plot. In dealing with samples of unknown structure, this is clearly a disadvantage, since experimental spectra will typically have signal-to-noise ratios lower than those shown in the examples in this chapter.

### Intentionally Folded
### INADEQUATE Spectra

To overcome the problems associated with limited digitization in the $F_1$ frequency domain, we have recourse to two, alternative solutions. First, we may simply perform more experiments, with the consequent penalty in time imposed by the inherently low sensitivity of the INADEQUATE experiment. The second and more viable choice is to intentionally fold the double quantum or $F_1$ frequency domain by altering the incrementation of the evolution time, $t_1$.

For a given $F_2$ spectral width, the normal range of double quantum frequencies permissible will be twice that of the $F_2$ range. Normally, this would require the incrementation of the evolution time, $t_1$, as a function of the $F_2$ dwell time/2 as in the spectrum shown in Fig. 5-4. Under these conditions, all possible responses will be within the $F_1$ frequency range and the skew diagonal will pass through the data matrix as shown above. Quite obviously, if only 32 or perhaps 64 increments of the evolution time are to be employed, this will result in very coarse digitization in $F_1$, since in the present example for a simple

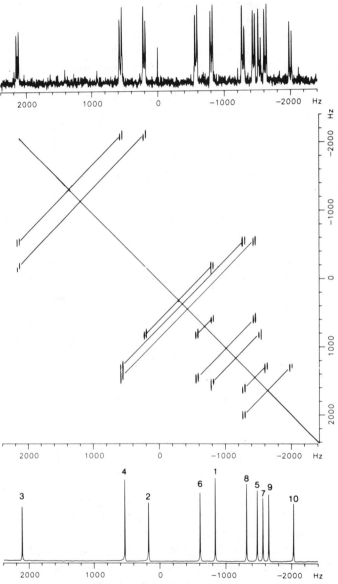

*Fig. 5-5.*　　Autocorrelated INADEQUATE spectrum of menthol **(5-1)** recorded using the pulse sequence shown in Fig. 5-3. The conditions used in recording this spectrum were identical to those of the previous example save the requirements for the incrementation of the evolution time, $t_1$, which results in an effective doubling of the digitization in $F_1$.

aliphatic compound this amouts to ±4800 Hz. Furthermore, there are areas of the data matrix that cannot possibly contain a response, eg, the lower left and upper right regions of the spectrum shown in Fig. 5-4. To reduce the problem of coarse digitization and unused regions of the data matrix, we may intentionally fold the spectrum in $F_1$. We shall examine two spectra obtained in this fashion.

Using the pulse sequence shown in Fig. 5-2a, the spectrum presented in Fig. 5-6 was acquired with the evolution time, $t_1$, incremented as a function of the $F_2$ dwell time. The obvious result of this change is the halving of the $F_1$ frequency range to ±2400 Hz. From the discussion above, we will recall that the $F_1$ frequencies of some of the responses exceeded 2400 Hz, which will lead to folding in $F_1$. In particular, the response correlating C3 with C4 had an $F_1$ frequency of 2760 Hz. As a result of the folding in $F_1$, we now observe the C3-C4 response folded 360 Hz from the bottom edge of the $F_1$ frequency range and appearing in the previously unused lower left quadrant of the data matrix. At the other extreme of the $F_1$ frequency range, the C8-C9 and C8-C10 responses are folded into the upper right quadrant of the data matrix as shown in Fig. 5-6.

Carrying the case of folding in $F_1$ still further, we may increment the evolution time, $t_1$, as a function of the $F_2$ dwell time x 2, which will give a frequency range of ±1200 Hz in $F_1$. Relative to the initial starting point in Fig. 5-4, we have, at this point, quadrupled the relative digitization in $F_1$ in the spectrum shown in Fig. 5-7. Relative to the spectrum shown in Fig. 5-4, any response with an $F_1$ frequency > ±1200 Hz will be folded. In this case both the C3-C4 and C3-C2 responses will fold as well as all responses beyond C1-C6, which has a frequency of -1400 Hz.

Other than requiring a slight amount of additional book-keeping of response locations, folding in the $F_1$ frequency domain imposes no penalties on the user. The principles governing where a response will appear are the same regardless of whether a spectrum is unfolded, folded once or even twice.

## Assembling the Carbon Skeleton
## of Cedrol Using INADEQUATE Data

Having considered menthol in some detail, it is appropriate to examine the assemblage of a carbon skeleton from ¹³C-¹³C INADEQUATE data. Rather than beginning with a structure, let us instead start from the INADEQUATE spectrum shown in Fig. 5-8, recorded using the conventional pulse sequence shown in Fig. 5-2a. The empirical formula of the compound $C_{15}H_{25}O$ is consistent with a molecule containing three rings. Beginning with the carbon resonating furthest downfield at 72.0 ppm (1), which is consistent with a 3° alcohol, we have a convenient starting point. Note that there are three

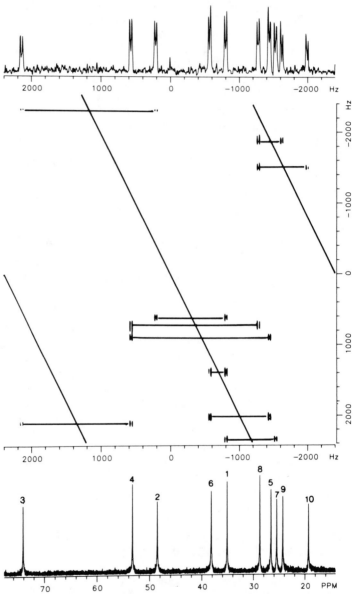

*Fig. 5-6.* INADEQUATE spectrum of menthol (**5-1**) recorded in which the evolution time, $t_1$, was incremented as a function of the $F_2$ dwell time. This modification halves the $F_1$ frequency range relative to that in the spectrum shown in Fig. 5-4. Consequently, the skew diagonal and responses along it, which are outside of the frequncy range, fold as shown here and discussed in the text.

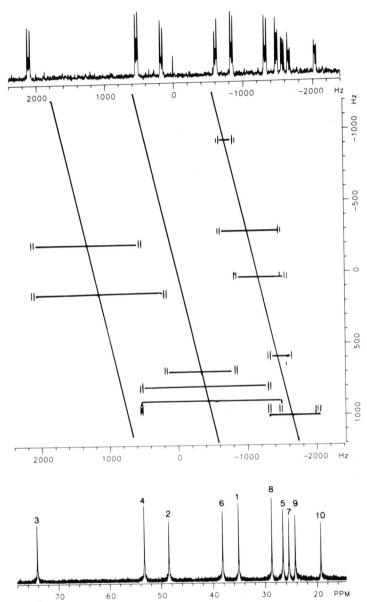

*Fig. 5-7.*     INADEQUATE spectrum of menthol (**5-1**) recorded in which the evolution time, $t_1$, was incremented as a function of the 2 x $F_2$ dwell time. This modification halves the $F_1$ frequency range relative to that in the spectrum shown in Fig. 5-6. Folding is more pronounced than in the spectrum in Fig. 5-6, digital resolution in $F_1$ is higher and more efficient usage is made of the data matrix.

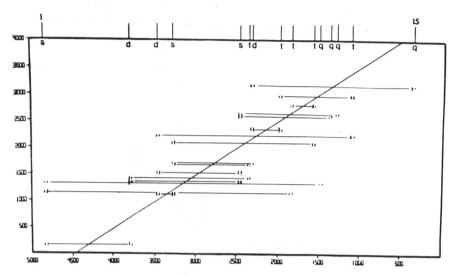

*Fig. 5-8.*     75 MHz $^{13}$C-$^{13}$C INADEQUATE spectrum of cedrol. (J. N. Shoolery, *J. Nat. Prod.*, **47**, 226 (1984) with permission).

connectivities associated with this resonance. Referring to the carbon spectrum above the contour plot, carbons are labeled in ascending numerical order from left to right; multiplicities determined from an APT spectrum are shown below the spectrum (s = C; d = CH; t = CH$_2$; q = CH$_3$, respectively). Hence the connectivities to C2, C9 and C11 correspond to methine, methylene and methyl carbons, respectively. (Refer to Fig. 5-9a.) Translating this first set of connectivities into molecular structure information, we generate the fragment **5-2**.

9           11

$H_2C$      $CH_3$

1
2   OH

CH

**5-2**

      We may continue to assemble the atom-to-atom connectivities using any of the leads from C1, but let us continue from the methine carbon, C2. Hence we observe that C2 correlates with C5 and C6

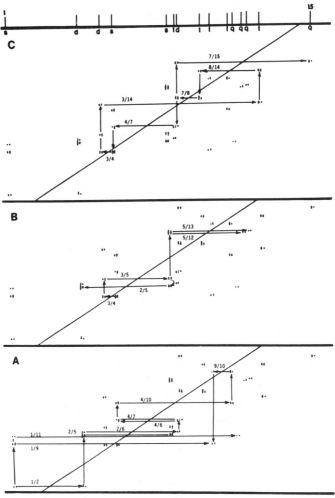

*Fig. 5-9.*     Connectivity networks identifying the cedrane carbon skeleton. (A) Beginning from the tertiary alcohol carbon resonating furthest downfield (1), the structural fragment shown by **5-2** is established as a starting point, with 2, 9 and 11 identified as nearest neighbors. Further connectivities shown identify 2, 5, 6, 4, 3, 7 and 10, affording the structural fragment represented by **5-3**. (B) Connectivities linking 3 and 5 and identifying the *gem*-dimethyl resonances, thereby affording the structural fragment represented by **5-4**. (C) Connectivities linking 3 and 7 via 14 and 8 and identifying the final methyl group (15), completing the structure shown by **5-5**.

which correspond to a quaternary and methylene carbon, respectively. Here again, a choice in direction must be made in terms of which branch to follow. Selecting the methylene carbon, C6, we note that it correlates with the C4 doublet downfield. Although it is quite difficult to see because of the very close double quantum frequencies for the two correlations, C4 correlates with C3, C7 and C10, which correspond to methine, methine and methylene carbons, respectively. Here again, given a branchpoint, we must follow one of the two connectivities. Practically, that from C10 is easier to follow as it correlates only with C9, thus completing a six membered ring as shown by **5-3**. The complete set of connectivities just described is shown in Fig. 5-9a.

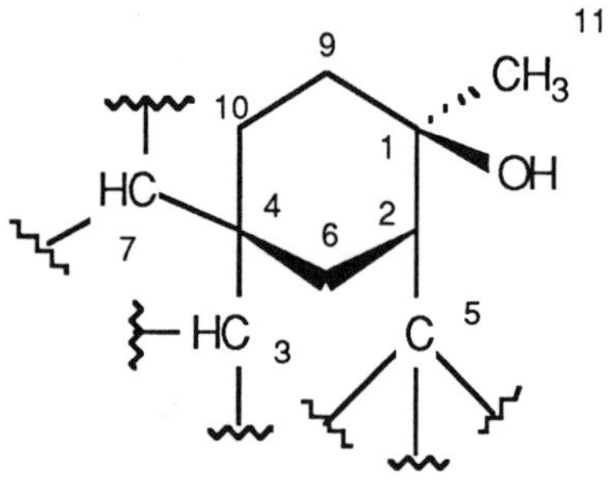

**5-3**

To continue the elucidation of the carbon skeleton, it is convenient to backtrack slightly to C4, referring to Fig. 5-9b. Using the C3 resonance as the next point of departure, we find quite quickly that C3 correlates with C5, thus completing another ring. The final connectivities from C5 correlate it with C12 and C13, constituting a *gem*-dimethyl pair. Thus, we have assembled the portion of the skeleton shown by **5-4.**

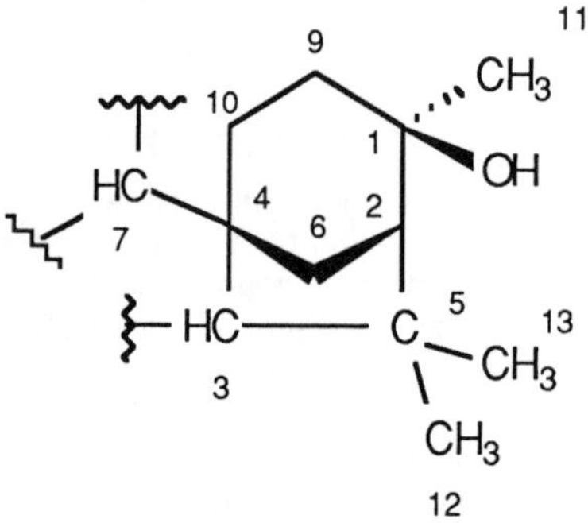

**5-4**

Continuing from the C3 methine carbon, the other connectivity carries us first to the C14 methylene and then on to the C8 methylene before finally establishing the identity of the third ring with the connectivity between C8 and C7. The only remaining connectivity to be discussed at this point is that between C7 and the Cl5 methyl. This final set of connectivities is illustrated by Fig. 5-9c.

**5-5**

In conclusion, without any prior knowledge, we have just deduced the cedrane carbon skeleton (**5-5**) using the ¹³C-¹³C INADEQUATE spectrum shown in Figs. 5-8 and 5-9. Cedrol, selected here as an example, fortuitously had no connectivity breaks due to heteroatoms and/or a ¹³C-¹³C AB pairing. This will assuredly not always be the case. Rather, breaks in the connectivity network can occur with reasonable frequency. The final compound discussed below, benzo[3,4]phenanthro[1,2-*b*]thiophene (**5-6**), illustrates some of the typically encountered problems. It is also useful to note that in such cases, recourse to other two-dimensional NMR experiments such as long range heteronuclear chemical shift correlation (see Chapter 3) or heteronuclear relayed coherence transfer (see Chapter 4) is available. When used in a concerted fashion, these experiments are generally capable of surmounting any difficulties or shortcomings of any of them singly.

### The INADEQUATE Spectrum of Benzo[3,4]phenanthro[1,2-*b*]thiophene

The assignment of the ¹³C-NMR spectrum of benzo[3,4]-phenanthro[1,2-*b*]thiophene (**5-6**) represents the sort of challenge more typical of that encountered. Specifically, the ¹³C-¹³C INADEQUATE spectrum shown in Fig. 5-10 contains breaks in the connectivity network due to heteroatoms and AB ¹³C-¹³C pairings and due to compromise relaxation considerations.

Despite the fact that the data was collected on a 500 MHz spectrometer at an observation frequency for ¹³C of 125.76 MHz, not all

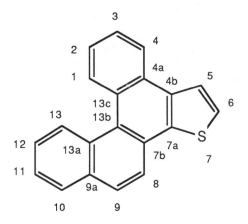

**5-6**

of the possible connectivities are observed.   In particular, the $T_1$ relaxation times for resonances in the interior of the benzo[c]phenanthrene nucleus (C13a-c) range upward of 10 sec.   The spectrum shown in Figs. 5-10 and 5-11 was recorded using an interpulse delay of 2.7 sec with 30 x 8K blocks of data, which were zero-filled to 64 x 4K. Clearly, when relaxation times are in the range of 10+ sec, a 2.7 sec recycle time will be very ineffective in providing adequate relaxation when the quaternary carbons in the helical turn are considered, especially when it is generally recommended that the interpulse delay be set as a function of 1.5 x $T_1$!  Hence, we would expect to see breaks in the connectivity network for this reason.  Second, there is one strongly coupled AB pair and one pair that would probably best be categorized as AM.  In the former case, there is no trace of a response correlating C1 and C13c, which have resonance frequencies of 128.96 and 128.84 ppm, respectively.  In the latter case, the response correlating C11 with C12 is identified only by the inner resonances of the doublet, the outer resonances being well below the noise threshold in the spectrum.  Here, C11 and C12 resonate at 125.43 and 126.25 ppm, respectively.  Finally, there will also be a break in the case of the thiophene ring caused by the intervening sulfur atom.

Overall, based on the limitations invoked by relaxation, chemical shifts and heteroatom location, the best that can be achieved from the INADEQUATE spectrum shown in Fig. 5-10 alone is the structural fragment represented by **5-7**.  While a substantial piece of information, this alone would be insufficient to unequivocally "pin down" the structure of the molecule.  However in the original report by Martin and co-workers[19] the INADEQUATE spectrum shown in Fig. 5-10 was used in conjunction with a long range optimized proton-carbon spectrum recorded using the Freeman-Morris sequence (see Fig. 3-10) that afforded a total assignment of $^{13}$C-NMR spectrum.

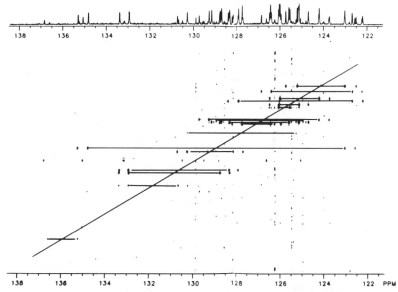

*Fig. 5-10.*    125.76 MHz $^{13}C$-$^{13}C$ INADEQUATE spectrum of benzo[3,4]phenanthro-[1,2-*b*]thiophene (**5-6**) recorded in deuterochloroform. The data was taken as 30 x 8K complex points and processed to afford a final matrix consisting of 64 x 4K points. Total acquisition time was 19 hours.

**5-7**

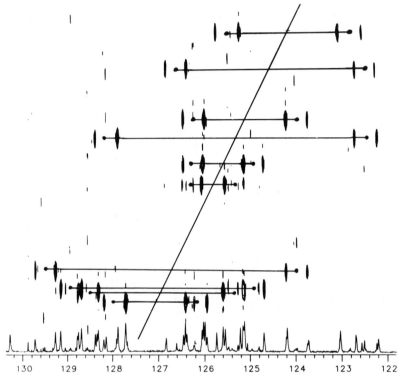

*Fig. 5-11.*    Expansion of the 125.76 MHz $^{13}$C-$^{13}$C INADEQUATE spectrum of benzo[3,4]phenanthro[1,2-*b*]thiophene **(5-6)** in the congested region from 120 to122 ppm.

## Parameter Selection for Two-Dimensional INADEQUATE

As has been discussed above, the double selection modification of Lallemand and coworkers[15] should be used with either the method of Mareci and Freeman[14] or the method of Turner.[12] Thus, it is necessary to calibrate pulse angles of 90°, 135° and 180°, see Chapter 1 for a discussion on pulse widths.  In addition, it is advisable to use a composite pulse of the form 90°(x)180°(y)90°(x) in place of the 180° pulse as discussed in Chapter 1. The $\tau$ delay should be set equal to  1/4 x $^nJ_{CC}$, where n ≥ 1 although,  due to sensitivity considerations, only one bond couplings are usually optimized for. Here, for survey conditions, values of $\tau$ ranging from 6.0 msec for aliphatic compounds to 5.0 msec for aryl compounds should be used in the absence of measured $^{13}$C-$^{13}$C coupling constants.  As presented above, due to the evolution of the two-dimensional INADEQUATE pulse sequence, the considerations dealing with quadrature detection have been fully discussed. As such, the experiment benefits from being set up with equal spectral widths in both $F_1$ and $F_2$. The $F_1$ dimension should be digitized with 1-8K depending on the density of responses in the spectrum; for the $F_2$ dimension, the number of blocks that have been used in the literature varies from 30 points to 512 points depending on, respectively, whether individual slices from the matrix are not needed or are necessary for structural considerations. Finally, consideration to the interpulse delay must be made. For nonprotonated carbons, long relaxation times can cause the loss of responses from the matrix. It is therefore necessary to allow sufficient time for relaxation to occur, usually 2-5 times the longest $T_1$; in some instances it may be desirable to add a relaxation reagent such as $Cr(acac)_3$ or $Gd(FOD)_3$. Either of these reagents may be used in low concentrations (0.01-0.05 molar) to effectively reduce the long $T_1$'s of nonprotonated carbons. Care must be taken when using one of these reagents since there is the chance for the chemical shifts to change due to the paramagnetic nature of these compounds.

## A Brief Survey of Applications of the INADEQUATE Experiment

One of the first practical uses of the two-dimensional INADEQUATE experiment was the solution of the structure of a photodimer reported by Freeman, Frenkiel and Rubin.[20] In this report the two-dimensional INADEQUATE experiment demonstrated its useful-ness as a means of obtaining structural information not available by other means. In addition, in this report a modification was employed that allowed one to obtain $^{13}C$-$^{13}C$ connectivities exclusively to non-protonated carbons. Essentially the modification removes the responses from carbons not coupled to a nonprotonated carbon through the combination of an additional delay of $1/2^1J_{CH}$ after the last pulse before acquisition and not using proton decoupling during acquisition. It was this modification that allowed the determination of the the structure of the photodimer of cholesta-4,6-diene-3-one.

Brown, Pai and Naik[21] used two-dimensional INADEQUATE to obtain positive structural proof of nonadienes produced by hydroboration. Their choice for using this experiment was influenced by the fact that they had large amounts of soluble materials and thus, sensitivity was not a problem. Berger and Zeller[22] utilized two-dimensional INADEQUATE to measure all the $^{13}C$-$^{13}C$ couplings in azulene. In their work they optimized the $\tau$ in the excitation scheme for couplings of 3, 5, 7 and 57 Hz, corresponding to couplings over 4, 3, 2 and 1 bonds, respectively. Thus, they generated two-dimensional matrices for these correlations and were able to unequivocally determine both direct and long range coupling constants between carbons. The assignment of vitamin $D_3$ was reported by Kruk, Jans and Lugtenberg[23] using the two-dimensional INADEQUATE pulse sequence. This experiment has also been used to assign several tricyclo[6.3.0.0$^{2,6}$]-undecane derivatives.[24] The $^{13}$C spectral assignments of a variety of alkaloids including harmane,[25] panamine[14] and omorsina[26] have also been completed using the INADEQUATE technique.

In a series of papers by Martin and coworkers, the two-dimensional INADEQUATE experiment has been utilized to assign the $^{13}$C NMR spectra of complex polynuclear heteroaromatics. The two-dimensional INADEQUATE experiment was first used in the assignment of phenanthro[1,2-b]thiophene[27] in combination with data from other two-dimensional NMR experiments to assign the proton and carbon spectra. Subsequently a number of applications in the assignment of the carbon spectra of more complex polynuclear aromatic systems have been reported.[19,28-30] In a related direction, Craik, Hall and Munro[31] have employed the INADEQUATE experiment in conjunction with heteronuclear relayed coherence transfer to unequivocally assign the aromatic carbon resonances of a series of imipramine (dibenzazepine) analogs.

Perhaps one of the most powerful demonstrations of the capability of $^{13}$C-$^{13}$C double quantum INADEQUATE spectroscopy is in the elucidation of the carbon skeletons of new natural products. There are a growing number of such applications contained in the literature. For example, Kikuchi et al.[32] have used the technique to deduce the carbon skeleton and obtain a complete set of $^{13}$C NMR assignments for the novel cyclopropyl containing steroid cyclonervilasterol. Dekker and co-workers[33] have utilized the INADEQUATE technique to determine the carbon skeleton of jaherin, a new daphnane diterpene. Shoolery[34] has used cedrol as a model compound to illustrate the power of the INADEQUATE technique in assembling natural products carbon skeletons. In an unrelated study, Waegell and coworkers[35] have also applied the INADEQUATE technique to a series of cedrane analogs. Faure and coworkers[36] have reported a total assignment of the carbon NMR spectrum of α-longipinene based on an INADEQUATE spectrum. Further examples are contained in the work of Kikuchi and co-workers,[37] who used INADEQUATE in the determination of the structure of new clerodane diterpenes, and Benn and Mynott,[38] who have used the technique in the structure determination and $^{13}$C spectral assignment of corrinoids.

A number of other applications of the INADEQUATE technique are contained in the primary literature and the list is continually growing. Clearly, it is beyond the scope of this brief chapter segment to exhaustively review this broad group of papers. Those cited above are intended, rather, to provide a very brief overview of some of the applications that have appeared.

## USING PROTON DETECTION TO
## INCREASE THE SENSITIVITY OF
## THE INADEQUATE TECHNIQUE

Although the methods discussed above, such as modified read pulses and double selection, can improve the sensitivity of the two-dimensional INADEQUATE experiment, the increase in sensitivity is modest to say the least. Several methods that rely on proton polarization to enhance the sensitivity of the detected carbon signal have been reported.[39,40] These methods, however, still fall short of a major breakthrough in increasing the sensitivity. One method that does increase the sensitivity of the experiment drastically is the proton detected method described recently by Keller and Vogele.[41]

Problems with the technique include incomplete suppression of the residual signal of protons not modulated by the carbon double quantum coherence and failure to detect the double quantum domain in quadrature. Although these problems are formidable, they can be overcome or ingnored, since the gain in sensitivity more than compensates for the other problems. One problem that must be overcome involves the major modifications to the spectrometer that must be completed beforehand. The modifications arise from the necessity to broadband decouple as well as observe on the proton channel. The pulse sequence of Keller and Vogele is shown in Fig. 5-12.

The pulse sequence shown in Fig. 5-12 is essentially an INADEQUATE followed by a reverse INEPT.[42] The sequence has been dubbed INSIPID, for INadequate Sensitivity Improvement by Proton Indirect Detection. The originators of this sequence suggest that a decoupling scheme such as WALTZ or MLEV be used.[43]

With the information gleaned from the discussions of the preceding sections, it is possible to consider several modifications to improve this sequence. First, saturation of the proton signals due to non-$^{13}$C coupled protons can be accomplished by utilizing a BIRD pulse (see Chapters 1 and 3) during the interpulse delay. The use of the BIRD during the interpulse delay was discussed in Chapter 3 in the section on proton detected heteronuclear chemical shift correlations. Second, quadrature detection of the double quantum frequency domain could presumably be obtained by increasing the flip angle of the final excitation pulse and the read pulse to 135°. This plausible modification comes from the work of Lallemand[15] and has been considered in this chapter.

## ASSIGNMENT OF QUATERNARY
## CARBONS USING CARBON RELAYED
## PROTON-CARBON SPECTROSCOPY

Kessler, Bermel and Griesinger[44] have devised a modification of the INEPT-INADEQUATE[39] experiment designed to aid in the

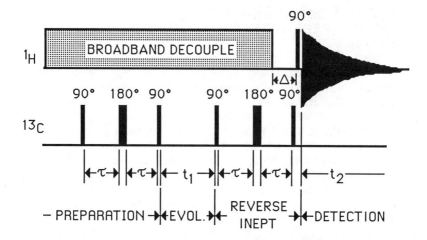

*Fig. 5-12.* Proton detected double quantum pulse sequence developed by Keller and Vogele.[41] Preparation begins with proton presaturation and sufficient time for carbon relaxation. The fixed delays, $\tau$, during the preparation and reverse INEPT periods are optimized as a function of $1/4J_{CC}$ in the usual sense. The interval $\Delta$ is optimized as a function of $1/2(^1J_{CH})$ to $1/4(^1J_{CH})$.

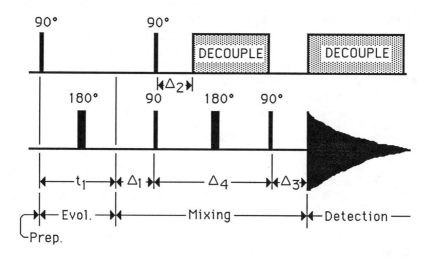

*Fig. 5-13.* Carbon relayed proton-carbon correlation experiment proposed by Kessler and coworkers[44] as an means of assigning quaternary carbon resonances.

assignment of quaternary carbon resonances. The pulse sequence for their experiment, which utilizes first transfer of magnetization from proton to carbon and then transfer to an adjacent carbon via carbon-carbon polarization transfer, is shown in Fig. 5-13.

Basically, the first segment of the experiment up through the end of the $\Delta_1$ interval operates like a conventional heteronuclear chemical shift correlation pulse sequence (see Chapter 3). Magnetization terms evolve and are then transferred to carbon by the pair of 90° pulses at the end of $\Delta_1$. During the subsequent $\Delta_2$ interval, which is contained within $\Delta_4$, HC terms are refocused and then locked when the decoupler is gated on. During the $\Delta_4$ interval, $^1J_{CC}$ coupling evolves, and chemical shift terms are refocused by the 180° $^{13}$C pulse midway through the interval. The final 90° $^{13}$C pulse provides the necessary carbon-carbon polarization transfer. To suppress strong signals from the protonated carbons, an additional delay, $\Delta_3$, is inserted prior to acquisition, which is optimized as a function of $1/2(^1J_{CH})$ to allow the protonated carbon terms to become antiphase, at which point they cancel when the decoupler is switched on. During the $\Delta_3$ interval there is also partial refocusing of the $^{13}$C-$^{13}$C coupling, which makes phasing of the spectrum an impossibility that must be circumvented by using sinusoidal multiplication and an absolute value calculation prior to presentation.

With regard to sensitivity, Kessler and coworkers[44] claim it to be better than the INEPT-INADEQUATE experiment of Sørensen et al.[39] Some potential applications have been demonstrated in Kessler's work but it remains to be seen if there will be others. It should be noted, however, that Lee and Morris[45] have proposed an extension of the sequence shown in Fig. 5-13, so there is assuredly some interest in this technique.

## REFERENCES

1.    A. Bax, R. Freeman and S. P. Kempsell, *J. Am. Chem. Soc.*, **102**, 4849 (1980).

2.    W. P. Aue, E. Bartholdi and R. R. Ernst, *J. Chem. Phys.*, **64**, 2229 (1976).

3.    A. Wokaun and R. R. Ernst, *Chem. Phys. Lett.*, **52**, 407 (1977).

4.    A. Wokaun and R. R. Ernst, *Mol. Phys.*, **36**, 317 (1978).

5.    G. Bodenhausen, R. L. Vold and R. R. Vold, *J. Magn. Reson.*, **37**, 93 (1980)

6.    S. Vega and A. Pines, *J. Chem. Phys.*, **66**, 5624 (1977).

7.    L. Müeller, *J. Am. Chem. Soc.*, **101**, 4481 (1979).

8.    G. D. Mateescu and A. Valeriu, in *Magnetic Resonance: Introduction, Advanced Topics and Applications to Fossil Energy*, L. Petrakis and J.P. Fraissard, Eds., D. Reidel, Boston, 1984.

9.    A. Bax, R. Freeman, T. A. Frenkiel and M. H. Levitt, *J. Magn. Reson.*, **43**, 478 (1980).

10.   A. Bax, R. Freeman, and T. A. Frenkiel, *J. Am. Chem. Soc.*, **102**, 2102 (1980).

11.   D. L. Turner, *Mol. Phys.*, **44**, 1051 (1981).

12.   D. L. Turner, *J. Magn. Reson.*, **49**, 175 (1982).

13.   A. Bax and T. H. Mareci, *J. Magn. Reson.*, **53**, 360 (1983).

14.   T. H. Mareci and R. Freeman, *J. Magn. Reson.*, **48**, 158 (1982).

15.     D. Piveteau, M. A. Delsuc, E. Guittet and J. Y. Lallemand, *Magn. Reson. Chem.* , **23**, 127 (1985).
16.     D. L. Turner, *J. Magn. Reson.*, **53**, 259 (1983).
17.     M. H. Levitt and R. R. Ernst, *Mol. Phys.*, **50**, 1109 (1983).
18.     J. L. Marshall, *Carbon-Carbon and Carbon-Proton NMR Couplings: Applications to Stereochemistry and Conformational Analysis*, Methods in Stereochemical Analysis, Vol. 2, VCH, New York, 1983.
19.     J. G. Stuart, M. J. Quast, G. E. Martin, V. M. Lynch, S. H. Simonsen, R. N. Castle, J. L. Dallas, B. K. John and L. F. Johnson, *J. Heterocyclic Chem.*, **23**, 1215 (1986).
20.     R. Freeman, T.A. Frenkiel and M.R. Rubin, *J. Am. Chem. Soc.*, **104**, 5545 (1982).
21.     H. C. Brown, G. G. Pai and R. G. Naik, *J. Org. Chem.*, **49**, 1072 (1984).
22.     S. Berger and K.-P. Zeller, *J. Org. Chem.*, **49**, 3725 (1984).
23.     C. Kruk, A. W. H. Jans and J. Lugtenberg, *Magn. Reson. Chem.*, **23**, 267 (1985).
24.     J. L. Jurlina, A. J. Ragauskas and J. B. Stothers, *Magn. Reson. Chem.*, **23**, 689 (1985).
25.     D. H. Welti, *Magn. Reson. Chem.*, **23**, 872 (1985).
26.     N. S. Bhacca, M. F. Balandrin, A. D. Kinghorn, T. A. Frenkiel, R. Freeman and G. A. Morris, *J. Am. Chem. Soc.*, **105**, 2538 (1983).
27.     G. E. Martin, S. L. Smith W. J. Layton, M. R. Wilcott,III, M. Iwao, M. L. Lee, and R. N. Castle, *J. Heterocyclic Chem.*, **20**, 1367(1983).
28.     M. J. Musmar, M. R. Wilcott,III, G. E. Martin, R. T. Gampe, Jr., M. Iwao, M. L. Lee, R. E. Hurd, L. F. Johnson and R. N. Castle, *J. Heterocyclic Chem.*, **20**, 1661(1983).
29.     M. J. Musmar, G. E. Martin, M. L. Tedjamulia, H. Kudo, R. N. Castle and M. L. Lee, *J. Heterocyclic Chem.*, **21**, 929 (1983).
30.     M. J. Musmar, A. S. Zektzer, G. E. Martin, R .T. Gampe, Jr., M .L. Tedjamulia, R. N. Castle and R. Hurd, *Magn. Reson. Chem.*, **24**, 1039 (1987).
31.     D. J. Craik, J. G. Hall and S. L. A. Munro, *Chem. Pharm. Bull.*, **35**, 188 (1987).
32.     T. Kikuchi, S. Kadota, S. Matsuda and H. Suchara, *Tetrahedron Lett.*, 2505 (1984).
33.     T. G. Dekker, T. G. Fourie, E. Matthee, F. O. Snyckers and W. Ammann, *S. Afr. J. Chem.*, **40**, 74 (1987).
34.     J. N. Shoolery, *J. Nat. Prod.*, **47**, 226 (1984).
35.     P. Brun, J. Casanova, M. S. Raju, B. Waegell, E. Wenkert, J. P. Zahra, *Magn. Reson. Chem.*, **25**, 619 (1987).
36.     R. Faure, E. J. Vincent, E. M. Gaydou and O. Rakotonirainy, *Magn. Reson. Chem.*, **24**, 883 (1986).
37.     T. Kikuchi, K. Tsubono, S. Kadota, H. Kizu, Y. Imoto and T. Tomimori, *Chem. Lett.*, 987 (1987).
38.     R. Benn and R. Mynott, *Angew. Chem., Intl Ed., Engl.,* **24**, 333 (1985).
39.     O.W. Sørensen, R. Freeman, T. Frenkiel, T .H. Mareci and R. Schuck, *J. Magn. Reson.*, **46**, 180 (1982).
40.     S. W. Sparks and P. D. Ellis, *J. Magn. Reson.*, **62**, 1 (1985).
41.     P. J. Keller and K. E. Vogele, *J. Magn. Reson.*, **68**, 389 (1986).
42.     R. Freeman, T. H. Mareci and G. A. Morris, *J. Magn. Reson.*, **42**, 341(1981).
43.     A. J. Shaka, J. Keeler and R. Freeman, *J. Magn. Reson.*, **53**, 313 (1983).
44.     H. Kessler, W. Bermel and C. Griesinger, *J. Magn. Reson.*, **62**, 573 (1985).
45.     K. S. Lee and G. Morris, *J. Magn. Reson.*, **70**, 332 (1986).

# CHAPTER 6

## APPLICATIONS PROBLEMS

### INTRODUCTION

Mastery of any technique or procedure comes with practice. The same is quite true of two-dimensional NMR spectroscopy. Although there are numerous examples scattered through this book, competence comes only after working through problems for oneself. Thus, this chapter contains a number of problems. Initially, the problems are based on known structures and on spectral assignment rather than structure elucidation. Further, the early problems are also based on the use of single or perhaps two experiments while later problems require the concerted application of frequently numerous experiments to deduce complex structures.

The approach to the problems is essentially topical following the content of Chapters 2-5 initially. Later problems, as difficulty increases, begin to combine techniques leading finally to unknown molecular structures. The problems chosen for inclusion in this chapter by no means represent ideal data in each case. Early problems are generally typified by fairly clear spectra. Later problems, by intent, incorporate spectra that are typical of what is often encountered in the "real world."

### HOMONUCLEAR CORRELATION PROBLEMS

#### 6.1. Assignment of the Proton NMR Resonances of a Simple Dinaphthothiophene Derivative

Isomeric dinaphthothiophenes have been of interest to a number of investigators as potential constituents of solvent refined coal liquids. One such compound is dinaphtho[1,2-b;2',3'-d]thiophene (6-1) which was originally prepared by Castle and coworkers.[1] Subsequent preparation of **6-1** by Klemm and coworkers by an alternative route gave proton NMR chemical shifts that differed somewhat from those originally reported.[2] The total assignment of the proton NMR spectrum of **6-1** was finally reported by Johnston, Martin and Castle.[3] Hence for this first problem we have the task of unequivocally assigning the spectrum of a molecule of well established structure.

The COSY spectrum of **6-1** recorded at 300 MHz is shown in Fig. 6-1. The half of the COSY data matrix below the diagonal was processed using sinusoidal multiplication prior to both Fourier transformations. The data shown above the diagonal were processed using a 2.5 Hz Gaussian multiplication prior to both Fourier transformations. Some ancillary information that may be useful in making the total assignment was the observation of a 12.7 % nOe between the proton furthest downfield at 8.68 ppm and the doublet centered at 8.31 ppm. No additional information should be necessary to assemble an unequivocal set of proton resonance assignments for **6-1**.

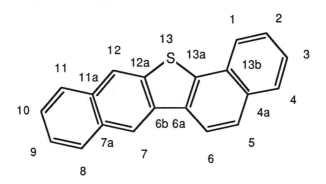

**6-1**

## 6.2.  Assignment of the Proton NMR Spectrum
of the Sesquiterpene Annulatol-A

The marine kingdom provides a diverse assortment of natural product structures. Recently, Cardellina and coworkers[4] reported the isolation of a group of brominated sesquiterpene alcohols from the calcareous green alga *Neomeris annulata*. One of these, annulatol-A (**6-2**), provides an interesting assignment challenge.

All of the resonances in the 300 MHz spectrum of **6-2** resonate between 3.6 and 0.4 ppm. The COSY spectrum is shown in Fig. 6-2 and was acquired as 300 x 1K complex points. The data were processed using sinusoidal multiplication with zero filling to 512 x 512 points prior to symmetrization. The zero quantum (ZQCOSY) spectrum of **6-2** acquired using a 5 Hz optimization is shown in Fig. 6-3. The data were acquired as 380 x 1K complex points with the evolution time

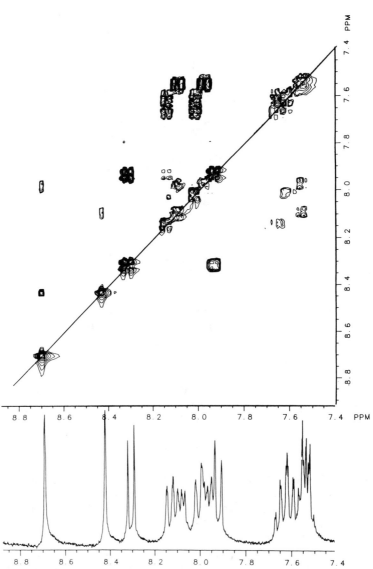

*Fig. 6-1.*     COSY spectrum of **6-1** recorded in deuterochloroform at 300.068
MHz.    Data presented above the diagonal were processed using
sinusoidal multiplication prior to both Fourier transformations
followed by symmetrization while below the diagonal data were
subjected to a 2.5 Hz Gaussian multiplication prior to both Fourier
transformations and symmetrization.

incremented as a function of the dwell time giving equal frequency ranges in both $F_1$ and $F_2$.

Develop arguments using either the COSY or ZQCOSY quantum data to assign as many resonances as possible in the proton NMR spectrum of **6-2**.

**6-2**

## SIMULTANEOUS USE OF HOMO- AND HETERONUCLEAR CORRELATION

### 6.3   Assignment of the Proton NMR Spectrum of the Alkaloid α-Yohimbine with Simultaneous Verification of $^{13}C$-NMR Assignments

The indole alkaloid α-yohimbine (**6-3**) presents a somewhat more complex assignment problem than those preceding it. The $^{13}C$-NMR spectral assignments for **6-3** have been reported by Wenkert and coworkers[5] and are shown on the structure below.

The COSY spectrum is shown in Fig. 6-4 as both sinusoidally and Gaussian processed data. The proton-carbon heteronuclear chemical shift correlation spectrum (HC-COSY) is presented as a contour plot in Fig. 6-5 and as a stack plot of the aliphatic carbon chemical shift slices in Fig. 6-6. The objective of this problem is to illustrate the usage of the carbon NMR chemical shift assignments of yohimbine (**6-3**) proposed by Wenkert and coworkers[5] as a means of correlating proton and carbon chemical shift information. Subsequently, using the homonuclear connectivity information contained in the COSY spectrum presented in Fig. 6-4, verify the aliphatic carbon resonance assignments made by Wenkert. It should be possible, with a little diligence, to link all of the aliphatic carbons and protons into a single heteronuclear connectivity network. In the process, you will obviously generate a

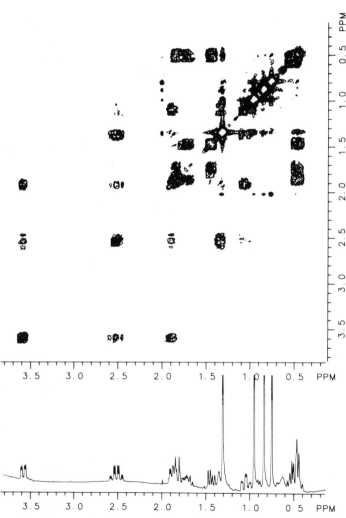

*Fig. 6-2.*     COSY spectrum of annulatol-A (**6-2**) recorded in deuterochloroform at 300.068 MHz. The data were acquired as 300 x 1K complex points and were processed using a sinusoidal multiplication prior to both Fourier transformations with zero filling to 512 points prior to the second.

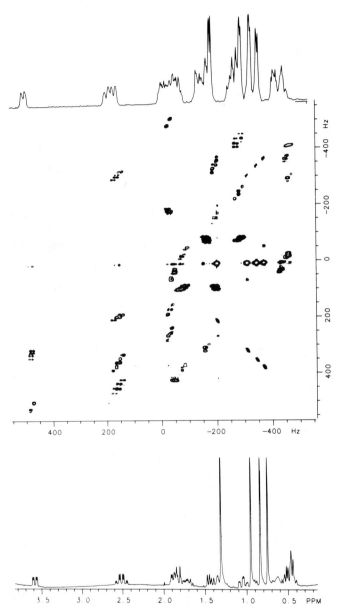

*Fig. 6-3.* ZQCOSY spectrum of annulatol-A (**6-2**) acquired using the Müller pulse sequence shown in Fig. 2-40. The fixed delay during the excitation interval, τ , was optimized for 5 Hz. The data were acquired as 384 x 1K complex points and were processed using sinusoidal multiplication prior to both Fourier transformations with zero filling to 512 points prior to the second.

complete set of proton resonance assignments for this reasonably complex alkaloid.

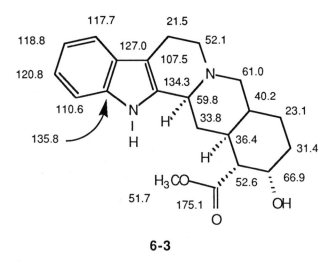

**6-3**

## 6.4   Identifying Heteronuclear Spin Systems of a Polynuclear Aromatic Using a Combination of Heteronuclear Chemical Shift Correlation and Heteronuclear Relayed Coherence Transfer

Polynuclear aromatic compounds can represent a substantial assignment challenge because of the numbers of resonances that can appear within restricted chemical shift ranges. Fortunately, however, the proton-proton coupling networks in such molecules are usually fairly straightforward. In that regard, this and the following problem are complementary. First, in this problem, we shall consider the means available for identifying component resonances of heteronuclear spin systems. In the following problem, we will consider a total assignment of a complex polynuclear aromatic.

Benzo[b]phenanthro[4,3-d]thiophene (**6-4**) is a helical poly-nuclear aromatic compound that provides a good introductory challenge in the identification of heteronuclear spin systems.[6] The proton-carbon heteronuclear correlation spectrum is shown in Fig. 6-7 while the heteronuclear relayed coherence transfer spectrum is shown in Fig. 6-8. Using the former to identify direct responses, thereby providing a starting point in the latter, it should be readily possible to identify all of the component proton and carbon resonances of the four heteronuclear spin systems contained in the structure of **6-4**. The proton spin systems in this case can be classified as follows: there are two ABMX spin systems, one AM and one AB spin system.

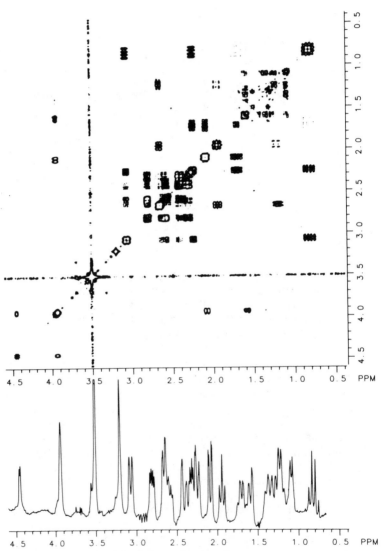

*Fig. 6-4.*          COSY spectrum of α-yohimbine (**6-3**) recorded in deuterochloroform
at an observation frequency of 300.068 MHz. The transmitter was
located in the middle of the aliphatic region to optimize digitization.
The data were collected as 384 x 1K complex points and were
processed using a sinusoidal multiplication prior to both Fourier
transformations with zero filling to 512 points prior to the second to
give the spectrum shown.

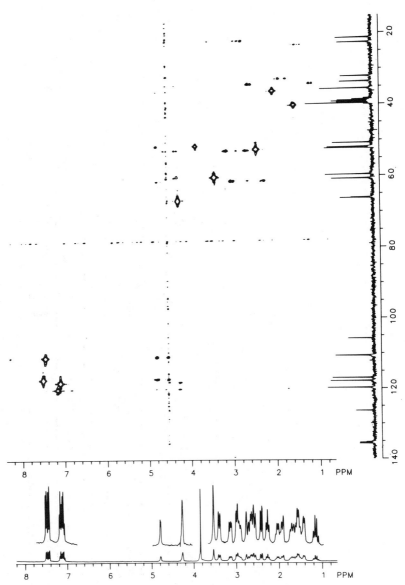

*Fig. 6-5.*    HC-COSY spectrum of α-yohimbine (**6-3**) recorded in deuterochloroform at observation frequencies of 300.068 and 75.459 MHz for [1]H
and [13]C, respectively. The data were recorded as 256 x 1K complex
points and were processed using an exponential multiplication prior to
both Fourier transformations.

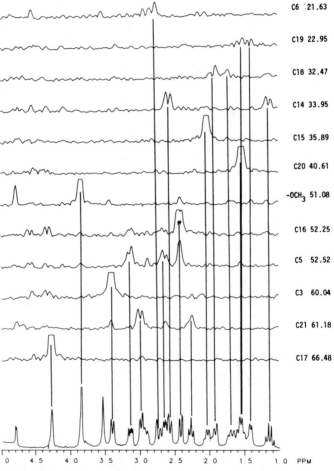

*Fig. 6-6.*     Stack plotted slice summary of aliphatic carbon responses from the HC-COSY spectrum of α-yohimbine (**6-3**). Chemical shifts are specified to the left of each individual trace.

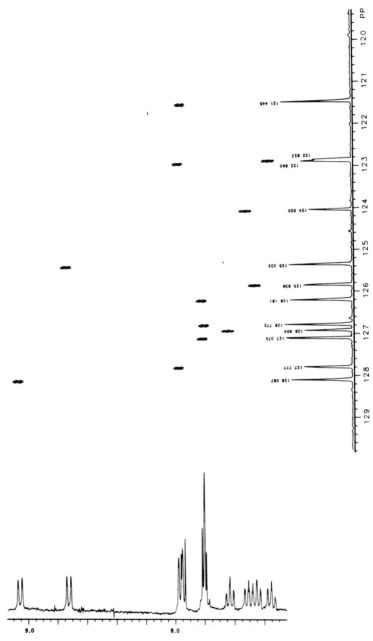

*Fig. 6-7*   HC-COSY spectrum of benzo[*b*]phenanthro[4,3-*d*]thiophene (**6-4**) recorded in deuterochloroform at observation frequencies of 300.068 and 75.459 MHz for ¹H and ¹³C, respectively. The data were recorded as 256 x 1K complex points and were processed using an exponential multiplication prior to both Fourier transformations.

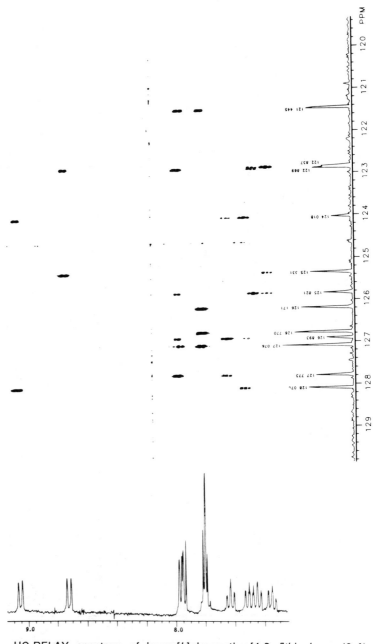

*Fig. 6-8.*     HC-RELAY spectrum of benzo[*b*]phenanthro[4,3-*d*]thiophene (**6-4**) recorded in deuterochloroform. The data were acquired and processed in a fashion identical to that described with Fig. 6-7. Magnetization was relayed assuming a 7 Hz vicinal proton-proton coupling.

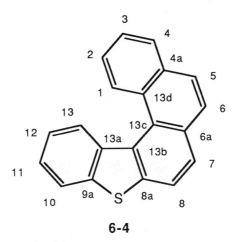

**6-4**

## 6.5 Assigning the Proton and Carbon-NMR Spectra of the Complex Helicene Phenanthro[3,4:3',4']phenanthro-[2,1-*b*]thiophene Using Long Range Heteronuclear Chemical Shift Correlation

Using what we have learned about subgrouping resonances of heteronuclear spin systems from the previous problem, let us now consider a more complex problem, the total assignment of the proton and carbon NMR spectra of the complex polynuclear heteroaromatic, phenanthro[3,4:3',4']-phenanthro[2,1-*b*]thiophene (**6-5**).

**6-5**

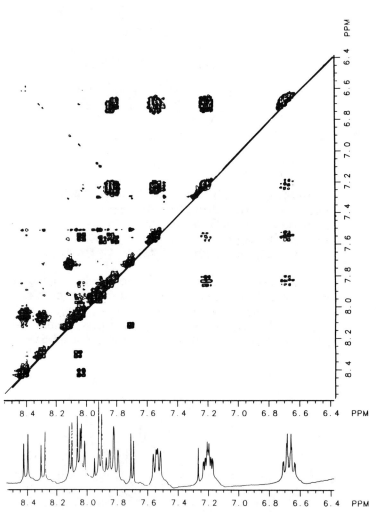

*Fig. 6-9.*     Composite COSY/HOHAHA spectrum of phenanthro[3,4:3',4']-phenanthro[2,1-*b*]thiophene (**6-5**) recorded in deuterochloroform at 300.068 MHz. Both spectra were recorded as 256 x 512 points and were processed using a sinusoidal multiplication and zero filling prior to both Fourier transformations to give a final matrix comprised of 512 x 512 points. The HOHAHA portion of the spectrum utilized an MLEV-17 sequence, which was applied for 10 msec flanked by 2.5 msec trim pulses. Acquisition times were 3 and 12 hours, respectively.

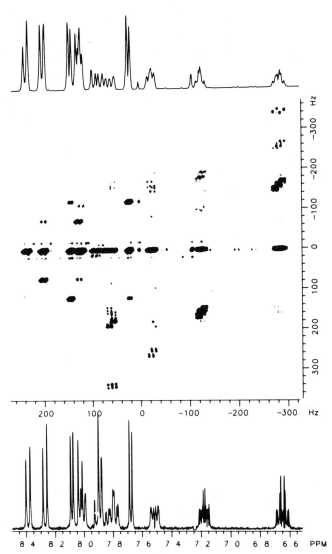

*Fig. 6-10.* ZQCOSY spectrum of phenanthro[3,4:3',4']phenanthro[2,1-*b*]thiophene
(**6-5**) recorded in deuterochloroform at 300.068 MHz. The pulse
sequence of Müller was employed (see Fig. 2-40) with the fixed delay,
τ, optimized for a 7 Hz vicinal coupling. The data were taken as 384 x
1K complex points and were processed using a sinusoidal multiplication
prior to both Fourier transformations. Shift axes labeled in ppm are
relative to TMS, those labeled in hertz are relative to the transmitter.

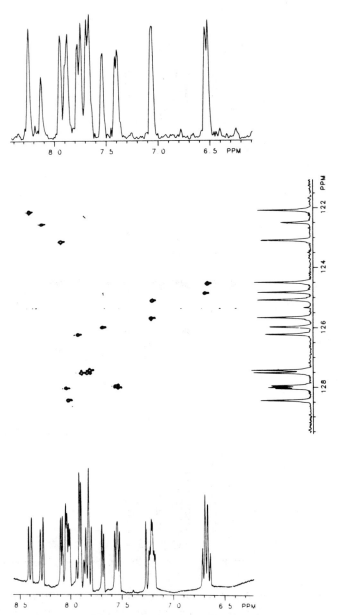

Fig. 6-11.    HC-COSY spectrum of phenanthro[3,4:3',4']phenanthro[2,1-*b*]-
thiophene (**6-5**) recorded in deuterochloroform as 256 x 1K complex
points.    The data were processed using exponential multiplications
prior to both Fourier transformations.    The contour plot is flanked by
high resolution proton and carbon spectra.

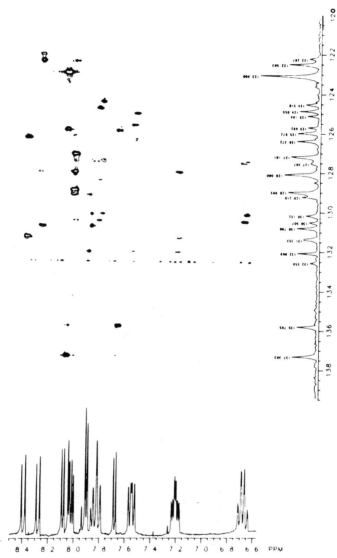

*Fig. 6-12.*    Long range optimized HC-COSY spectrum of phenanthro[3,4:3',4']-
phenanthro[2,1-*b*]thiophene (**6-5**) recorded in deuterochloroform as
256 x 1K complex points.  The pulse sequence employed was that
shown in Fig. 3-10 with no provision made for one bond modulation
decoupling.  The data were processed using exponential multiplications
prior to both Fourier transformations.  The contour plot is flanked by
high resolution proton and carbon spectra.

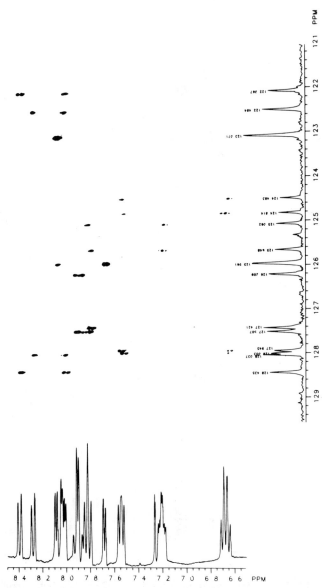

*Fig. 6-13.* HC-RELAY spectrum of phenanthro[3,4:3',4']phenanthro[2,1-*b*]-thiophene (**6-5**) recorded in deuterochloroform as 256 x 1K complex points. The data were recorded using the pulse sequence shown in Fig. 4-20 with the magnetization relay optimized for a 7 Hz vicinal coupling. The data were processed using exponential multiplications prior to both Fourier transformations. The contour plot is flanked by high resolution proton and carbon spectra.

The COSY/HOHAHA composite spectrum of **6-5,** taken from the work of Johnston and Martin,[7] is presented in Fig. 6-9. As an alternative, the proton zero quantum spectrum of the molecule is presented in Fig. 6-10.[8] The direct proton-carbon heteronuclear chemical shift correlation spectrum of **6-5** is presented in Fig 6-11; the long range optimized spectrum in Fig. 6-12 and the heteronuclear relayed coherence transfer spectrum in Fig. 6-13.[9]

## MOLECULAR STRUCTURE DETERMINATION

Having gone through five problems that have entailed the assignment and/or verification of spectral assignments for a number of molecules, a degree of familiarity with the techniques should be felt. At this point it is appropriate to make the transition from assigning the spectra of compounds of known structure to one of deducing the structure while making the assignments. This area represents one of the most fundamental applications of two-dimensional NMR spectroscopy. It is also the area in which the techniques discussed in the previous five chapters have had the greatest impact. It is now quite possible to undertake assignment problems that would have been impossible only a few short years ago. It is also possible to avoid complex chemical degradation schemes that irretrievably destroy precious samples. Hence, the problems that follow will deal with structure elucidation and will make use of virtually all of the techniques that we have presented in this monograph in one way or antoher.

### 6.6    Elucidation of the Structure
### of a Marine Sesquiterpene

A sample of a chloroform soluble sesquiterpene of marine origin was received and subjected to detailed analysis by two-dimensional NMR techniques. The compound gave an elemental analysis that corresponds to the molecular formula $C_{15}H_{25}OBr$. The only proton two-dimensional NMR experiment performed was a ZQCOSY experiment whose spectrum is shown in Figs. 6-14 and 6-15 (the latter plotted at a much deeper contour level). The compound exhibited an infrared absorption at 3620 cm[-1] indicating a non-hydrogen-bonded hydroxyl group, which, in conjunction with the proton resonance at 2.95 ppm (1H, d) and a carbon resonance at 69.8 ppm, could be construed as a secondary alcohol. This contention is supported by the loss of the resonances at 2.95 and 69.8 ppm in the proton and carbon spectra, respectively, and the appearance of a carbonyl resonance in the carbon spectrum resonating at 211.0 ppm when the material was subjected to Jones oxidation.

Other data obtained on the compound included a direct heteronuclear chemical shift correlation spectrum, which is presented in Fig. 6-16, and a proton (reverse) detected long range heteronuclear multiple quantum coherence spectrum shown in Figs. 6-17 and 6.18.

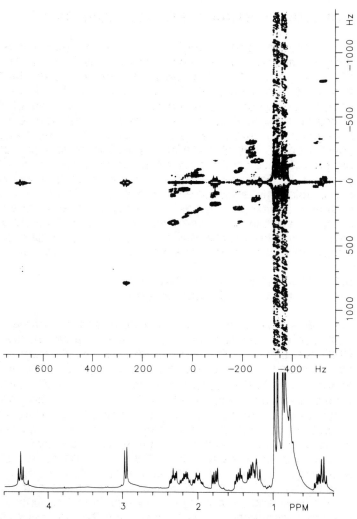

*Fig. 6-14.* ZQCOSY spectrum of a marine sesquiterpene recorded in deuterochloroform at an observation frequency of 300.068 MHz. The data were recorded using the Müller pulse sequence shown in Fig. 2-40 with the fixed delay, τ , optimized for 5 Hz. The data were acquried as 384 x 1K complex points and were processed using a sinusoidal multiplication prior to both Fourier transformations.

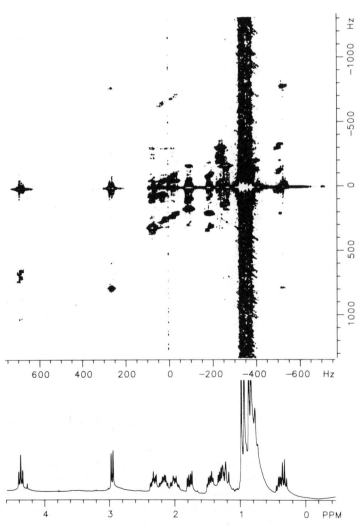

*Fig. 6-15.*    ZQCOSY spectrum taken from the same data set as Fig. 6-14 but plotted at a deeper contour level.

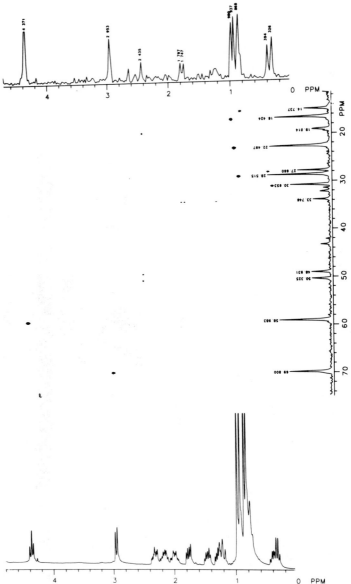

*Fig. 6-16.*    HC-COSY spectrum of a marine sesequiterpene recorded using the pulse sequence shown in Fig. 3-10. The durations of the delays, $\Delta_1$ and $\Delta_2$ were optimized as $1/2(^1J_{CH})$ and $1/3(^1J_{CH})$, where an assumed average one bond coupling of 135 Hz was utilized. The data were recorded as 200 x 1K complex points and were processed using exponential multiplication prior to both Fourier transformations and zero filling to 256 points prior to the second. The contour plot shown is flanked by high resolution proton and carbon reference spectra.

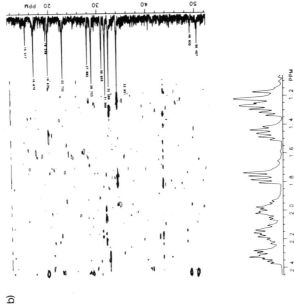

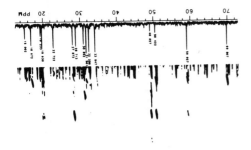

b)

a)

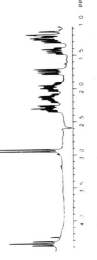

*Fig. 6-17* (a) Long range heteronuclear chemical shift correlation spectrum recorded using the proton detected long range heteronuclear multipler quantum coherence technique (HMBC, see Fig. 3-61). The low pass J-filter was optimized for 135 Hz; the long range transfer delay Δ, was optimized for 8.3 Hz. The data were recorded using 384 x 1K complex points and were processed as described in Chapter 3. High resolution proton and carbon spectra are plotted flanking the contour plot, which shows responses downfield of the methyl proton region. (b) Segment of the spectrum shown in (a) plotted at a much deeper contour level to show responses for broad multiplets.

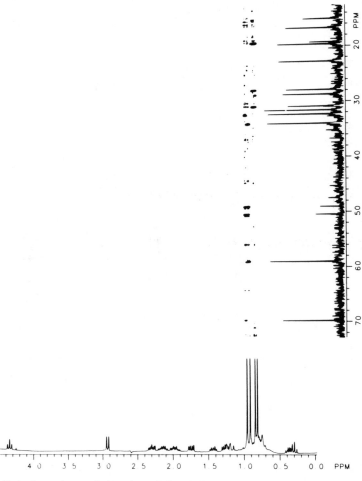

*Fig. 6-18.*        Data from the methyl region of the HMBC spectrum presented in Fig.
6-17.

The latter shows only responses arising from the methyl groups while the former shows the balance of the spectrum. The objective of this problem is to assemble the structure of this novel sesquiterpene while simultaneously assigning the proton and carbon NMR spectra.

## 6.7 ELUCIDATION OF THE STRUCTURE OF A COMPLEX ALKALOID USING INADEQUATE

Thus far we have considered the use of homonuclear proton-proton and heteronuclear proton-carbon connectivity information for the elucidation of structures in the preceding problems. The power of these techniques in inarguable. However, when sufficient samples of material are available, the carbon-carbon double quantum INADEQUATE experiment provides experimental data which is equally if not more powerful. Indeed, the INADEQUATE spectrum imposes no requirements for resolution in the proton spectrum, that are at least marginally necessary if the long range proton-carbon heteronuclear chemical shift correlation techniques are to be successfully employed.

To acquaint the student with the utilization of INADEQUATE data, the spectrum of an alkaloid with a molecular formula $C_{20}H_{33}N$ is shown in Fig. 6-19. Resonances are labeled "a" through "t" above the spectrum. The data were acquired at 50.3 MHz. Chemical shifts, $^{13}C$-resonance multiplicities and the directly attached proton shifts from a heteronuclear shift correlation spectrum are contained in Table 6-1.

Table 6-1. $^{13}C$-NMR chemical shifts, resonance multiplicities and the dhemical shift of the directly attached proton for an alkaloid with a molecular formula $C_{20}H_{33}N_3$.

| | | | |
|---|---|---|---|
| a | 68.66 | d | 2.63 |
| b | 68.00 | d | 3.61 |
| c | 64.98 | d | 1.82 |
| d | 61.58 | t | 2.53, 2.82 |
| e | 56.78 | d | 2.77 |
| f | 55.41 | t | 2.12, 2.87 |
| g | 50.96 | t | 2.36, 2.68 |
| h | 38.68 | t | 1.14, 1.45 |
| i | 37.57 | s | --- |
| j | 35.30 | d | 1.65 |
| k | 33.85 | t | 1.33, 1.54 |
| l | 32.42 | t | 0.98, 1.65 |
| m | 32.25 | d | 1.81 |
| n | 30.56 | t | 1.07, 1.79 |
| o | 25.38 | t | 1.64 |
| p | 25.27 | t | 1.60 |
| q | 24.94 | t | 1.31, 1.72 |
| r | 24.40 | t | 1.90 |
| s | 23.80 | t | 1.69, 2.17 |
| t | 20.74 | t | 1.65, 1.92 |

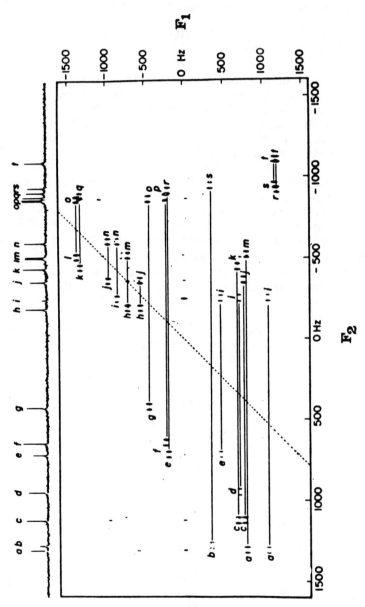

*Fig. 6-19.*        $^{13}C$-$^{13}C$ double quantum INADEQUATE spectrum of an alkaloid recorded at 50.3 MHz. (Used with permission, R. Freeman and T. H. Mareci, *J. Magn. Reson.*, **48**, 158 (1982).)

Assemble as much of the carbon skeleton as possible using the connectivity information contained in the INADEQUATE specturm shown in Fig. 6-19. Students should also be careful to recall that a break in the carbon-carbon connectivity network can occur when a pair of carbon resonances takes on AB rather than AX character. Pairings of this type in this problem would include the following pairs of resonances, if they happened to be adjacent to one another: l and m; p and q; and q and r.

## 6.8    Establishing the Structure of a Complex Alkaloid Using Proton-Proton, Proton-Carbon and Long Range Proton-Carbon Connectivity Data

An alkaloid was isolated from a plant source and gave an elemental analysis consistent with the molecular formula $C_{20}H_{22}N_2O_2$. The infrared spectrum was not particularly remarkable but gave absorptions consistent with an amide carbonyl and an N-H stretch. The COSY spectrum of the compound is shown in Fig. 6-20; the proton-carbon heteronuclear shift correlation spectrum was acquired using the method of Bax[10] with a BIRD pulse located midway through the evolution time (see Fig. 3-25) and is shown in Fig. 6-21; the long range heteronuclear correlation spectrum was acquired using the method described by Zektzer, John and Martin[11] which employed a BIRD pulse midway through the $\Delta_2$ refocusing interval (see Fig. 3-51). The long range correlation data are presented as the complete spectrum in Fig. 6-22 and as an expansion of the aliphatic region of the carbon spectrum in Fig. 6-23.

## 6.9    Determination of the Structure of a Dinaphthothiophene Isomer of the Compound Considered in Problem 6.1

Investigation of solvent refined coal liquids resulted in the isolation of a number of dinaphthothiophene analogs whose structures had to be determined. One such compound was an isomer of the compound whose proton NMR spectrum was assigned in Problem 6.1. Data available include: the COSY spectrum shown in Fig. 6-24, in which data obtained following sinusoidal multiplication prior to both Fourier transformations are shown above the diagonal and 2.5 Hz Gaussian multiplied data are shown beneath the diagonal; and the 7 Hz optimized zero quantum spectrum shown in Fig. 6-25 which was recorded using the Müller pulse sequence shown in Fig. 2-40. Heteronuclear chemical shift correlation data include: a proton-carbon heteronuclear correlation spectrum with semiselective broadband homonuclear decoupling in

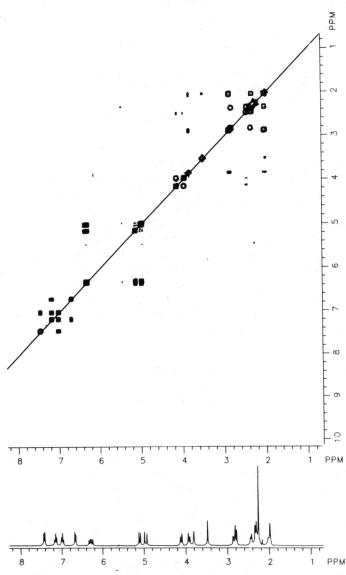

*Fig. 6-20.*    COSY spectrum of an alkaloid recorded in deuterochloroform at 300.068 MHz. The data were acquired as 256 x 1K complex points and were processed using sinusoidal multiplication prior to both Fourier transformations with zero filling to 512 points prior to the second to give a final matrix consisting of 512 x 512 points, which was symmetrized prior to plotting.

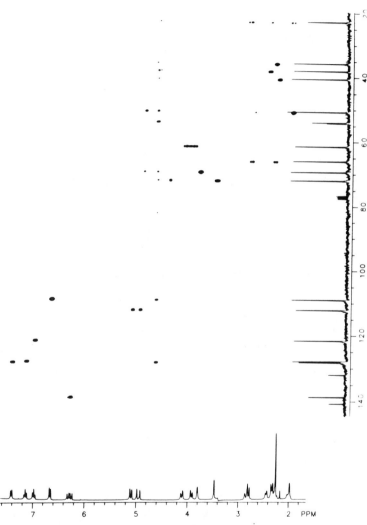

*Fig. 6-21*    HC-COSY spectrum of an alkaloid recorded using semiselective
decoupling as provided by the pulse sequence shown in Fig. 3-25; the τ
delay in the BIRD pulse was set for an assumed 145 Hz one bond
coupling.   The data were taken as 200 x 1K complex points and were
processed using an exponential multiplication prior to both Fourier
transformations with zero filling to 512 points prior to the second.
High resolution proton and carbon spectra are plotted flanking the
contour plot.

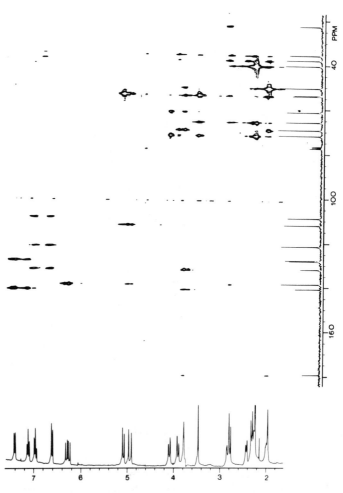

*Fig. 6-22.*     Long range optimized HC-COSY spectrum of an alkaloid using the pulse
sequence shown in Fig. 3-51 to provide one bond modulation
decoupling.    Data acquisition conditions were the same as those
specified in Fig. 6-21 except that the long range delays were
optimized for an assumed 10 Hz coupling.

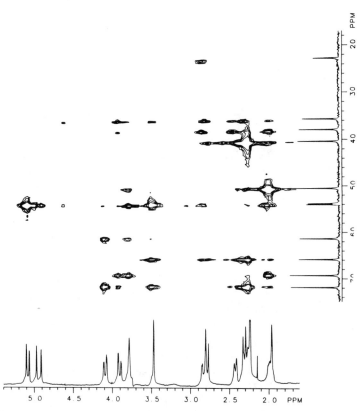

*Fig. 6-23.*     Expansion of the aliphatic region of the long range heteronuclear
chemical shift correlation spectrum shown in Fig. 6-22.

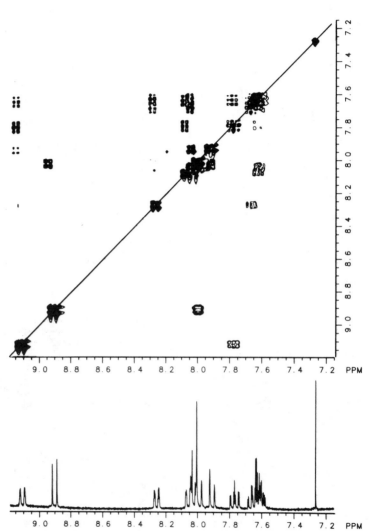

*Fig. 6-24.*        COSY spectrum of a dinaphthothiophene analog acquired as 256 x 1K complex data points. The data above the diagonal were processed using sinusoidal multiplication prior to both Fourier transformations with zero filling prior to the second to afford a matrix consisting of 512 x 512 complex points; the data below the diagonal were treated in the same fashion except that a 2.5 Hz Gaussian multiplication was performed in place of the sinusoidal multiplication.

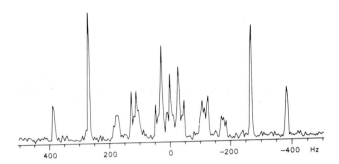

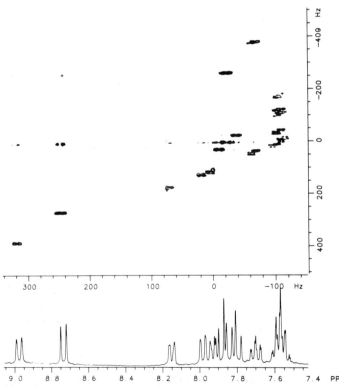

*Fig. 6-25.*    ZQCOSY spectrum of a dinaphthothiophene analog acquired as 384 x 1K
complex points using the Müller pulse sequence shown in Fig. 2-40 in
which the fixed delay during the evolution period was optimized for 7
Hz.   Chemical shift axes labeled in ppm are downfield of TMS, those
labeled in hertz are relative to the transmitter.

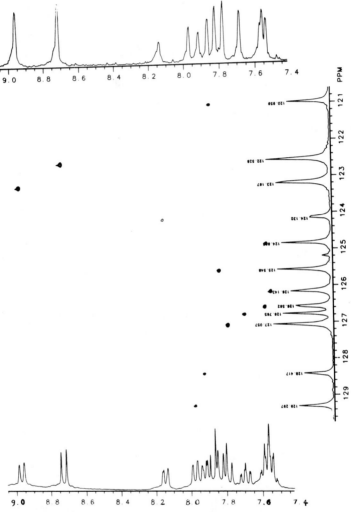

Fig. 6-26.    HC-COSY spectrum of a dinaphthothiophene analog recorded using the
              semiselective decoupling pulse sequence shown in Fig. 3-25 in which
              the  delay in the BIRD pulse was optimized for an average 165 Hz one
              bond coupling.  The data were acquired as 200 x 1K complex points and
              were processed using an exponential multiplication prior to both
              Fourier transformations with zero filling to 256 points prior to the
              second.   The contour plot is flanked by high resolution proton and
              carbon reference spectra.

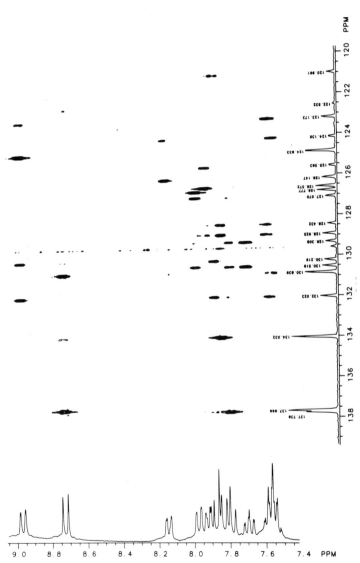

*Fig. 6-27.* Long range HC-COSY spectrum of a dinaphthothiophene analog recorded using the pulse sequence shown in Fig. 3-51. Parameters and processing considerations were identical those described in Fig. 6-26 except that the long range magnetization transfer delays were optimized for 10 Hz.

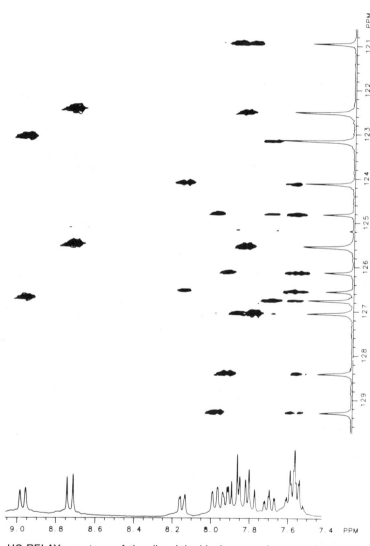

*Fig. 6-28.*     HC-RELAY spectrum of the dinaphthothiophene analog recorded using the pulse sequence shown in Fig. 4-20. Parameters and processing considerations were identical to those described in Fig. 6-26 except that the magnetization relay utilized an assumed 7 Hz vicinal proton-proton coupling.

both frequency domains shown in Fig. 6-26 (see Fig. 3-25); a long range optimized heteronuclear chemical shift correlation spectrum shown in Fig. 6-27, recorded using the pulse sequence shown in Fig. 3-51 with a 10 Hz optimization; and a heteronuclear relayed coherence transfer spectrum, presented in Fig. 6-28, recorded using the pulse sequence shown in Fig. 4-20. Using the data given, determine the structure of this dinaphthothiophene isomer and totally assign the proton and carbon NMR spectra.

## REFERENCES

1.    M. L. Tedjamulia, Y. Tominaga, R. N. Castle and M. L. Lee, *J. Heterocyclic Chem.*, **20**, 1143 (1983).
2.    L. H. Klemm, M. P. Stevens, L. K. Tran and J. Sheley, *J. Heterocyclic Chem.*, **25**, in press (1988).
3.    M. D. Johnston, Jr., G. E. Martin and R. N. Castle, *J. Heterocyclic Chem.*, **25**, in press (1988).
4.    D. E. Barnekow, J. H. Cardellina, II, A. S. Zektzer and G. E. Martin, *J. Am. Chem. Soc.*, **110**, in press (1988).
5.    E. Wenkert, C. J. Chang, H. P. S. Chawla, D. W. Cochran, E. W. Hagaman, J. C. King and K. Orito, *J. Am. Chem. Soc.*, **98**, 3645 (1976).
6.    M. J. Musmar, A. S. Zektzer, G. E. Martin, R. T. Gampe, Jr., M. L. Lee, M. L. Tedjamulia, R. N. Castle and R. E. Hurd, *Magn. Reson. Chem.*, **25**, 1039 (1987).
7.    M. D. Johnston, Jr. and G. E. Martin, *Magn. Reson. Chem.*, submitted (1988).
8.    A. S. Zektzer, G. E. Martin and R. N. Castle, *J. Heterocyclic Chem.*, **24**, 879 (1987).
9.    A. S. Zektzer, J. G. Stuart, G. E. Martin and R. N. Castle, *J. Heterocyclic Chem.*, **23**, 1587 (1986).
10.   A. Bax, *J. Magn. Reson.*, **53**, 517 (1983).
11.   A. S. Zektzer, B. K. John and G. E. Martin, *Magn. Reson. Chem.*, **25**, 752 (1987).

# CHAPTER 7

## SOLUTIONS TO PROBLEMS

### INTRODUCTION

Working any problem is a useful exercise only if the student is able to check his or her work. Detailed solutions to the problems given in Chapter 6 are provided. The greatest benefit will, of course, be derived from working the problems prior to examining the solutions. However, it is also useful to work on a given problem until a snag is encountered, at which point considering part of the solution is helpful. At that point, once sufficient information is gleaned from the solution to get beyond the point of difficulty, it is suggested that the student attempt to complete the problem before proceeding any further with the discussion of it in this chapter.

### SOLUTION TO PROBLEM 6.1

### Assignment of the Proton NMR Resonances of a Simple Dinaphthothiophene

#### Identification of Resonances in Individual Spin Systems

The assignment of the 300 MHz proton NMR spectrum of dinaphtho[1,2-b:2',1'-d]thiophene (**7-1**, hereinafter) shown in Fig. 6-1 requires first that resonances be subgrouped into individual homonuclear spin systems.[1] From simple inspection of the structure of the molecule, it should be clear that H7 and H12 will appear nominally as singlets. It might be further argued that H7 should resonate downfield of its H12 counterpart because of additional deshielding from the aromatic ring across the bay. This surmise, while a useful initial

418

argument, can cause problems without substantiation. Hence, we shall put it aside for potential later use.

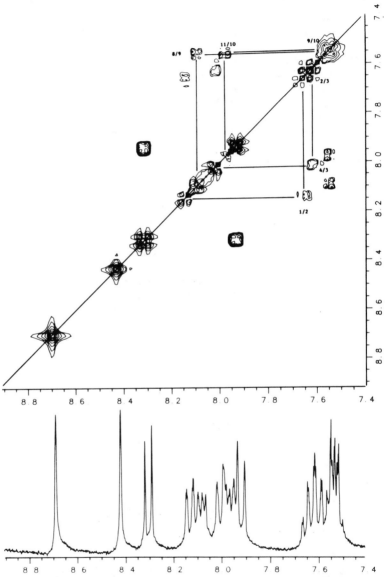

*Fig. 7-1.*   COSY spectrum of **7-1** identical to that shown in Fig. 6-1 except that the entire spectrum shown was processed with a 2.5 Hz Gaussian multiplication.  The connectivities for H1-H4 are shown below the diagonal and those for H8-H11 above the diagonal.  The process utilized to establish these connectivities is described in the text.

Beyond the two singlets, we can beneficially begin to use the data contained in the COSY spectra shown in Fig. 7-1. We would expect the spectrum of **7-1** to contain two four spin systems and a two spin system. To use the COSY data, it is most convenient to begin from the data contained above the diagonal, which were processed using a 2.5 Hz Gaussian multiplication prior to both Fourier transformations. In particular, this method of data processing "washes out" long range couplings, which are frequent in the COSY spectra of polynuclear aromatics, thereby making the identification of vicinal neighbors more straightforward.

Utilizing the strategy just proposed, we note that the doublet resonating at 8.31 ppm has an off-diagonal response correlating it with the doublet resonating at 7.92 ppm. There is no further off-diagonal response correlating the doublet resonating at 7.92 ppm with any other resonance. Hence, we have identified the components of the two spin system comprised of H5 and H6.

Continuing, we must identify the individual component resonances of the two four spin systems, which is a more challenging undertaking. A convenient starting point for this task is provided by the pair of resonances located between 8.05 and 8.15 ppm. In the Gaussian processed data (Fig. 7-1), the proton of the pair resonating downfield exhibits a correlation to the complex multiplet centered at about 7.62 ppm. In particular, the proton resonating at 8.13 ppm correlates with the resonance centered at 7.59 ppm. A near off-diagonal response correlates the proton at 7.59 ppm with its vicinal neighbor just downfield, which resonates at 7.65 ppm. Finally, the proton at 7.65 ppm correlates with a proton resonating at 8.04 ppm. At this point, although we have just identified the constituent spins of one of the four spin systems, we still do not know which one, nor do we have the spins ordered relative to the other spin systems in the molecule. That is to say that the orientation of the spin system, if it were to be H1-H4, could have H1 as either the proton at 8.13 or 8.04 ppm in the absence of any additional information.

The final spin system components to be identified are those in the H8-H11 four spin system. From Fig. 6-1, we note that the proton resonating at 8.09 ppm is correlated to the complex multiplet resonating furthest upfield int he 2.5 Hz Gaussian processed data. Specifically, the proton resonating at 8.09 ppm correlates with the proton resonating at 7.51 ppm. There is next, although difficult to discern, a correlation between the proton at 7.51 and that at 7.55 ppm, the latter, in turn, finally coupled to a proton resonating at 7.96 ppm. This completes the identification of the second four spin system, which would be best classified as an ABXY spin system on the basis of the chemical shifts of the constituent protons. Once again, as with the case of the other four spin system, it is possible that the spin system could be misordered in the absence of additional information.

## Orienting Spin Systems
## Relative to One Another

Having identified all of the constituents of the individual spin systems contained in **7-1**, we must next complete the assignment by finally orienting the spin systems relative to one another. To accomplish this final step in the assignment process, some additional information must be gleaned from the data shown in Fig. 7-2.

Perhaps the simplest strategy in completing the assignment is to examine the spectrum for additional responses correlating resonances in different spin systems to one another. Clearly, this information will be contained only in the presentation of sinusoidally multiplied data. Here we note a wealth of long range coupling information. In particular, both of the singlets resonating downfield, which would correspond to H7 and to H12, exhibit a long range coupling. First, they are correlated with each other. Next, the proton resonating furthest downfield, at 8.68 ppm, exhibits a long range coupling to the proton resonating at 7.96 ppm. This coupling pathway could arise from either a *peri* coupling or from a five bond epi-zig-zag coupling. In similar fashion, the proton resonating at 8.42 ppm exhibits a long range coupling to the proton resonating at 8.09 ppm. Here again, this could again be either a *peri* coupling or a five bond epi-zig-zag coupling.

Before making specific assignments, let us examine the information currently in hand. The long range couplings between the resonances at 8.68 and 8.42 ppm correlate the H7 and H12 protons to the four spin ABXY spin system with terminal spins of 8.09 and 7.96 ppm. Thus, at this point we have circumstantial evidence that the other four spin system is linked to the two spin system comprised of H5 and H6. Let us, however, examine the spectrum for specific confirmation of this attribution. Here, confirmation is found in the form of a long range coupling between the proton resonating at 7.92 ppm and the proton resonating at 8.13 ppm. Thus, for a third time we must decide between a *peri* coupling, which in this case would be between H4 and H5, and a five bond epi-zig-zag coupling between H1 and H5. Usefully, long range couplings have been reviewed by Bartle, Jones and Matthews.[2] Typical *peri* couplings in aromatic systems range between 0.20 and 0.50 Hz, while in contrast five bond epi-zig-zag couplings range up to 1.1 Hz. On the basis of the evolution time, it is quite unlikely that it would be possible to observe *peri* coupling responses in the spectra recorded for **7-1**. Rather, the much larger five bond epi-zig-zag coupling responses would be seen preferentially. Thus, the response correlating the resonances at 7.92 (H5) and 8.13 ppm establishes the identity of the latter as H1, consequently orienting the H1-H4 four spin system relative to the two spin system. Hence, assignments for H1-H4 are as follows: H1, 8.13 ppm; H2, 7.59 ppm; H3, 7.65 ppm and H4 8.04 ppm. H5, as noted above, is clearly assigned as the proton resonating at 7.92 ppm, and its AX partner, H6, is assigned as the proton resonating at 8.31 ppm.

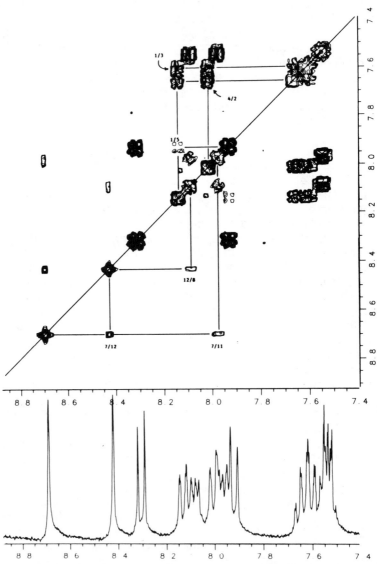

Fig. 7-2.    COSY spectrum of **7-1** processed using sinusoidal multiplication prior to both Fourier transformations. Several long range couplings are observed in the spectra and are discussed in the text.

We are left now with the task of orienting the H8-H11 spin system relative to the H7 and H12 resonances. Here, because both H7 and H12 are long range coupled to one of the terminal spins of the four spin system, we do not have any point from which to argue. While it is essentially certain that both of the long range couplings in question are five bond epi-zig-zag we must unequivocally differentiate H7 from H12 if the assignment is to be completed. Here, no information contained within either of the two-dimensional NMR spectra in Fig. 6-1 provides any assistance. Rather, we must finally resort to the ancillary information provided with the problem, specifically the 12.7% nOe observed between the proton resonating at 8.68 ppm and the doublet resonating at 8.31 ppm. By examining **7-1**, we see that H6 and H7 are in close proximity to one another and can account for the sizable nOe observed. In contrast, there is no way that H12 can exhibit an nOe to either of the spins of the two spin H5-H6 system.

Given the means of assigning H7 as the proton resonating at 8.68 ppm, we consequently establish the assignment of H11 as the proton long range coupled to H7 via a five bond epi-zig-zag coupling, simultaneously orienting the four spin system relative to the H7 and H12 resonances. Final assignments for **7-1** are given in Table 7-1.

Table 7-1.     Proton resonance assignments for dinaphtho[1,2-*b*:2',1'-*d*]thiophene (**7-1**) in deuterochloroform at 300 MHz.

| Position | Chemical Shift (δ) |
|----------|--------------------|
| H1 | 8.13 |
| H2 | 7.59 |
| H3 | 7.65 |
| H4 | 8.04 |
| H5 | 7.92 |
| H6 | 8.31 |
| H7 | 8.68 |
| H8 | 8.09 |
| H9 | 7.55 |
| H10 | 7.51 |
| H11 | 7.96 |
| H12 | 8.42 |

## SOLUTION TO PROBLEM 6.2

## Assigning the Ring A
## Protons of Annulatol-A

## Using the COSY Data

The spectra of annulatol-A (**7-2**) shown in Figs. 6-2 and 6-3 recorded at 300 MHz are reasonably well resolved.[3] The COSY spectrum presented in Fig. 6-2 shows a pair of correlations from the proton resonating furthest downfield at 3.55 ppm, which may be assigned as H1.  Correlations are observed to two protons, one resonating at 2.50 ppm and the other at 1.90 ppm which is part of the cluster of three protons resonating between 1.70 and 1.90 ppm.  Clearly, the two protons to which correlations are observed must be H2a/b.

Continuing for the proton at 2.50 ppm, we note that it exhibits an off-diagonal response correlating it with the proton resonating at 1.90 ppm corresponding to the geminal coupling.  Further correlations are noted to a proton resonating at 1.35 ppm, submerged beneath a methyl resonance, and a very weak response to a proton resonating at 1.05 ppm.  Given the structure of **7-2,** these must be the H3a/b resonances, although the coupling to the latter is very ill defined from the spectrum shown in Fig. 7-3.  Finally, progressing from the proton resonating at 1.90 ppm, we can convince ourselves that it exhibits couplings to the same two protons, although the intensity of the off-diagonal responses is opposite those of the proton resonating at 2.50 ppm.

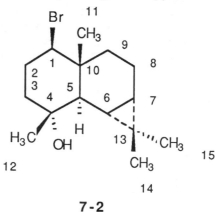

**7-2**

One of the problems inherent in the usage of COSY spectral data is highlighted in the preceding paragraph.  Specifically, whenever we have several protons with closely similar chemical shifts, it can become quite difficult to track connectivity networks.  Let us contrast the difficulties with using COSY data with the same connectivity network in the zero quantum spectrum.  Rather than referring back to Fig. 6-3, however, we shall instead refer to Fig. 7-4.

## Using the Zero Quantum Data

Beginning again from the H1 proton resonating at 3.55 ppm ($F_2$ = +490 Hz), we note correlations in $F_1$ at +340 and +520 Hz, which correlate H1 with H2a/b.  The H2a/b geminal correlation is indicated by the dashed line ($F_1$ = ±200 Hz).  H2a (2.50 ppm), aside from the geminal correlation response ($F_1$ = +200 Hz), also exhibits responses at $F_1$ = +370 and +455 Hz correlating H2a with H3a and H3b resonating at 1.35

and 1.05 ppm, respectively. Similarly, H2b (1.90 ppm) exhibits responses at $F_1$ = +175 and +260 Hz correlating it to H3a and H3b, respectively.

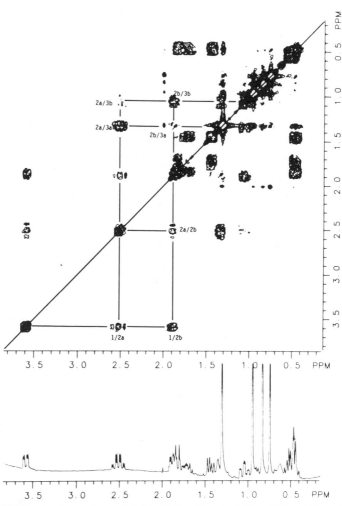

*Fig. 7-3.* COSY spectrum of annulatol-A (**7-2**). Connectivities linking H1 to the H2a/b anisochronous methylene protons are shown below the diagonal. Further connectivities linking H2a and H2b to the anisochronous H3a/b methylene protons are presented above the diagonal.

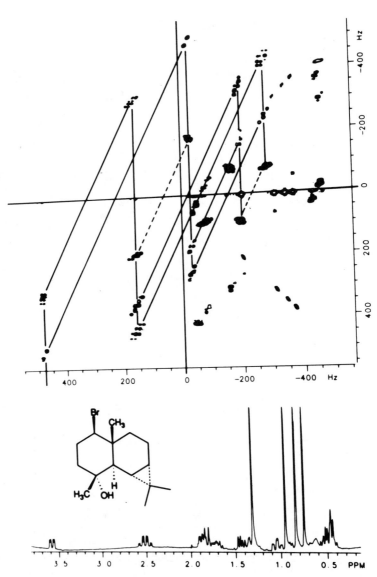

*Fig. 7-4.*    ZQCOSY spectrum of annulatol-A **7-2** showing connectivities in ring
A.    Vicinal connectivities are denoted by solid lines, geminal
connectivities are denoted by dashed lines.

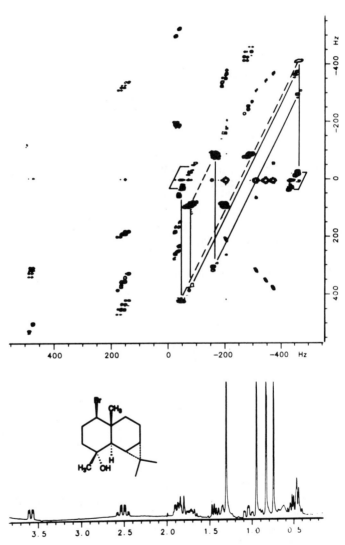

*Fig. 7-5.* ZQCOSY spectrum of annulatol-A (**7-2**) showing connectivities in ring B. Vicinal correlations are denoted by solid lines, geminal couplings are denoted by dashed lines.

It is useful to note that while the off-diagonal responses correlating H2a (2.50 ppm) to H3b (1.05 ppm) and H2b (1.90 ppm) to H3a (1.35 ppm) were weak and possibly questionable, the exact opposite is true in the 5 Hz optimized zero quantum spectrum shown in Fig. 7-4. Here there is no question about the connectivity pathways even though H2b resonates in a cluster of three resonances.

## Assigning the Ring B
## Protons of Annulatol-A

No unequivocal, convenient starting point can be identified from which the assignment of the proton resonances in ring B of annulatol-A (7-2) can be begun. During the elucidation of the structure by Cardellina and coworkers,[3] a long range correlation was observed between the 11-methyl proton resonance and a proton resonating at 1.82 ppm, which was assigned as H9a in the proton detected long range heteronuclear multiple quantum spectrum. Given this entry point, the assignment of the ring B protons may be successfully undertaken. These connec-tivities are shown in Fig. 7-5 and the assignments are summarized in Table 7-2. However, since the assignment of the ring B protons could not be done from the data provided in Chapter 6, we will not discuss the assignment further here. Rather, the interested reader is referred to the account of Cardellina and coworkers for further details.[3]

## SOLUTION TO PROBLEM 6.3

## Assignment of the Proton NMR Spectrum
## of the Alkaloid α-Yohimbine with
## Simultaneous Verification of the
## $^{13}$C-NMR Assignments

## Locating a Starting Point

The problem to be dealt with in the case of α–yohimbine (7-3) is somewhat more complex than the preceding problems. In particular, one source of difficulty is in the location of a starting point. Any number of strategies may be invoked. However, a reasonable strategy is to pick a resonance in the proton spectrum that is well resolved and therefore assured of exhibiting connectivities that are irrefutable. In the case at hand, referring to the COSY spectrum shown in Fig. 6-4, such a starting point is provided by the quartet resonating furthest upfield in the proton spectrum. From the proton carbon spectrum shown either as a contour plot in Fig. 6-5 or as a slice summary in Fig. 6-6, we note that this proton is directly bonded to a carbon resonating at 33.95 ppm. Using Wenkert's[4] $^{13}$C-NMR assignments shown on 6-3 and also in Table 7-3, we observe that this resonance corresponds to the 14 position (see 7-3 for numbering scheme).

Table 7-2.     Proton resonance assignments of annulatol-A (**7-2**) made using a concerted application of two-dimensional NMR techniques.[3]

| Position | δ | Position | δ |
|---|---|---|---|
| 1 | 3.55 | 8 | 1.73 |
| 2 | 2.50 | | 1.40 |
| | 1.90 | 9 | 1.85 |
| 3 | 1.35 | | 0.45 |
| | 1.05 | 11 | 1.30 |
| 5 | 0.50 | 12 | 0.85 |
| 6 | 0.57 | 14 | 0.95 |
| 7 | 0.55 | 15 | 0.75 |

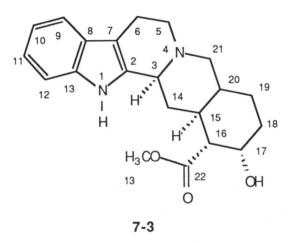

**7-3**

Conveniently, a position such as 14 is useful in that there are limited connectivities to vicinal neighbors. Here we can be reasonably certain of establishing at least a fragment consisting of probably three or more carbons and their directly attached protons.

Alternative strategies for determining a starting point might include using the methine proton at the 17 position. In the absence of other assignment information, H17 has a reasonably unique chemical shift, which can then be used to identify the directly attached carbon via Fig. 6-5. Other similar approaches may also be used, but rather than continuing to describe alternatives, let us redirect our attention to the problem at hand and use the H14 resonance as a point of departure.

## Establishing a Heteronuclear
## Connectivity Network

Beginning from the carbon resonating at 33.95 ppm, which corresponds to that assigned to C14 by Wenkert and coworkers,[4] we

Table 7-3.    Aliphatic proton and carbon shift assignments for α–yohimbine (7-3).

| | δ (ppm) | | | |
|---|---|---|---|---|
| | This | | | |
| Resonance | Problem | Wenkert[4] | Resonance | δ (ppm) |
| C3 | 60.04 | 59.8 | H3 | 3.33 |
| C5 | 52.52 | 52.1 | H5a | 2.93 |
| | | | H5b | 2.45 |
| C6 | 21.63 | 21.5 | H6a | 2.71 |
| | | | H6b | 2.54 |
| C14 | 33.95 | 33.8 | H14a | 2.37 |
| | | | H14b | 0.94 |
| C15 | 35.89 | 36.4 | H15 | 1.82 |
| C16 | 52.25 | 52.6 | H16 | 2.21 |
| C17 | 66.48 | 66.9 | H17 | 4.06 |
| C18 | 32.47 | 31.4 | H18a | 1.72 |
| | | | H18b | 1.47 |
| C19 | 22.95 | 23.1 | H19a/b | 1.21 |
| C20 | 40.06 | 40.2 | H20 | 1.35 |
| C21 | 61.18 | 61.0 | H21a | 2.78 |
| | | | H21b | 2.07 |
| C23(OMe) | 51.08 | 51.7 | H23 | 3.63 |

must next turn our attention to the COSY spectrum shown initially in Fig. 6-4.    Rather than referring back, we shall instead begin to utilize the annotated COSY spectrum shown in Fig. 7-6.

Having a tentative assignment based on the proton-carbon correlation spectrum of the furthest upfield proton (0.94 ppm) as H14, we may begin to utilize the homonuclear correlation information contained in the COSY spectrum. It will be noted here that the H14b resonance shows correlations to three other protons via its off-diagonal responses. Using the stacked slice summary from the heteronuclear correlation spectrum (see Fig. 7-7), we may quickly identify the H14a proton resonating at 2.37 ppm. The geminal correlation response, 14a/b, is labeled accordingly in Fig. 7-6. Clearly then, the other two off-diagonal responses emanating from the H14b proton must provide correlations to the vicinal neighbors. It is at this point that we actually begin to establish the heteronuclear connectivity network essential to verifying the carbon NMR assignments.

The correlation to the doublet resonating at 3.19 ppm is shown to be directly bound to the carbon resonating at 60.04 ppm, which Wenkert has assigned as C3. We note also that the proton now identified as H3 also correlates vicinally with H14a in addition to some other correlations, which will become important later in this problem. Finally, the weaker correlation linking H14b to a proton resonating at 1.82 ppm establishes the vicinal connectivity between H14b and H15, the latter also vicinally correlated to H14a. Thus, to this point, we have established the structural fragment defined by **7-4**. In addition, the proton-proton connectivities from the COSY spectrum are supportive of the carbon NMR assignments made by Wenkert for C3, C14 and C15.

**7 - 4**

Continuing from the H15 resonance observed at 1.82 ppm, we next note an off-diagonal response correlating it with a proton resonating as a double doublet at 2.21 ppm (connectivities shown above the diagonal, Fig. 7-6). From **7-3** we could, at this point, be observing a correlation to either H16 or the H20 proton, both of which are vicinal to H15. Turning to the slice summary shown in Fig. 7-7 we note that the proton in question is directly bound to a carbon resonating at 52.25 ppm, which Wenkert has assigned as H16. By careful inspection of the COSY spectrum we note a very weak response correlating H15 to a proton resonating at 1.35 ppm, which, as we shall subsequently see, is indeed H20.

Given the location of H16, we note that H16 exhibits only one other off-diagonal response correlating it with a proton resonating at 4.06 ppm, which is directly bound to the carbon resonating at 66.48 ppm, which is assigned as C17. The H17 methine proton, in turn, has three sets of off-diagonal correlations that are ultimately assignable to H18a (1.72 ppm), H18b (1.47 ppm) and the strongly coupled H19a/b AB system (centered at 1.21 ppm). Of these connectivities, the strongest is observed to the proton resonating at 1.72 ppm corresponding to H18a. That this is the vicinally coupled proton is confirmed by the data processed using a Gaussian multiplication (see Fig. 6-4).

Hence, through the series of correlations just described, we have extended the structural fragment represented by **7-4** to the larger fragment defined by **7-5**. It should also be noted that all of the carbon assignments made by Wenkert[4] and coworkers, to this point, have been in agreement with the proton-proton connectivity network, underscoring the complementary nature of the data provided by the HH- and HC-COSY experiments.

Continuing the assignment of the proton spectrum, we will now refer to Fig. 7-8. From the H18a/b resonance we also observe correlations to H19a/b. From the H19a/b protons we also note a correlation to the H20 methine proton, which we had not previously identified in a reliable fashion, the C20 carbon resonating at 40.06 ppm in complete agreement with the assignment of Wenkert[4] at 40.2 ppm (see Fig. 7-7). Finally, correlations are observed from H20 to the

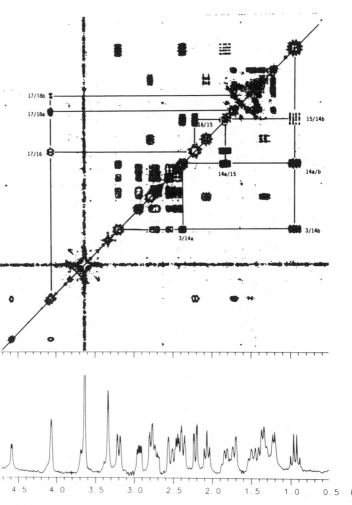

Fig. 7-6.     COSY spectrum of α-yohimbine (**7-3**) recorded at 300 MHz in deuterochloroform. The data were processed using a sinusoidal multiplication prior to both Fourier transformations. Connectivities emanating from H3 are shown below the diagonal represented by **7-4**. Connectivities continuing from H15 through H18a/b are shown above the diagonal as shown by **7-5**.

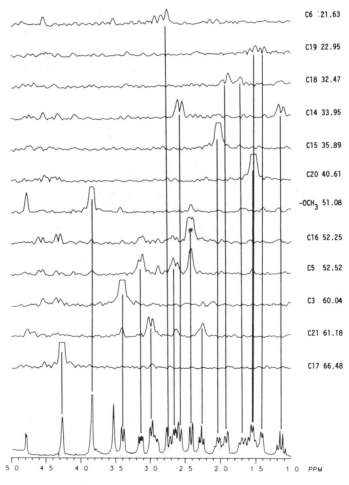

C6 ⁻21.63
C19 22.95
C18 32.47
C14 33.95
C15 35.89
C20 40.61
-OCH₃ 51.08
C16 52.25
C5 52.52
C3 60.04
C21 61.18
C17 66.48

5 0    4 5    4 0    3 5    3 0    2 5    2 0    1 5    1 0  PPM

*Fig. 7-7.*     Stacked slice summary showing responses for the aliphatic carbons of
α-yohimbine (**7-3**) taken from the HC-COSY spectrum shown in Fig.
6-5.  Carbon chemical shifts are specified with the individual slices.

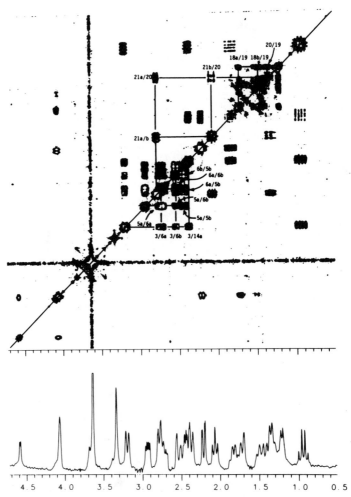

*Fig. 7-8.*     COSY spectrum of α-yohimbine (**7-3**) processed in a fashion identical to that shown in Fig. 7-6. Correlations shown above the diagonal provide additional correlations linking H18 through H21 as shown by **7-6**. Correlations shown below the diagonal correlate H3 with protons across the piperidine ring as shown by **7-7**.

**7-5**

anisochronous H21a/b resonances, resonating at 2.78 and 2.07 ppm, respectively. From Fig. 7-7, the C21 carbon resonates at 61.18 ppm, again in agreement with the assignment of Wenkert.[4] Aside from the H15-H20 connectivity, which was not readily or reliably distinguished in the COSY spectrum, we have now assembled a rather large, contiguous structural fragment as represented by **7-6** leaving only the 5 and 6 positions of the piperidine ring to be assigned.

**7-6**

Still referring to Fig. 7-8, let us return to consider the H3 resonance at 3.33 ppm a final time. In the discussion above, we noted that H3 exhibited off-diagonal response that correlated it with H14a/b and another pair of protons, which we left undefined at that time. At this point, to complete the assignment of the aliphatic proton and carbon resonances of α-yohimbine (**7-3**), we must redirect our attention to those responses. Since we have assigned all of the protons in the aliphatic region of the spectrum save those for at the 5 and 6 positions, H3 must clearly correlate with two of the remaining protons via the off-diagonal responses observed at 2.71 and 2.54 ppm in Fig. 7-8. Returning to Fig. 7-7, we note that these responses correspond to an anisochronous geminal methylene pair and that further, the carbon to which they are attached resonates at 21.63 ppm. Clearly the chemical shift of the directly bonded carbon is inconsistent with a nitrogen bearing methylene. Indeed, we note that Wenkert[4] has assigned the resonance at 21.5 ppm in his study as C6 and that at 52.1 ppm as C5. Thus, the responses from H3 correlate it long range with the H6a/b pair. The final sets of connectivities correlating H3 with the protons at the 5 and 6 positions are shown beneath the diagonal in Fig. 7-8. Building from the structural fragment shown by **7-6** we have now constructed essentially the entire aliphatic portion of the structure of α-yohimbine, as show by **7-7.** Final assignments for the aliphatic proton resonances of **7-3** are given in Table 7-3 accompanied by a comparison of the carbon chemical shifts in this region of the molecule. In the case of the latter, all of the assignments made by Wenkert[4] at 25.2 MHz are borne out in the two-dimensional NMR spectra utilized in solving this problem.

## SOLUTION TO PROBLEM 6.4

**Identifying Heteronuclear Spin Systems
of a Polynuclear Aromatic Using a
Combination of Heteronuclear Chemical
Shift Correlation and Heteronuclear
Relayed Coherence Transfer**

### Locating a Starting Point

The first task before we may begin to make assignments requires the location of a starting point in the molecular structure. Using the proton-carbon heteronuclear correlation spectrum shown in Fig. 6-8, we note that the proton resonating furthest downfield (9.06 ppm) correlates with the carbon resonating furthest downfield at 128.07 ppm. This provides us with the necessary starting point from which to proceed with the identification of the constituents of the first spin system.

### Assigning the Two Proton,
### Four Spin Systems and Their
### Heteronuclear Partners

Beginning from the proton resonating downfield at 9.06 ppm, referring to Fig. 7-9, we note that at the chemical shift of the directly attached carbon, magnetization has been relayed to a proton resonating upfield at 7.51 ppm. This constitutes the vicinal neighbor proton of our

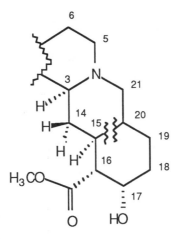

**7-7**

starting proton and the correlation between the two protons is designated by an arrow. Following the vertical arrow in the figure, we next locate the chemical shift of the direct proton-carbon correlation, 7.51/123.97 ppm. At this carbon chemical shift we observe responses at three proton chemical shifts: 9.06, 7.63 and 7.51 ppm. Quite obviously the 9.06 ppm response represents magnetization relayed back to the starting point, 7.51 ppm is the directly attached proton and finally, the proton resonating at 7.63 ppm represents magnetization transferred to the next vicinal neighbor. Following this correlation, the directly coupled pair 7.63/126.86 ppm is easily identified and confirmed from the heteronuclear correlation spectrum shown in Fig. 6-7. Here again responses from three protons are observed in the heteronuclear relay spectrum shown in Fig. 7-9: 7.97, 7.63 and 7.51 ppm. As in the previous case we may identify the downfield proton of the trio resonating at 7.97 ppm as the response to the vicinal neighbor while those at 7.63 and 7.51 ppm represent the direct and previous protons, respectively. Following the connectivity arrows shown in Fig. 7-9 a step further, we locate the direct carbon pairing 7.97/127.74 ppm. Here there is only one relayed response, that back to the preceding proton resonating at 7.63 ppm. Thus, through this series of operations, we have established the complete identities of one of the heteronuclear spin systems associated with the proton ABMX spin system of the compound. These are shown on structure **7-8**.

It must be emphasized at this point that from the information elucidated thus far using the heteronuclear relay spectrum there is no way in which the spin system we have identified could be specifically assigned as shown in the structure. Rather, to locate and orient the spin system identified to this point relative to the rest in the structure, we must

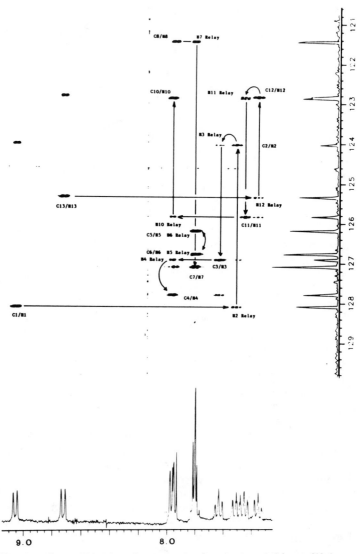

Fig. 7-9.    Heteronuclear relayed coherence transfer spectrum of benzo[b]phen-
anthro[4,3-d]thiophene (7-8) recorded in deuterochloroform at
observation frequencies of 300.068/75.459 MHz for $^1H/^{13}C$,
respectively. Direct responses are identified using the heteronuclear
chemical shift correlation spectrum shown in Fig. 6-8.

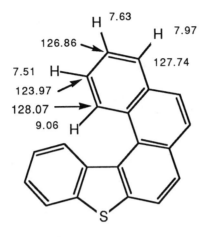

**7-8**

have additional information that unequivocally links one or more of the spins in the system to another. There are several ways of accomplishing this task. In their original work, Martin and coworkers[5] utilized carbon-carbon double quantum coherence (see Figs. 5-10, 5-11 for a similar example). This task could also be accomplished using long range heteronuclear chemical shift correlation or via long range proton-proton connectivity information. The former method is utilized in the following problem. The latter requires the observation of appropriate long range proton-proton coupling responses. There are several ways of observing the requisite responses. One may employ a long range optimized COSY experiment (see Chapter 2) or alternatively, it may be possible to employ a HOHAHA spectrum (see Chapter 4) as recently demonstrated by Johnston and Martin.[6] Any or all of these methods operate by linking resonances of one spin system to the adjacent spin system via carbon-carbon, carbon-proton or proton-proton coupling pathways.

Returning to the problem at hand, the constituent resonances of the other ABMX proton spin system can be identified using Fig. 7-9 beginning from the proton resonating as a doublet at 8.73 ppm. The directly bound carbon resonates at 125.33 ppm. Here, following the connectivity arrows in the figure, the vicinal neighbor is identified as the proton resonating furthest upfield at 7.36 ppm, which is directly bound to the carbon resonating at 122.79 ppm. Responses for proton magnetization transferred to this carbon include protons resonating at 8.73, 7.45 and 7.36 ppm. The response of interest in this case is the proton resonating at 7.45 ppm, which is directly bound to the carbon observed at 125.82 ppm, this pair constituting the third pair of spins in the system. Responses are observed at 7.97, 7.45 and 7.36 ppm. Of these, that of interest is the proton resonating at 7.97 ppm, which is directly bound to the carbon resonating at 122.83 ppm, which completes the identification of the second proton four spin system and the associated heteronuclear spins.

## Assigning the Two Proton, Two Spin Systems and Their Heteronuclear Partners

After the assignment of the two four spin systems, the major portion of the work required in this problem is behind us. All that remains are the two proton, two spin systems. The first, comprised of H7 and H8 resonating at 7.79 and 7.94 ppm, respectively, is straightforward and easily dispensed with. Connectivities are shown in Fig. 7-9. Beginning from the H8 direct response 7.94/121.45 ppm, we have only a single relay response to the proton at 7.79, H7 to consider. The direct response in this case is located at 7.79/127.08 ppm and completes the assignment of this spin system.

The remaining two spin system is somewhat more interesting. This system, comprised of H5/H6 resonating at 7.80/7.77 ppm is a strongly coupled AB pair. Here the heteronuclear relay experiment that we have been employing experiences difficulty in that the relay responses are essentially superimposed over the direct proton/carbon responses, thereby making the identification of these spins more difficult. The best means of identifying these protons as a vicinal pair would probably be provided by a proton double or zero quantum coherence spectrum (see Chapter 2). Alternatively, in the original work, the C5/C6 pairing was identified from the carbon-carbon double quantum spectrum.

## Conclusions

Overall, the heteronuclear relayed coherence transfer experiment, given the direct proton-carbon pairings, provides a powerful tool for the identification of heteronuclear spin systems in molecules where there are reasonably uniform vicinal proton-proton coupling constants. Examples would include polynuclear aromatics[5] and saccharides.[7] Although it is certainly possible to apply the heteronuclear relay experiment to other classes of molecules, the results obtained will be less predictable due to the variation in vicinal coupling constants. In such cases, the heteronuclear relayed Hartmann-Hahn experiment proposed by Bax, Davis and Sarkar[8] (see Fig. 4-28) would probably be a better choice. Unfortunately, at this point in time, there are no examples of the application of this experiment contained in the literature.

## SOLUTION TO PROBLEM 6.5

## Total Assignment of the Proton and Carbon NMR Spectra of the Complex Helicene Phenanthro[3,4:3',4']phenanthro[2,1-*b*]thiophene

### Where to Start?

Faced with a problem of the complexity of the present one, there is the perennial question of where to begin. In the case of the benzo[*b*]phenanthro[4,3-*d*]thiophene (**7-8**) discussed immediately above, we had the luxury of two downfield shifted bay region protons, which would provide an excellent starting point for the assignment. In the present case, the helical nature of the molecule has undoubtedly shifted the bay region protons upfield into the aromatic "envelope," thereby making their identification somewhat more difficult. The assignment is made even more problematic by the fact that the two four spin systems are extensively overlapped. Hence, we must find some other point of departure.

While it is certainly not as convenient to begin from the center of the helix, working our way outward, it can be done and does provide a cleaner point of entry into the problem. Thus, the three doublets resonating furthest downfield at 8.40, 8.28 and 8.09 ppm provide us with viable entry points. As will be noted from the COSY portion of the composite spectrum shown in Fig. 7-10, each of these three protons is a component of a different two spin system, all of them AX in character. Arbitrarily, let us select the resonance at 8.40 ppm as our starting point.

### Establishing a Heteronuclear Connectivity Network from the Proton Resonating at 8.40 ppm

Given the correlations in the COSY spectrum shown in Fig. 7-10, the proton-carbon heteronuclear chemical shift correlation spectrum (see Fig. 6-11) immediately provides the identities of the directly attached protons. Similar although redundant information is provided from the heteronuclear relay spectrum shown in Fig. 7-11. While these operations do identify all of the constituents of three heteronuclear spin systems, they do not carry the process any further forward. Thus, we must consider using additional data if the starting point that we have chosen is to be of any utility.

The next point of attack requires the utilization of the long range optimized heteronuclear chemical shift correlation spectrum shown in Fig. 7-12. From this spectrum, we note that the proton resonating at 8.40 ppm exhibits two strong correlations to quaternary carbons resonating at 127.51 and 131.32 ppm and a rather weak correlation to a protonated carbon resonating at 122.07 ppm. There is also a "spot," for lack of a better word, in the contour plot at the chemical shift of the quaternary carbon resonating at 135.76 ppm. We shall come back to this latter

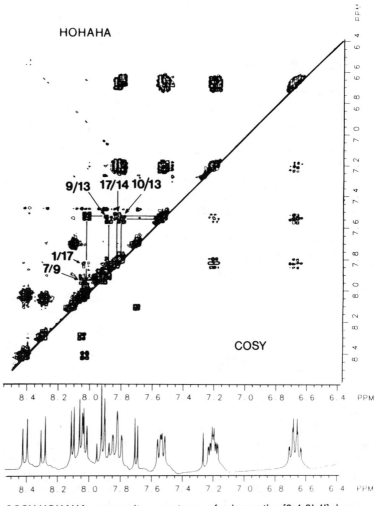

*Fig. 7-10*     COSY/HOHAHA composite spectrum of phenanthro[3,4:3',4']phen-
anthro[2,1-*b*]thiophene (**6-5**) recorded in deuterochloroform at
300.068 MHz.

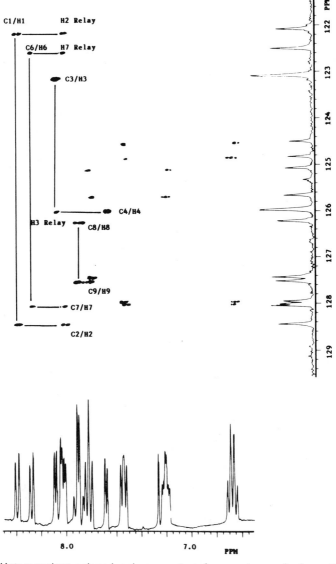

*Fig. 7-11.* Heteronuclear relayed coherence transfer spectrum of phenanthro-[3,4:3',4']phenanthro[2,1-*b*]thiophene (**6-5**). Direct correlation responses are identified using the heteronuclear chemical shift correlation spectrum shown in Fig. 6-11. Connectivities are shown only for the four two spin systems in the spectrum.

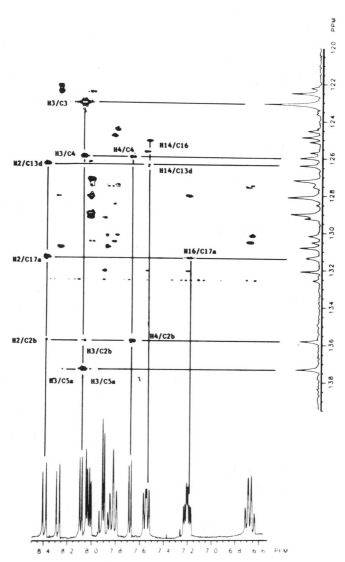

*Fig. 7-12*    Long range optimized heteronuclear chemical shift correlation spectrum of phenanthro[3,4:3',4']phenanthro[2,1-*b*]thiophene (**6-5**). The experiment was optimized for an assumed 10 Hz long range correlation using the pulse sequence shown in Fig. 3-10 without provision for one bond modulation decoupling.

piece of information later. Given the information above, in conjunction with the connectivity information provided by the COSY and heteronuclear relay spectra, we note that the protonated carbon to which the weak response was observed corresponds to the directly attached carbon. Coupled with the knowledge that the three bond heteronuclear couplings are generally substantially larger than their two bond counterparts, we may assemble a tentative structural fragment as shown by **7-9**.

8.03/128.44

**7-9**

In the structural fragment (**7-9**), the location of the two protons as part of a two spin system is provided by the COSY data and cross-confirmed by the heteronuclear relay experiment (Fig. 7-11). You will also note that the identity of the two quaternary carbons has not been affixed as yet. This is because they could very easily be interchanged at this point. We must progress further in the assignment before we can make tentative or more definite assignments.

## Expanding the Connectivity Network Outward

Having established a "beachhead," we must next begin to expand the connectivity network outward to generate a larger structural fragment. The most logical way in which to accomplish this task is to use the correlations to the two quaternary carbons that we already have from the proton resonating at 8.40 ppm.

Let us turn our attention first to the "spot" referred to above at the chemical shift of quaternary carbon resonating at 135.76 ppm. If we pull the appropriate slice from the data matrix at this chemical shift, which is shown in Fig. 7-13, we note that there is indeed a response there correlating the proton resonating at 8.40 ppm with the carbon resonating at 135.76 ppm, which was barely visible at the threshold of the contour plot. This particular response is quite important in that the quaternary carbons resonating furthest downfield in polynuclear aromatics of the

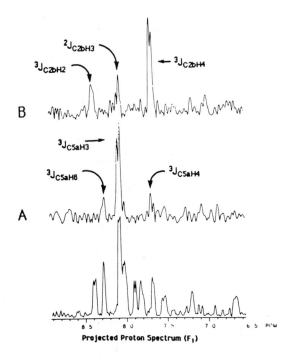

*Fig. 7-13.*     Selected slices taken from the long range hetrornuclear chemical shift correlation spectrum presented as a contour plot in Fig. 7-12.

type in this problem are generally the carbons common to the thiophene ring. If this can be demonstrated, then we have located the fragment shown by **7-9** in proximity to the thiophene ring. This is an important step in the completion of the total assignment!

From the slice shown in Fig. 7-13 and the contour plot shown in Fig. 7-12, we note that the quaternary carbon in question is correlated wtih both members of one of the two spin systems that we identified above, specifically the protons resonating at 8.09 and 7.69 ppm. Furthermore, these protons also correlate with the quaternary carbon resonating furthest downfield at 137.25 ppm, which is in turn coupled to a proton resonating at 8.28 ppm, which is a component of the third two spin system (see Fig. 7-13).

Although the next statement represents somewhat of a quantum leap in deduction, it is substantiated by one additional piece of information, which we will consider in a moment. Given the set of connectivities we have just developed, the protons resonating at 8.03 and 8.40 can be assigned as H1 and H2, respectively. The subsequent correlations link H2 to the thiophene protons, H3 and H4, which resonate at 8.09 and 7.69 ppm, respectively, through the quaternary carbon resonating at 135.76 ppm. Finally, the proton resonating at 8.28 ppm may be assigned as H6 through its mutual correlation with H5 to the quaternary carbon resonating at 137.25 ppm.

To substantiate the massive structural fragment proposed in the preceding paragraph, let us consider one additional correlation to the quaternary carbon resonating at 131.32 ppm which was correlated to the proton resonating at 8.40 ppm. We note that this carbon also correlates with one of the two protons contained in the multiplet centered at about 7.20 ppm. From the triplet structure of the two proton multiplets contained in this overlapped pair, they must both be component resonances of the four spin systems. Thus, if the proton resonating at 8.40 ppm is correlated to a quaternary carbon resonating at 131.25 ppm, which is in turn correlated to a proton of a four spin system then we establish the location of the three two spin systems above as suggested. These correlations are shown by fragment **7-10**.

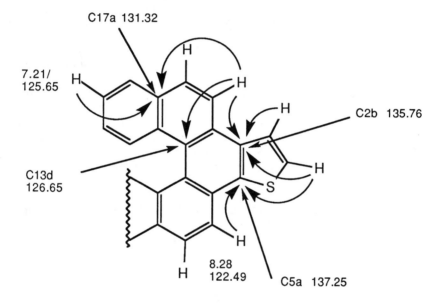

**7-10**

At this point, we have dispensed with the assignments of two of the three quaternary carbons to which H2 resonating at 8.40 ppm is coupled: C17a resonating at 131.32 ppm and C2b resonating at 135.76 ppm. The remaining quaternary carbon to which H2 may be coupled could either by C2a via a two bond coupling or C13d via a three bond coupling, the latter being the more probable coupling pathway. As we shall see below, the latter is the correct choice, as is indicated in structure **7-10**.

Given the identity of the three two spin systems and their orientation relative to one another in the molecular framework, we may direct our attention to the completion of the assignments of the four spin system comprised of H14-H17.

## Assigning H14-H17

Using the heteronuclear relay data once again, connectivities within the two four spin systems are illustrated in Fig. 7-14. As will be noted, the correlations linking H15 to H16 and H11 to H12 are absent in the relay spectrum. Thus, several spectra must be employed to complete the assignment of H14-H17 and their corresponding carbon resonances.

To this point in the assignment, we have not utilized the HOHAHA data presented above the diagonal with the COSY data in Fig. 7-10. It is now, however, germane to begin to use the information contained in this spectrum. Having identified H1 as the proton resonating at 8.03 ppm, we note a response correlating H1 with a proton resonating at 7.44 ppm that correlates with a directly bonded carbon (see Fig. 6-11) resonating at 128.00 ppm. The coupling pathway highlighted in the HOHAHA spectrum by this response could either be a four bond *peri* coupling between H1 and H17 or a five bond epi-zig-zag coupling between H1 and H14. Since we have located H16 as the proton resonating at 7.21 ppm from the long range data discussed above, it is a simple matter to identify the proton resonating at 7.44 ppm from the relay data presented in Fig. 7-14. If the proton resonating at 7.44 is indeed H17, it should exhibit a response in the relay spectrum to the H16 proton (7.21 ppm). Conversely, if the proton exhibits a response elsewhere, then its identity is confirmed as H14. In this case, the latter applies and we thus note that H14 couples to H15, which resonates at 6.70 ppm, C15 resonating at 124.81 ppm. Also using Fig. 7-14, we note that H16 correlates with a proton resonating at 7.84 ppm, which is thus assigned as H17, C17 resonating at 127.51 ppm. It is also useful to note that a *peri* coupling response is observed in the HOHAHA spectrum which correlates H1 with H17. Finally, in the HOHAHA spectrum we also observe a response correlating H14 and H17.

Given the identification and orientation of the components of the H14-H17 four spin system we can complete the assignment of the remaining quaternary carbon resonances in the "northern" region of the hexihelicene core of the molecule. From the long range optimized data, the connectivities shown in **7-11** can easily be deduced. Thus, H14, resonating at 7.44 ppm, is long range coupled to carbons resonating at 125.07 (C16), 131.32 (C17a) and 126.35 (C13b). H15 is long range coupled to carbons resonating at 130.48 (C13e); H16 is long range coupled to the carbon resonating at 131.32 (C17a); and finally H14 is long range coupled to the quaternary carbon resonating at 130.48 (C13e). All of these assignments are summarized in Table 7-4.

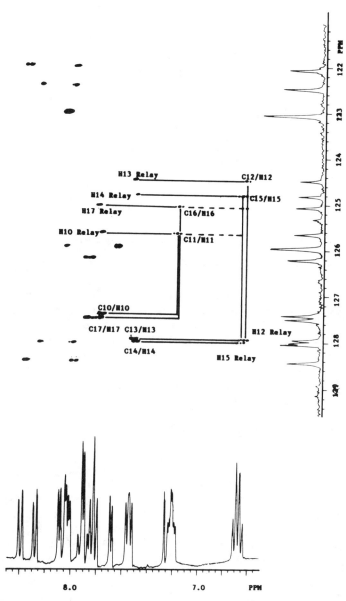

*Fig. 7-14.*    Heteronuclear relayed coherence transfer spectrum of phenanthro-
[3,4:3',4']phenanthro[2,1-*b*]thiophene (**6-5**) showing connectivities
in the two four spin systems.

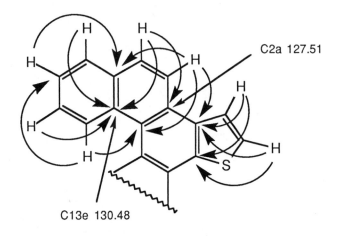

**7-11**

## Assigning the Remainder of the Spectrum

Given the assignments for half of the hexihelicene core and the thiophene ring plus the assignments of H6 via long range coupling to C5a and H7/C7 via the relay experiment, we can easily develop arguments for the assignment of the balance of the spectrum.

Beginning from H6 resonating at 8.28 ppm, we observe long range couplings to carbons resonating at 137.25 ppm, which was previously assigned as C5a, to a protonated carbon at 128.04 ppm, which is assigned as C7 and confirmed by the heteronuclear relay spectrum shown in Fig. 7-11; a strong response to a protonated carbon resonating at 122.49 ppm, which is the directly bonded proton and finally, a quaternary carbon resonating very far upfield at 122.16 ppm. The latter response must be assigned as the C13c quaternary carbon and is unique both in its chemical shift and in the fact that H6 will be the only proton to which it can be long range coupled.

Continuing to the long range couplings of H7, we observe couplings to carbons resonating at 129.20, 128.96, 126.04 and 122.49 ppm. The latter two are protonated carbons corresponding to the directly bonded C7 and C6, respectively. Responses to the quaternary carbons are more interesting in that they extend the connectivity network still further. The latter is a unique response, the carbon in question coupling to no other proton in the molecule. This is a likely occurrence only in the case of the C5b carbon. The response to the quaternary carbon resonating at 129.20 ppm is of greater interest in that it is also coupled to a proton resonating at 7.93 ppm, which is part of an AB spin system, which must hence be H8/H9. Thus, the carbon resonating at 129.20 ppm is assigned as C13b. It is also interesting to note that H7

(8.04 ppm) is correlated with H9 in the HOHAHA spectrum shown in Fig. 7-10. These coupling pathways are illustrated by **7-12.**

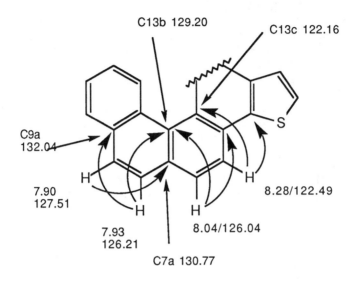

**7-12**

**7-13**

The final set of correlations that must be considered to complete the assignment hinges on C9a and C13a. The former we have already identified, and it resonates at 132.04 ppm. The latter we would expect to exhibit correlations to H9, H10 and H12. Indeed, these connectivities are observed and correlate the requisite three protons with the quaternary carbon resonating at 130.14 ppm, which is thus assigned as C13a. It will be recalled from the figure showing connectivities within the two four spin systems that the central spins of both four spin systems failed to give observable connectivities (see Fig. 7-14). The terminal spins were, however, linked to their respective vicinal neighbors and, based on the long range connectivities into the C9a and C13a carbons, the assignment can be completed as shown by the connectivities with **7-13**.

Table 7-4.    Proton and carbon NMR chemical shift assignments of phenanthro-[3,4:3',4']phenanthro[2,1-*b*]thiophene (**6-5**).

| Position | $\delta^1H$ (ppm) | $\delta^{13}C$ (ppm) |
|:---:|:---:|:---:|
| 1 | 8.03 | 128.44 |
| 2 | 8.40 | 122.07 |
| 2a | | 127.51 |
| 2b | | 135.76 |
| 3 | 8.09 | 123.10 |
| 4 | 7.69 | 125.97 |
| 5a | | 137.25 |
| 5b | | 128.96 |
| 6 | 8.28 | 122.49 |
| 7 | 8.04 | 128.04 |
| 7a | | 130.77 |
| 8 | 7.93 | 126.21 |
| 9 | 7.90 | 127.51 |
| 9a | | 132.04 |
| 10 | 7.81 | 127.42 |
| 11 | 7.22 | 125.65 |
| 12 | 6.68 | 124.48 |
| 13 | 7.56 | 127.94 |
| 13a | | 130.14 |
| 13b | | 129.20 |
| 13c | | 122.16 |
| 13d | | 126.35 |
| 13e | | 130.48 |
| 14 | 7.44 | 128.00 |
| 15 | 6.70 | 124.81 |
| 16 | 7.21 | 125.07 |
| 17 | 7.84 | 127.51 |
| 17a | | 131.32 |

## SOLUTION TO PROBLEM 6.6

### Elucidation of the Structure of
### A Marine Sesquiterpene

#### Finding a Place to Start

Given the set of spectra provided in Chapter 6, perhaps the best first undertaking would be to prepare a table of chemical shifts, both proton and carbon before going any further. This operation is easily completed using the heteronuclear chemical shift correlation spectrum shown in Fig. 6-16. This tabulation is presented in Table 7-5.

Table 7-5.   Proton and carbon chemical shift pairings.   Carbon resonance multiplicities from an APT spectrum are designated s = quaternary; d = methine; t = methylene; and q = methyl.

| δ C (ppm) | δ H (ppm) | |
|---|---|---|
| 69.8 (d) | 2.95 | |
| 59.0 (d) | 4.35 | |
| 50.3 (s) | | |
| 48.9 (s) | | |
| 33.7 (t) | 1.75 | 1.23 |
| 32.5 (t) | 2.35 | 1.31 |
| 31.5 (t) | 2.15 | 2.05 |
| 30.7 (d) | 0.30 | |
| 28.5 (q) | 0.88 | |
| 27.7 (d) | 0.35 | |
| 22.5 (q) | 0.95 | |
| 19.5 (q) | 0.82 | |
| 19.0 (s) | | |
| 16.4 (q) | 1.00 | |
| 14.7 (t) | 1.48 | 0.75 |

Given the structural information provided with the problem and the proton-carbon concordance presented, there are two plausible starting points. The first uses the hydroxyl bearing carbon resonating at 69.8 ppm as a starting point; the second utilizes the bromo methine carbon resonating at 59.0 ppm as a starting point. As we shall see in the ensuing solution of this problem, it is actually necessary to employ both of these starting points. Hence we shall begin from the hydroxyl methine carbon.

#### Assembly of a Structural Fragment
#### Beginning from the Hydroxyl Methine

Using the hydroxyl methine proton resonating at 2.95 ppm as a starting point, we note from the ZQCOSY spectrum shown in Fig. 7-15

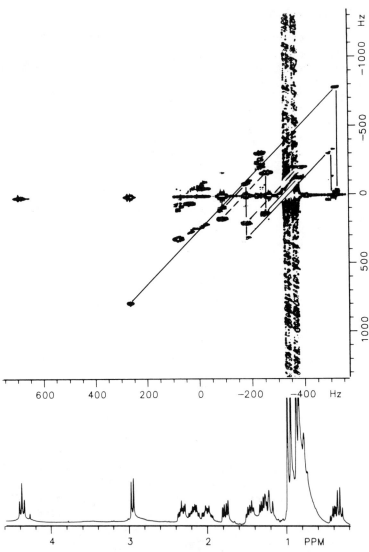

*Fig. 7-15.*          ZQCOSY spectrum of an unknown marine sesquiterpene acquired using
a 5 Hz optimization of the fixed delays in the pulse sequence shown in
Fig. 2-40. The data were taken as 512 x 1K complex points and were
processed using a sinusoidal multiplication prior to both Fourier
transformations. Shifts in hertz are relative to the transmitter;
shifts in ppm are downfield of tetramethylsilane.

that the proton resonance is a simple doublet. This observation allows us to imply that the hydroxymethine carbon is flanked on one side by a quaternary carbon. Hence, we will begin to assemble a structural fragment that can be extended in only one direction. Thus, we first establish a connectivity to the proton resonating at 0.30 ppm via the ZQCOSY spectrum shown in Fig. 7-15 using the responses located at ±775 Hz relative to $F_1$ = 0 Hz and the requisite $F_2$ shifts of the two protons. Given this chemical shift, it is quite probable that the molecule whose structure we are working on contians a cyclopropane ring somewhere in its structure. The fact that the proton resonating at 0.30 ppm is also correlated with a proton resonating at 0.35 ppm further confirms the cyclopropane suggestion. Indeed, from the information available to us thus far, we may draw the hypothesis that we are dealing with a cyclopropane ring fused to the framework of the molecule and further that the other available position on the cyclopropane is disubstituted. This latter contention is based on the fact that there is no methylene carbon paired with protons that would have chemical shifts consistent with a cyclopropane origin. To this point in the development of our structural problem, we have assembled the fragment defined by **7-14**.

**7-14**

By examining the multiplet for the proton resonating at 0.35 ppm, we note that although it is partially overlapped by its cyclopropyl counterpart resonating just upfield, there is more structure to the multiplet, which is suggestive of a pair of anisochronous methylene protons as vicinal neighbors. Returning to the ZQCOSY spectrum shown in Fig. 7-15, we note a further connectivity of the proton resonating at 0.35 ppm located at $F_1$ = -300 Hz. There is also a very weak trace of a possible response located at $F_1$ = -130 Hz. The former correlates with a proton resonating at 1.48 ppm, while the $F_1$ = +130 Hz correlation would locate a response at about 0.8 ppm in the band of $t_1$ noise associated with the methyl groups in the spectrum shown in Fig. 7-15. Returning to Table 7-5, we note that the methylene carbon resonating at 14.7 ppm has protons with chemical shifts of 1.48 and 0.75 ppm, which correspond to the connectivities observed in the ZQCOSY spectrum shown in Fig. 7-15. Now, if we direct our attention to the ZQCOSY spectrum shown in Fig. 7-16 which was optimized for 5 Hz, we observe the two responses just discussed with much better intensity as a function of the different optimization. Thus, we have confirmed our assignment of the anisochronous geminal methylene protons on the carbon flanking the cyclopropyl moiety.

Continuing, the protons at 0.75 and 1.48 ppm are correlated by a geminal coupling (dashed line), and each correlates further to protons resonating at 1.75 and 1.23 ppm the latter directly bound to the carbon resonating at 33.7 ppm. The final pair of anisochronous methylene protons are also geminally correlated (dashed line). It will be noted that the methylene proton resonating at 1.75 ppm appears as an apparent doublet of doublets due to the very weak coupling to the proton resonating at 0.75 ppm, which was not resolved. In turn, the proton resonating at 1.23 ppm also has the appearance of a doublet of doublets for similar reasons. Hence, since all of the couplings in both of these proton multiplets can be accounted for by protons we have already identified, we have reached the end of the connectivity network for this structural fragment, at least as definable by the ZQCOSY data the we have to work from. This leads to the conclusion that the next carbon must be, once again, a quaternary carbon. Consequently, the final structural fragment, which has been assembled from the hydroxyl methine starting point, is given by **7-15**.

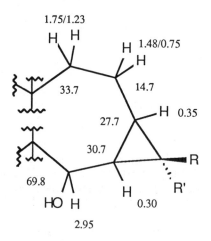

**7-15**

## Assembly of a Structural Fragment
## Beginning from the Bromo Methine

Our second potential starting point for the elucidation of the structure of the sesquiterpene was the overlapped double doublet from the bromo methine carbon, the proton in question resonating at 4.35

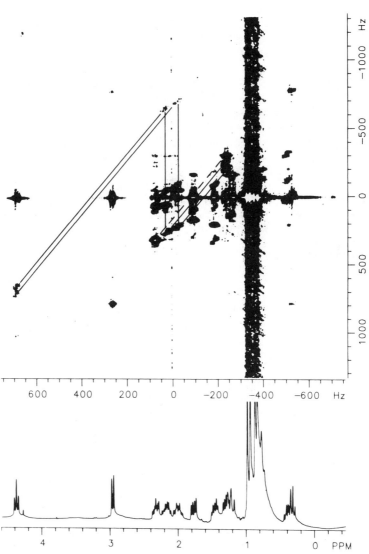

*Fig. 7-16.* ZQCOSY spectrum identical to that shown in Fig. 7-15 plotted at a significantly deeper contour level.

ppm. Using the structure of the multiplet in this case, which has a pair of equivalent vicinal couplings, we must conclude that either the bromo methine proton is flanked by methine protons on either side or by a methylene with a pair of anisochronous protons. Returning to the ZQCOSY spectrum shown in Fig. 7-16, we note that the proton at 4.35 ppm exhibits connectivities to two other protons with $F_1$ frequencies of ±655 and ±700 Hz. The responses correlate the methine proton with protons resonating at 2.15 and 2.05 ppm, respectively. From Table 7-5 we will note that both of these protons are directly bound to a methylene carbon resonating at 31.5 ppm, the latter of the two choices presented above. Hence, the bromo methine carbon must reside at the terminus of a structural fragment that may be assembled on the basis of proton homonuclear couplings flanked by a quaternary carbon on one side and the methylene carbon just identified on the other.

Continuing, we note in the ZQCOSY spectrum that the protons resonating at 2.15 and 2.05 ppm are also linked by a connectivity (dashed line) which is consistent with their geminal nature. Next, both of the methylene protons just identified are correlated with a proton resonating just downfield at 2.35 ppm and a proton resonating substantially upfield at 1.31 ppm. The protons at 2.35 and 1.31 ppm are also correlated with one another (dashed line) and are identified as another anisochronous methylene pair from the proton-carbon spectrum shown in Fig. 6-16 and data in Table 7-5. No further connectivities can be established from either of these assignments and we have hence reached another end point in our expanding connectivity network. This structural fragment must thus also be linked to a quaternary carbon, which allows us to now consider the second structural fragment represented by **7-16**.

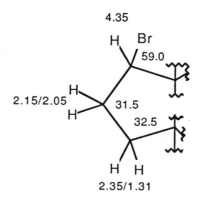

**7-16**

## What Pieces Are Left?

Having assembled two major pieces of the structure of our unknown sesquiterpene, it is now appropriate to take stock of the structural components that are left in our tally list in Table 7-5. Quite obviously, we have done nothing to assign any of the methyl resonances and we have only tentatively attributed the location of one of the quaternary carbons. In total, there are four methyls to be positioned in the molecular structure and three quaternary carbons. From the data given with the problem in Chapter 6, the next obvious step in the elucidation of the structure is to utilize the information contained in the proton detected long range heteronuclear multiple quantum spectrum shown in Figs. 6-17 and 6-18 (see Chapter 3 for a discussion of this experiment if necessary).

Before we consider the data, however, let us see what we have in terms of overall structural possibilities. First, since there are only two remaining quaternary carbon resonances (assuming that we are correct in our assumption that the third must be contained in the cyclopropyl moiety), these must be common to the termination points of both **7-15** and **7-16**. Hence, we at this point may conclude that we are dealing with a 5-7-3 ring system. Since the structural fragments represented by **7-15** and **7-16** may be linked in either of two possible orientations, there are, at this point, two structural possibilities, those represented by **7-17** and **7-18**. Hence, our final task is to determine unequivocally which of these possible structures is the correct one.

**7-17**                    **7-18**

## Linking the Structural Framents With Long Range Heteronuclear Connectivity Information from the HMBC Experiment

The segment of the HMBC spectrum shown in Fig. 7-17 provides correlations to the multiplets downfield of the methyl region; correlations to the methyls are shown in Fig. 7-18. Beginning from the hydroxyl methine proton resonating at 2.95 ppm, we note connectivities to three carbons that resonate at 30.7, 19.0 and 16.4 ppm, which correspond to a methylene, quaternary and methyl carbon, respectively. The identity of the carbon resonating at 30.7 ppm has been previously established as the vicinal neighbor carbon of the cyclopropyl moiety (see **7-14** and **7-15**). This information is hence not particularly valuable. The correlations to the quaternary and methyl carbons are somewhat more interesting as they represent new information. The former could be either a bridgehead carbon linking **7-15** and **7-16**, or alternatively, the quaternary cyclopropyl carbon whose existence was suggested above. Even more interesting is the correlation to the methyl carbon resonating at 16.4 ppm. This carbon must be one of the substituents of one of the quaternary carbons that link **7-15** and **7-16,** since the substituents of the cyclopropyl quaternary carbon would be too far removed to give a response to the hydroxyl methine.

Turning to connectivities involving the bromo methine proton resonating at 4.35 ppm, we observe only one connectivity to the methyl carbon resonating at 22.5 ppm. While useful information, it does not yet provide us with the means of discriminating between **7-17** and **7-18**.

The other information contained in the spectrum shown in Fig. 7-17 all arises from the proton resonating at 1.75 ppm, which is attached to the terminal methylene carbon in the structural fragment portrayed by **7-14**. Of paramount importance among these connectivities is the correlation with the bromo methine carbon resonating at 59.0 ppm. This is the first piece of information that links the fragments defined by **7-14** and **7-15** into a unified larger structure, shown by **7-19**. It must also be noted at this point that we have now also successfully differentiated between the two structural possibilities -- the correct structure of the sesquiterpene we are dealing with, at least circum-stantially, is that represented by **7-17,** although we must obtain final confirmation of this by linking all of the pieces together.

In addition to the key correlation to the bromo methine carbon, the proton resonating at 1.75 ppm correlates with the cyclopropyl methine carbon resonating at 27.7 ppm and both of the remaining quaternary carbons resonating at 50.3 and 48.9 ppm. All of these connectivities are shown by arrows on **7-19**. It must also be noted, however, that the two quaternary carbons, C4 and C10, have not been assigned at this point. There is no way that they can be differentiated from the connectivities to the H9 proton resonating at 1.75 ppm. Differentiation of the C4 and C10 resonances is, however provided by the connectivities of the H3 and H8 protons. The former, as shown by **7-19** correlates with C4, which resonates at 50.3 ppm, while the latter correlates with the quaternary resonating at 48.9 ppm, which must thus be C10. There are, in addition, numerous other connectivities that link the structural fragments shown by **7-14** and **7-15**, further serving to confirm the structure of the molecule. For example, H3 correlates with C6. H6 is correlated to the bridgehead methyl group, C12, which is in turn linked to C3. Thus, we are assured that the structure of the molecule has been correctly determined and the proton and carbon NMR spectra unequiv-

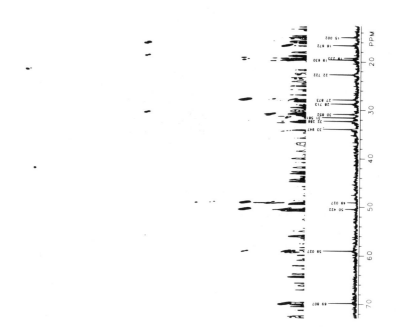

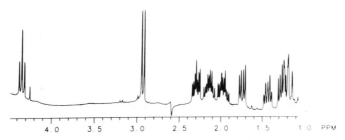

Fig. 7-17. Long range heteornuclear correlations observed in a proton detected long range heteronuclear multiple quantum (HMBC) experiment. Correlations are shown for the protons downfield of the methyl region.

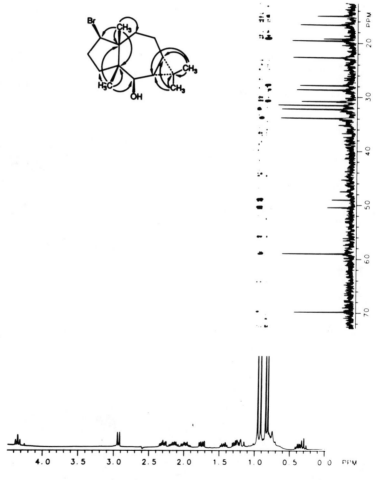

*Fig. 7-18.*          Long range heteronuclear correlations observed in a proton detected long range heteronuclear multiple quantum (HMBC) experiment. Connectivities are shown for the methyl groups.

ocally and totally assigned.   Final assignments for the molecule are
given in Table 7-6.

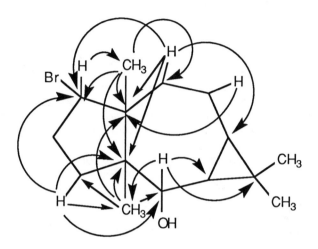

**7-19**

Table 7-6.   Proton and carbon chemical assignments for the sesquiterpene
neomerin (**7-17**) isolated from the calcareous alga *Neomeris annulata*.

| Position | $\delta$ C (ppm) | $\delta$ H (ppm) | |
|:---:|:---:|:---:|:---:|
| 1 | 59.0 | | 4.35 |
| 2 | 31.5 | 2.15 | 2.05 |
| 3 | 32.5 | 2.35 | 1.31 |
| 4 | 50.3 | | |
| 5 | 69.8 | | 2.95 |
| 6 | 30.7 | | 0.30 |
| 7 | 27.7 | | 0.35 |
| 8 | 14.7 | 1.48 | 0.75 |
| 9 | 33.7 | 1.75 | 1.23 |
| 10 | 48.9 | | |
| 11 | 22.5 | | 1.00 |
| 12 | 16.4 | | 0.95 |
| 13 | 19.0 | | |
| 14 | 28.5 | | 0.88 |
| 15 | 19.5 | | 0.82 |

## SOLUTION TO PROBLEM 6.7

### Assignment of the Structure of
### A Complex Alkaloid Using Carbon-
### Carbon INADEQUATE Data

### Where to Start?

Given that there are three nitrogen atoms in the molecular structure that we have been asked to assemble, the selection of a starting point is complicated somewhat. For want of anywhere better to start, let us begin from the methine carbon labeled "a," which resonates at 68.00 ppm furthest downfield in the spectrum shown in Fig. 7-19. On the basis of the chemical shift of this resonance, it is reasonably certain that it is directly attached to a nitrogen atom; the multiplicity then accounts for the attachment of two other carbons. By quickly inspecting the INADEQUATE spectrum shown in Fig. 7-19, this expectation is confirmed, by connectivities to resonances labeled "i" and "m," which resonate at 37.57 and 32.25 ppm, respectively. Resonances "i" and "m" are observed to be quaternary and methine, respectively, from the mutiplicity information contained in Table 6.1.

### Expanding the Connectivity
### Network Outward from Our
### Starting Point

Given resonance "a" as a starting point with neighboring resonances "i" and "m," let us begin to extend the connectivity network outward.

By examining the INADEQUATE spectrum shown in Fig. 7-19, we note that resonance "i," which is a quaternary carbon, is bound to three other carbons, "d," "e" and "n," which resonate at 61.58, 56.78 and 30.56 ppm, respectively, which are also identified as a methylene, methine and methylene respectively on the basis of the multiplicity information presented in Table 6-1.

In contrast, resonance "m," which is a methine carbon, exhibits only a connectivity to "h," a methylene carbon resonating at 38.68 ppm. The other empty valence position, based upon the chemical shift of "m", cannot be occupied by one of the nitrogen atoms. Hence, we have encountered the first potential break in the connectivity network. Referring back to the original presentation of the problem in Chapter 6, we note that resonances "l" and "m" were cited as having the potential to be AB rather than AX pairs that would, if they were adjacent to one another in the molecular structure, produce a break in the connectivity network. Hence, it would appear that we have encountered just such a case. Thus, although a connectivity has not been observed, we may tentatively locate "l" adjacent to "m" in the assembly of a structural fragment.

To this point, the structural fragment that we have assembled is depicted by **7-20**. The dashed line connecting positions "l" and "m" represents the inferred connectivity discussed above.

**7-20**

## Connectivities from "d"

To maintain an orderly expansion of the structural fragment from **7-20** let us consider extending the various "appendages" of the structural fragrment alphabetically. Thus, beginning with "d" we note that there is only the single connectivity to "i," which has already been identified. Hence, "d" must tentatively be a methylene carbon to which one of the nitrogen atoms is bound. This is also consistent with the chemical shift of "d," which resonates at 61.58 ppm.

## Connectivities from "e"

Expanding the connectivity network next from "e" affords us considerably more structural information than did the preceding effort. Thus, we observed that "e," a methine carbon, correlates with "r", a methylene carbon resonating at 24.40 ppm. As there is no other correlation to "e", the remaining valence site, on the basis of chemical shift, can probably be attributed to one of the nitrogen atoms.

From "r" we note in the portion of the spectrum folded in at the lower right that "r" correlates with "t" another methylene carbon that resonates at 20.74 ppm. The resonance labeled "t" correlates, in turn, with yet another methylene carbon labeled "s," which resonates at 23.80 ppm and is then correlated with the methine carbon labeled "b," which resonates at 68.00 ppm. No further connectivites from "b" are observed in the spectrum, which suggests that the remaining valence positions must be occupied by heteroatoms. This observation, in light of the chemical shift of "b" at 68.00 ppm, it somewhat surprising as we would normally expect a carbon atom attached to two heteroatoms to resonate somewhat further downfield. In any case, the structural fragment that we have assembled thus far is now shown by **7-21**.

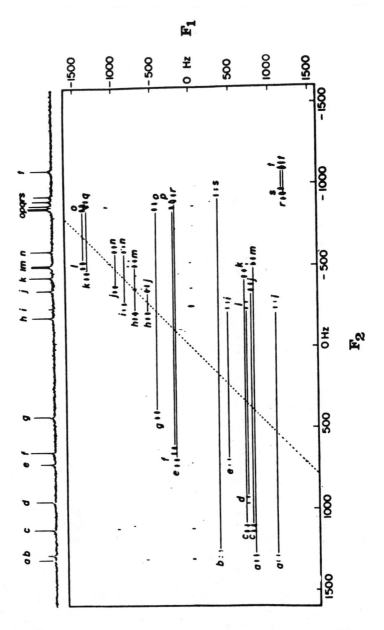

*Fig. 7-19.*        <sup>13</sup>C-<sup>13</sup>C INADEQUATE spectrum of an alkaloid recorded at 50.3 MHz. Chemical shifts and multiplicities are given in Table 6-1.

**7-21**

From even the most cursory examination of the structural fragment represented by **7-21,** we note that we have used five nitrogen atoms to account for multiplicity and chemical shift considerations while there are only three nitrogen atoms available to us from the molecular formula given with the problem in Chapter 6. Hence, we must begin to consider the formation of rings to condense the number of nitrogen atoms utilized in the structure. Before we undertake this operation, however, it is appropriate to complete the assembly of the carbon skeleton from the data contained in Fig. 7-19.

## Connectivities from "h"

Returning to our alphabetical protocol, let us next consider further connectivity information beginning from "h." First, we note that "h" is a methylene carbon and hence can have only one connectivity in addition to that to "m," which we have already established. Thus, we note that "h" is correlated with "j," a methine carbon resonating at 35.30 ppm.

Continuing from "j," we first observe a correlation to "c" a methine carbon resonating at 64.98 ppm which, on the basis of chemical shift considerations, is probably attached to a nitrogen atom. Second, we note that "j" is also correlated with "n," which we have previously identified as a methylene carbon attached to "i," which is shown in **7-21.** Consequently, the "j"-"n" correlation we have just observed completes the formation of a six-membered ring, giving us a bit more of the structure of the molecule.

Returning to "c" we note that it is correlated to a chain of four methylene carbons beginning with "k" resonating at 33.85 ppm, which is in turn linked to "q" resonating at 24.94 ppm. Next we observe that "q" exhibits no further correlation responses. Based on chemical shift considerations, "q" is obviously not linked to a heteroatom nor to a quaternary carbon, since the one quaternary carbon in the structure has been utilized in the form of "i." Thus, "q" may have a potential linkage to "p" resonating at 25.27 ppm, once again representing a break in the connectivity network due to an AB carbon-carbon pairing. Finally, "p" is linked to the fourth methylene carbon in the chain, "f," which resonates at 55.41 ppm. As there are no further correlations from "f," we can assume that it is attached to a nitrogen on the basis of chemical shift.

To this point, we have extended the connectivity network as far as we possibly can from "h" and now have expanded the structural fragment as shown by **7-22** where, once again, the dashed line indicates an inferred connectivity.

**7-22**

### Connectivities from "l"

The remaining point from which connectivities need to be traced is "l." Hence, we note that "l" is correlated with a methylene carbon, "o," which resonates at 23.38 ppm, which is in turn finally correlated with the methylene carbon "g" resonating at 50.96 ppm which exhibits no further connectivities. Based on chemical shift considerations, "g" is also a likely candidate for attachment to a heteroatom.

It is interesting to note in this case that all 20 of the carbon atoms have been linked into a single large structural fragment, which is depicted by **7-23**. The task remaining at this point is to try to link the various pendant chains from the central cyclohexyl unit into a single unified structure. It is also important to note that if a tally is taken of the numbers of hydrogens utilized in assembling **7-23**, we have accounted for 32 of the 33 possible hydrogen atoms, leaving one, which must therefore be attached to one of the nitrogen atoms.

**7-23**

## Linking the Structural Fragment
## into a Single Unified Structure

The single but nontrivial task remaining before us in the solution of this problem is that of tying the various appendages of our octopus-like structural fragment, **7-23**, back to give the requisite number of nitrogen atoms in a single structure. As depicted above, the structure requires potentially the influence of nitrogen at as many as eight carbons to account for chemical shift behavior. Clearly we have only three nitrogen atoms available to us and we are thus faced with the task of making some sense out of the structural fragment.

One thing to consider in starting the final task of assembling a structure is that alkaloids tend to be comprised largely of five- and six-membered rings with occasional larger rings such as seven and eight members. Smaller and larger rings than the range just specified are quite rare. Given this preamble, let us consider some of the possibilities.

One quite legitimate starting point entails the positioning of "b" between two nitrogen atoms. If indeed this is correct we must then have two nitrogen atoms located in reasonable proximity in the molecule. This requirement is satisfied if we utilize the nitrogen atoms linked to carbons "a" and "d" to provide the two nitrogens, with "b" positioned between them. This operation gives the structural fragment shown by **7-24** and reduces the number of "nitrogen candidate" locations in the molecule substantially. It should also be noted here that the nitrogens have now also been numbered and the two nitrogens just utilized were N1 and N3, respectively. Notice also that the structure has been redrawn relative to **7-23** and that in the operation that we have just described we have created two additional six membered rings, one a piperidine the other a 1,3-piperidine.

**7-24**

Given the rearranged drawing of the structural fragment, it should now be fairly obvious how the problem is finally solved. First, using the methylene chain appended to "m," we can make another piperidine ring once we realize that N1 and N2 are indeed the same nitrogen atom. Finally, we can construct two additional six-membered rings with a bridgehead nitrogen when we realize that N4, N5 and N6 are also one and the same. This completes the assembly of the structure which is shown by **7-25**, with the final detail being the location of the single remaining hydrogen atom. This final task represents no challenge as N3 still has one valence remaining open to which the hydrogen must be attached.

**7-25**

Assembling the structure of panamine,[9,10] as it was just presented, is straightforward only if the student realizes that the two nitrogens in proximity, N1 and N3, must also be the same two nitrogens with which carbon "b" is associated. Once this relationship is grasped, the number of plausible structures is substantially diminished, particularly if one restricts the possibilities to five- to eight-membered rings. Overall, the structure of the alkaloid just assembled is quite complex and beyond the reach of combined COSY/HC-COSY/long range HC-COSY experiments at medium field. This is in large part because the proton spectrum at 200 MHz, at which the work was originally done, is essentially unresolved. While the situation would improve at higher field, it would quite likely take the full resolution provided by a 600 MHz spectrometer to make this problem tractable with any technique short of the INADEQUATE experiment. Clearly, there are thus substantial indications for the use of the INADEQUATE experiment. However, since this is a demanding and very time consuming experiment when there are limited quantities of material available, it should probably be reserved for those cases where its use is absolutely essential.

## SOLUTION TO PROBLEM 6.8

### Establishing the Structure of a Complex Alkaloid Using Proton-Proton, Proton-Carbon and Long Range Proton-Carbon Connectivity Information

### Where to Begin?

Given the spectral data shown in Figs. 6-20 through 6-23 and the information about the infrared spectrum, we have a valuable clue. The amide and N-H stretching information, coupled with the pattern of proton resonances (see the reference spectrum plotted beneath the COSY contour plot, Fig. 6-20), it is reasonably clear that we have a 1,2-disubstituted aromatic ring. Further, the chemical shift of one of the protonated aromatic carbons above 110 ppm is suggestive of an indole nucleus. Typically, C7 in the indole nucleus resonates in this vicinity.[11] Since there are only two nitrogen atoms in the structure, given the infrared data, if the carbonyl is associated with the proposed indole nucleus, we must then be dealing with a spiro-oxindole alkaloid, **7-26**.

Given the premise of a spiro-oxindole nucleus from which to work, it is useful to see if this supposition can be confirmed using the data available to us before we go any further with the problem. Assuming the proton resonating at 6.65 ppm to correspond to the 7-position of the indole nucleus, we would expect to observe long range couplings to a protonated carbon and a quaternary carbon, which would be C5 and C3a, respectively using the numbering of indole. Examining the long range heteronuclear correlation spectrum presented in Fig. 6-22, we find that H7 is indeed long range coupled to two additional carbons, one of them protonated and resonating at 121.41 ppm, the other quaternary and resonating at 131.74 ppm. Given the identity of H5, the other proton resonance, which appears as an overlapped doublet of doublets (apparent triplet), must be H6. This coupling is confirmed by the COSY spectrum shown in Fig. 6-20. Next, we would expect the same long range coupling behavior of H6 as we have oberved for H5, which is indeed the case (referring to Fig. 6-22). Thus, H6 is long range coupled to a protonated carbon resonating at 127.97 ppm, which must be C4, and a quaternary carbon resonating at 140.52 ppm, which is consistent with what we would expect the chemical shift of C7a to be. Hence, in this simple series of operations we have managed to assign completely the phenyl ring of our anticipated indole nucleus.

7-26

As we continue, quite usefully we also should note that C3a exhibits a long range coupling to a proton resonating at 3.81 ppm, which provides us with a means of beginning to logically probe the aliphatic portion of the structure. Importantly, this same proton also couples to the carbonyl carbon resonating at 179.25 ppm, suggesting that it must be located within three bonds of the aliphatic proton in question. While this does not confirm that the carbonyl is a part of the spiro-oxindole nucleus that we are attempting to assemble, a proton located three bonds away from C3a would also be three bonds away from the carbonyl carbon located at the 2-position in oxindole! Furthermore, the chemical shift of the proton at 3.81 ppm and the data in Fig. 6-21 indicate that it is attached to an oxygenated methine carbon (69.24 ppm), which provides us with the location of the second oxygen in the structure of the alkaloid that we are determining. This structural information is embodied in structural fragment 7-27.

A further coupling from the proton resonating at 1.98 ppm to the carbonyl carbon plausibly establishes the identity of the other substituent attached to the spiro-center in 7-27. Thus, although we have not directly established the identity of the spiro-oxindole nucleus, we now have available very considerable evidence that supports this structural moiety.

Using data contained in the COSY spectrum shown in Fig. 6-20, we may next locate the vicinal neighbors of the two aliphatic protons just identified as coupling to the carbonyl. Assuming three bond

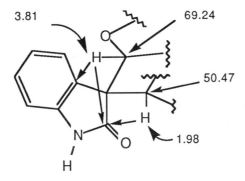

**7-27**

couplings to be the most prominent, we would thus anticipate that the vicinal neighbors would couple into the spiro-center thus confirming the identity of the spiro-oxindole nucleus. Thus, the proton resonating at 3.81 ppm is vicinally coupled to a pair of strongly anisochronous geminal methylene protons that resonate at 2.01 and 2.83 ppm (22.67 ppm, Fig. 6-21). The proton resonating at 1.98 ppm is vicinally couped to a methine proton resonating at 3.47 ppm (71.81 ppm, Fig. 6-21) the chemical shifts of this proton-carbon pairing suggestive of attachment to a nitrogen or oxygen. Since both of the oxygens available have been tentatively utilized, this leaves nitrogen as the preferred choice. Given the identity of these protons as the vicinal neighbors of the protons resonating at 3.81 and 1.98 ppm, we may begin to look for quaternary carbon resonances to which both of these protons couple as possible candidates for the spiro carbon of the indole nucleus.

Referring to Fig. 6-23, we find two resonances that are coupled to the requisite protons and are also quaternary: the carbons resonating at 53.96 and 53.82 ppm. Using slices extracted from the long range heteronuclear correlation spectrum, which are shown in Fig. 7-20, we note that the quaternary carbon resonating at 53.82 ppm is long range coupled to the exomethylene protons resonating at 4.95 and 5.09 ppm, while the former exhibits a coupling to the proton previously identified as its potential vicinal neighbor resonating at 3.81 ppm. Hence, we may assign the spiro-oxindole carbon as that resonating at 53.96 ppm, while the other quaternary carbon provides us with new information about further structural features, since it must probably be three bonds removed from both the proton resonating at 3.47 ppm and the anisochronous geminal methylene protons resonating at 2.01 and 2.83 ppm. This suggests the bond indicated by the dashed line in **7-28**.

To expand the structural fragment shown by **7-28** further, we must again return to the COSY spectrum shown in Fig. 6-20. Now we note that aside from the connectivity to the proton resonating at 3.47 ppm, the proton resonating at 1.98 ppm exhibits no other off-

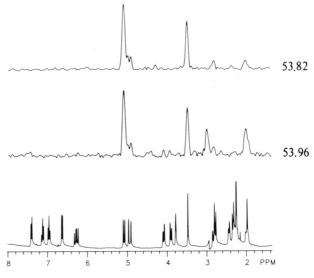

*Fig. 7-20*    Slices taken from the long range heteronuclear chemical shift correlation spectrum shown in Fig. 6-23 for the two aliphatic quaternary carbon resonances. Chemical shifts are labeled.

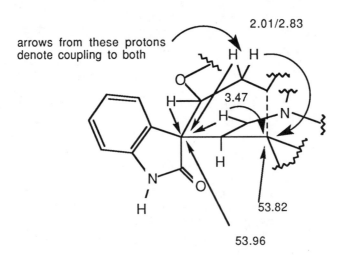

**7-28**

diagonal connectivity responses. This suggests that the proton resonating at 1.98 ppm is attached to a carbon linked to a quaternary carbon, which is also consistent with the inclusion of the dashed bond shown in **7-28**. The structural constituents attached to the quaternary center, however, cannot easily be traced from the proton resonating at

1.98 ppm that we have just been dealing with. Rather, it is now necessary to proceed from an alternative site in the structure.

Another logical starting point is provided by the anisochronous methylene protons resonating at 3.92 and 4.11 ppm, which are directly attached (Fig. 6-21) to a carbon resonating at 61.33 ppm. The chemical shift of the directly attached carbon is suggestive of attachment to oxygen. Referring back to **7-27**, we obviously have some substituent attached to the oxygen associated with the proton/carbon pair resonating at 3.81/69.24 ppm. Irrefutable support of the methylene carbon resonating at 61.33 ppm as the other ether carbon is afforded by the long range coupling between the proton resonating at 3.81 ppm and this carbon atom observed in Fig. 6-23. Conversely, the proton resonating at 3.92 ppm also exhibits a long range coupling to the carbon resonating at 69.24 ppm, further supporting the ether linkage we have just defined. Next, we note from the COSY spectrum shown in Fig. 6-20 that the anisochronous methylene protons are both coupled to a methine proton (see Fig. 6-21) resonating at 2.43 ppm, which is directly attached to a carbon resonating at 37.93 ppm. Usefully, we now also note long range coupling responses between the anisochronous methylene protons resonating at 3.92 and 4.11 ppm and the carbon resonating at 71.81 ppm, which bears the proton resonating at 3.47 ppm that we have been dealing with above. Since the long range coupling responses most often observed will be across three bonds, this implies an intermediary carbon (37.93 ppm) between the methylene carbon just identified and the methine carbon resonating at 71.81 ppm. Hence the bonds denoted by dashed lines in **7-29** are those inferred in the deductive process just described.

To this point, we have made substantial progress. Since natural products generally tend to contain oxygen heterocycles comprised of five, six, or seven members, we now may consider the size of the potential oxygen heterocycle that we are beginning to assemble atop the molecular structure shown in **7-29**. In the COSY spectrum shown in Fig. 6-20 we note that the proton resonating at 2.43 ppm is coupled to a proton that resonates at 2.30 ppm, which is coupled, in turn, to a proton resonating at 2.01 ppm. The latter we have previously identified as vicinal to the proton resonating at 3.81 ppm and also as being long range coupled to the quaternary carbon that resonates at 53.82 ppm. Given this set of observations, we have constructed a dihydropyran ring structure. Let us consider additional long range couplings that may further substantiate the structural fragment we have just inferred.

Using the proton resonating at 2.43 ppm as a point of departure, we observe that it exhibits a direct coupling in Fig. 6-23 and a coupling to a carbon resonating at 35.54 ppm that is directly bound to the proton resonating at 2.30 ppm. The carbon resonance associated with the proton resonating at 2.43 ppm also exhibits long range coupling responses to protons resonating at 2.01 and 2.83 ppm, which were previously identified as the anisochronous methylene protons vicinal to the oxygenated methine proton resonating at 3.81 ppm.

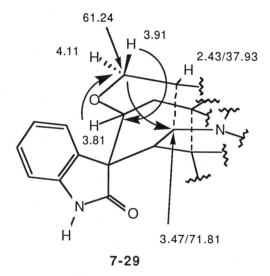

**7-29**

The carbon resonating at 35.54 ppm exhibits long range couplings to the proton resonating at 3.92 ppm, one of the ether methylene protons; a response to the proton resonating at 3.47 ppm; a response to the proton resonating at 2.01 ppm; finally, a response to a proton not previously dealt with, which resonates at 2.78 ppm and is associated with the methylene carbon reonating at 65.88 ppm. These correlations lead to the structure shown by **7-30**.

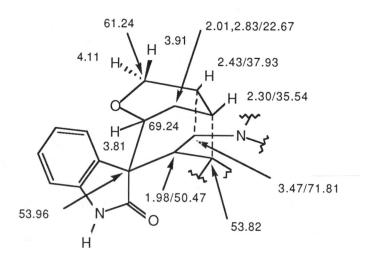

**7-30**

In concert, the long range couplings just enumerated confirm the closure of a tetrahydropyran ring and identify a methylene carbon long range coupled to the carbon resonating at 35.54 ppm. Continuing, we have only a few more pieces of the molecule left to assemble. The carbon just identified at 65.88 ppm is long range coupled to the proton resonating at 3.47 ppm whose directly bound carbon is coupled in turn, long range, to the N-methyl group, which is also coupled to the carbon at 65.88 ppm. This circular series of long range couplings supports the assembly of another and the final ring in the structure of the alkaloid that we are determining. This structure is shown by **7-31**.

To this point, we have accounted for the entire molecular structure with the exception of the two vinyl carbons and their three attendant protons. At this point, it is appropriate to note that the exomethylene protons resonating at 4.95 and 5.09 ppm both couple to the quaternary carbon resonating at 53.82 ppm, which now completes the structure of the molecule as shown by **7-32**.

The alkaloid that we have been dealing with is gelsemine. Historically, gelsemine was the first alkaloid whose [13]C-NMR spectrum was assigned, in a communication by Wenkert and coworkers.[12] More recently, Schun and Cordell[13] revised some of Wenkert's assignments on the basis of COSY and HC-COSY 2D-NMR data. The chemical shift assignments made in the process of deducing the structure of gelsemine are totally consistent with the assignments of Schun and Cordell. The COSY spectrum showing connectivities discussed above is shown in Fig. 7-21. The annotated long range heteronuclear chemical shift correlation spectra are shown in Figs. 7-22 and 7-23. Final chemical shift assignments are presented in Table 7-7.

**7-31**

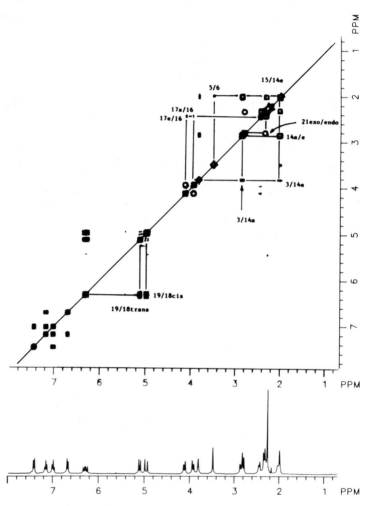

*Fig. 7-21.*     COSY spectrum of gelsemine (**7-32**) acquired as 512 x 1K complex points in deuterochlorform at 300.068 MHz. The data were processed using sinusoidal multiplication prior to both Fourier transformations and were symmetrized prior to plotting. Responses are labeled using the convention downfield/upfield.

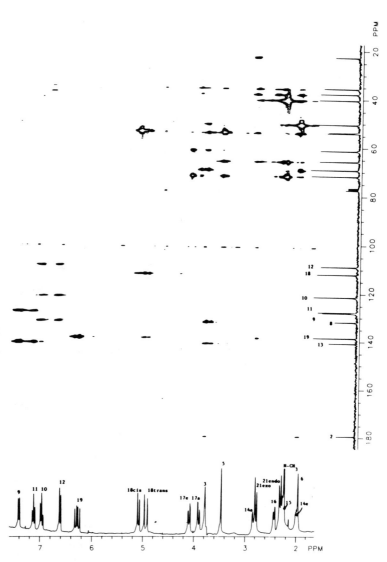

*Fig. 7-22.*    Long range heteronuclear chemical shift correlation spectrum of
gelsemine (**7-32**) recorded using the pulse sequence shown in Fig. 3-
51, which provides for one bond modulation decoupling. The long range
delays were optimized for 10 Hz. $^{13}$C-Responses in the vinyl and
aromatic region of the spectrum are labeled along the axis.

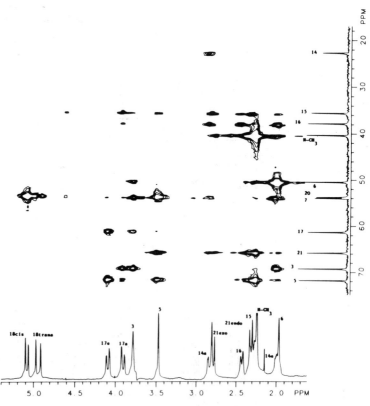

*Fig. 7-23.* Expansion of the long range heteronuclear chemical shift correlation spectrum shown in Fig. 7-22. Only the aliphatic region of the spectrum is plotted to show detail. Responses are labeled along the axis.

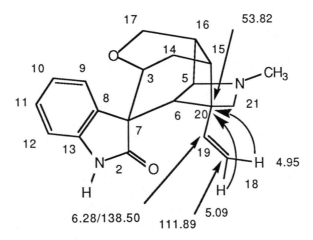

**7-32**

Table 7-7. Proton and carbon chemical shift assignments for the alkaloid gelsemine (**7-32**). The numbering scheme used is that shown on the structure in the text.

| Position | Proton Chemical Shift | Carbon Chemical Shift |
|---|---|---|
| 2 | | 179.25 |
| 3 | 3.81 | 69.24 |
| 5 | 3.47 | 71.81 |
| 6 | 1.98 | 50.47 |
| 7 | | 53.96 |
| 8 | | 131.74 |
| 9 | 7.43 | 127.97 |
| 10 | 6.97 | 121.41 |
| 11 | 7.15 | 127.70 |
| 12 | 6.65 | 108.80 |
| 13 | | 140.52 |
| 14 | 2.01    2.83 | 22.67 |
| 15 | 2.30 | 35.54 |
| 16 | 2.43 | 37.93 |
| 17 | 3.91    4.11 | 61.33 |
| 18 | 4.95    5.09 | 111.89 |
| 19 | 6.26 | 138.50 |
| 20 | | 53.82 |
| 21 | 2.32    2.78 | 65.88 |
| N-CH$_3$ | 2.24 | 40.40 |

## SOLUTION TO PROBLEM 6.9

### Determining the Structure of an Isomeric Dinaphthothiophene

#### Preliminary Structural Considerations

Before we embark on the determination of the structure, recall that we were told in the problem we are dealing with an isomer that is a dinaphthothiophene related to the compound considered in problem 6.1. Let's consider some structural possibilities. First, we know that there can be no linear isomers such as **6-1** since there are no proton singlets analogous to H7 and H12 in **6-1**. Hence, the compound that we are dealing with in this problem must have a nonlinear structure. Second, if we examine the COSY spectrum shown in Fig. 7-24, it is almost immediately apparent that the proton spectrum exhibits a degree of complexity inconsistent with either of the possible symmetric, non-linear structures. Hence, the most plausible starting structure is that shown by **7-33**.

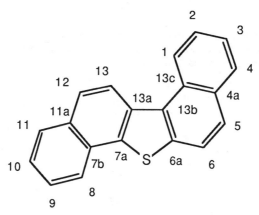

**7-33**

#### Interpreting the COSY Data

Examining the off-diagonal responses above the diagonal in the COSY spectrum shown in Fig. 7-24 we note that the proton resonating furthest downfield correlates with three other protons. In contrast, the sharper doublet just upfield correlates with only one other resonance. On this basis, since we have two protons in a bay region in **7-33** above, one a constituent of a four spin system, the other a member of a two spin system, we may tentatively assign these as H1 and H13, respectively. In the 2.5 Hz Gaussian processed data, we note that H1 exhibits a correlation to only one proton, which appears as a triplet resonating at 7.75 ppm in a clear region of the spectrum. Following this correlation

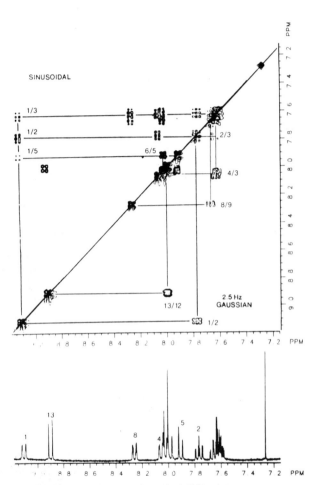

*Fig. 7-24.* COSY spectrum of a dilute solution of a dinaphthothiophene recorded in deuterochloroform at 300.068 MHz. The data were acquired as 256 x 512 complex points and were zero filled before each Fourier transformation during processing to give a final data matrix consisting of 512 x 512 points prior to symmetrization. The data above the diagonal were subjected to sinusoidal multiplication prior to both Fourier transformations; the data below the diagonal were processed using a 2.5 Hz Gaussian multiplication prior to both transformations.

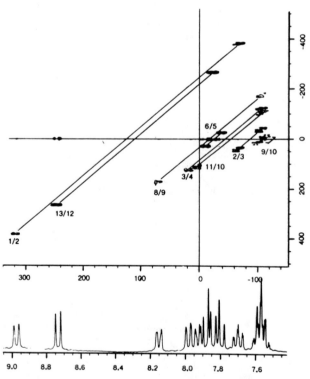

*Fig. 7-25.*     ZQCOSY spectrum of a solution of the dinaphthothiophene at the same concentration as that used for the heteronuclear correlation experiments. The spectrum was acquired as 256 x 512 complex points using the Müller ZQCOSY pulse sequence shown in Fig. 2-40, in which the fixed delay, τ, was optimized for a 7 Hz vicinal coupling. The data were processed with sinusoidal multiplication prior to both Fourier transformations. Axes labeled in ppm are downfield of TMS; axes labeled in hertz are relative to the transmitter.

further upfield, H2 is correlated to one of the three protons resonating as a part of a complex multiplet in the region of 7.6 ppm, and the connectivity cannot be satisfactorily followed beyond this point. Clearly at this point we have reason to resort to the ZQCOSY data shown in Fig. 7-25.

## Connectivity Information Provided by the ZQCOSY Experiment

Unlike the COSY spectrum shown in Fig. 7-24, correlations beyond H2 can be readily followed in the ZQCOSY spectrum shown in Fig. 7-25. Hence, H3 is traced as the "center" resonance of the three resonances furthest upfield. It should be noted that the ZQCOSY spectrum was also acquired at the same concentration used for the heteronuclear correlation experiments and that there is a significant shift in the position of some of the resonances in the ZQCOSY spectrum relative to their positions in the COSY spectrum. Hence, we shall utilize the chemical shifts in the former for all further discussion. Finally, H3 correlates with the final member of the four spin system, which resonates furthest downfield in the cluster in which it is contained, in direct contrast to what might be expected from the coupling responses in the COSY spectrum. Concentration induced shifts notwithstanding, the H4 resonance is clearly not the same proton to which H1 exhibits a correlation in the COSY spectrum. Indeed, this may be confirmed by examining a COSY spectrum run at higher concentration (not shown). Thus, the long range coupling in the COSY spectrum identifies H5 via a five bond epi-zig-zag coupling pathway and thus orients two of the spin systems relative to one another.

The balance of the connectivity information contained in the ZQCOSY spectrum is not remarkable other than to note that even the H9-H10 correlation between protons resonating at 7.462 and 7.494 ppm, respectively, is clearly distinguished in the ZQCOSY spectrum. In contrast, the off-diagonal responses correlating H9 with H10 in the COSY spectrum are obscured by the diagonal responses.

## Heteronuclear Correlation Data

The heteronuclear chemical shift correlation spectrum shown in Fig. 6-26 provides the identity of the directly attached carbon for each of the proton resonances identified in the COSY/ZQCOSY experiments. No further comment is warranted or necessary.

We are faced next with the problem of orienting the remaining four spin system relative to the H12-H13 two spin system. The latter is correctly oriented by virtue of the chemical shift of H13, which is in a bay region. Finally, we must also assign the quaternary carbon resonances to complete the assignment.

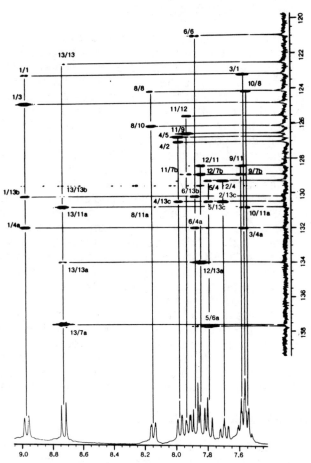

*Fig. 7-26.*      Long range optimized HC-COSY spectrum of **7-33** acquired using the
pulse sequence shown in Fig. 3-51, which provides for modulation
decoupling.   The BIRD pulse located midway through the $\Delta_2$ interval
was optimized for an assumed 165 Hz one bond coupling; long range
delays were optimized for 10 Hz.   The data were taken as 256 x 1K
complex points and were processed using an exponential multiplication
prior to both Fourier transformations.   High resolution proton and
carbon reference spectra are plotted flanking the contour plot.

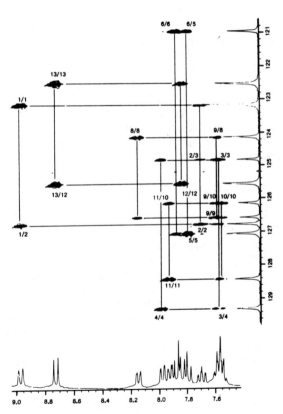

*Fig. 7-27.*     HC-RELAY spectrum of **7-33** acquired using the pulse sequence shown
in Fig. 4-20.    The data collection and processing were essentially
identical to "that" specified in Fig. 7-26 except that a 7 Hz vicinal
coupling was assumed for the magnetization relay portion of the
experiment.

## Using the HC-RELAY Data

To assemble the pieces of the last four spin system in correct order we must be very careful in our interpretation of the ZQCOSY correlation data; alternatively, we may use either the long range optimized heteronuclear correlation data shown in Fig. 7-26 or the HC-RELAY data presented in Fig. 7-27. The latter spectrum has the advantage of greater simplicity but is limited in that it provides no information about the identity of the quaternary carbon resonances. We shall, however, employ the HC-RELAY data to order the spins of the remaining four spin system.

Beginning from the proton resonating at 8.15 ppm, which is directly bound to the carbon resonating at 124.10 ppm, which we may tentatively assign as H8 on the basis of chemical shift considerations, we note first from the HC-RELAY spectrum shown in Fig. 7-27 that this proton relays to one of the heavily overlapped group of protons furthest upfield, specifically that resonating downfield in the group at 7.494 ppm. Going to the H9/C9 direct response (7.494/126.54 ppm), we note that H9 relays to H10, which resonates furthest upfield within the group at 7.462 ppm. Finally, from the H10/C10 direct response (7.462/126.12 ppm) we observe a relay response from the proton resonating at 8.063 ppm whose directly bound carbon resonate at 128.39 ppm. Hence through this simple series of operations we have correctly ordered the four proton and carbon resonances of the H8-H11 four spin system. This now leaves us only the task of assigning the eight quaternary carbon resonances.

## Assigning the Quaternary Carbons

Using the long range heteronuclear chemical shift correlation spectrum shown in Fig. 7-26, we may complete the spectral assignment given our knowledge of proton and protonated carbon resonance assignments. Simply on the basis of chemical shift considerations the two quaternary carbons resonating furthest downfield may be attributed to C6a and C7a. Hence, the resonance at 137.60 ppm may be assigned as C7a on the basis of a long range coupling to H13 while the resonance at 137.67 ppm may be assigned as C6a from its long range coupling to H5. These quaternary carbons are unusual in that they do not exhibit multiply redundant correlations because of their locations in the molecular structure. While it is conceivable that C7a could also exhibit a correlation to H8, this was not observed at optimizations of 5, 7.5 or 10 Hz.

In contrast to the single correlations observed to C6a and C7a, we may use our proton assignments to assign other quaternary carbons which exhibit multiple long range correlations. One such example is C13c. Using the location of H5, we observe a correlation to a carbon resonating at 130.46 ppm, which is reasonable for C13c. This assignment is irrefutably confirmed by correlations to H2 and H4 as illustrated by **7-34**. In similar fashion, C4a is assigned as the resonance at 131.96 ppm  by correlations to both H3 and H6; C7b is assigned at 128.88 ppm via correlations to H9, H11 and H12; C11a is assigned at 130.78 ppm through correlations to H8, H10 and H13; C13a is assigned as the resonance at 133.97 ppm on the basis of a single correlation to H12.  Most importantly, however, C13b exhibits correlations to both H1 and H13.  In the case of a molecule of unknown structure, this carbon would be of paramount importance in that it links together the two major segments of the molecule.  It is also worth noting that these correlations were observed only in the spectrum shown in Fig. 7-26 which was optimized for 10 Hz.  Reoptimization of the experiment for 5 or 7.5 Hz failed to give these responses in either case.

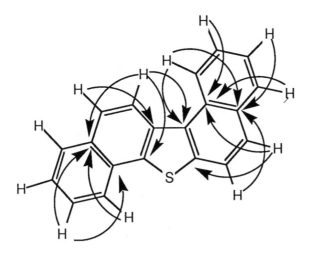

**7-34**

Final chemical shift assignments for **7-33** are given in Table 7-8. In determining the structure of this molecule, we were aided rather substantially by the knowledge that we were dealing with an isomer of the compound from the first problem.  Irrespective of this starting "boost," the method employed unequivocally yields a single unique structure for this set of data.[14]

## CONCLUSIONS

The problems that we have dealt with in Chapters 6 and 7 were predominantly taken from work conducted in the laboratories of the authors during the final months of the writing of this monograph.  They are not necessarily representative of the work typically done in that lab, nor will they necessarily reflect samples routinely encountered in any other laboratory.  We feel, however, that these problems are illustrative of some of the ways in which the techniques presented in this monograph can be combined to yield unequivocal structural information.

Table 7-8. Proton and carbon chemical shift assignments of dinaphtho-[1,2-*b*:1',2'-*d'*]thiophene (**7-33**) recorded in deuterochloroform.

| Position | $\delta^1H$ (ppm) | $\delta^{13}C$ (ppm) |
|----------|-------------------|----------------------|
| 1 | 8.885 | 123.13 |
| 2 | 7.609 | 126.74 |
| 3 | 7.482 | 124.80 |
| 4 | 7.893 | 126.27 |
| 4a | | 131.96 |
| 5 | 7.702 | 127.04 |
| 6 | 7.787 | 120.94 |
| 6a | | 137.67 |
| 7a | | 137.60 |
| 7b | | 128.88 |
| 8 | 7.838 | 124.10 |
| 9 | 7.494 | 126.54 |
| 10 | 7.462 | 126.12 |
| 11 | 8.063 | 128.39 |
| 11a | | 130.78 |
| 12 | 7.746 | 125.53 |
| 13 | 8.640 | 122.50 |
| 13a | | 133.97 |
| 13b | | 130.17 |
| 13c | | 130.46 |

The number of potential combinations of available two-dimensional NMR techniques that may be brought to bear on the solution of a given problem is quite substantial. In large part, it will depend upon the nature of the sample, the quantity of material available, spectral resolution and spectrometer time available to the investigator and the pulse sequence library. As each of these factors varies from one problem to the next, so probably will the approach used. Irrespective of this, the future is quite bright for the application of two dimensional NMR techniques in the elucidation of complex molecular structures. As additional new techniques are devised, structure elucidation strategies we cannot at present conceive may possibly become routine. Indeed, even during the course of writing this monograph, new techniques have become available that could not be included in the interest of a timely completion of the writing. We can only hope that readers will keep an open mind when considering the capabilities of new techniques and that they will find the challenge of using these experiments to solve their problems as stimulating as we have.

# REFERENCES

1.  M. D. Johnston, Jr., G. E. Martin and R. N. Castle, *J. Heterocyclic Chem.*, **25**, in press (1988).
2.  K. D. Bartle, D. W. Jones and R. S. Matthews, *Rev. Pure Appl. Chem.*, **21**, 2445 (1968).
3.  D. E. Barnekow, J. H. Cardellina, II, A. S. Zektzer and G. E. Martin, *J. Am. Chem. Soc.*, **110**, in press (1988).
4.  E. Wenkert, C.-J. Chang, H. P. S. Chawla, D. W. Cochran, E. W. Hagaman, J. C. King and K. Orito, *J. Am. Chem. Soc.*, **98**, 3645 (1976).
5.  M. J. Musmar, A. S. Zektzer, G. E. Martin, R. T. Gampe, Jr., M. L. Lee, M. L. Tedjamulia, R. N. Castle and R. E. Hurd, *Magn. Reson. Chem.*, **25**, 1039 (1987).
6.  M. D. Johnston, Jr. and G. E. Martin, *Magn. Reson. Chem.*, submitted (1988).
7.  P. Bigler, W. Ammann and R. Richarz, *Org. Magn. Reson.*, **22**, 109 (1984).
8.  A. Bax, D. G. Davis and S. K. Sarkar, *J. Magn. Reson.*, **63**, 230 (1985).
9.  T. H. Mareci and R. Freeman, *J. Magn. Reson.*, **48**, 158 (1982).
10. N. S. Bhacca, M. F. Balandrin, A. D. Kinghorn, T. A. Frenkiel, R. Freeman and G. A. Morris, *J. Am. Chem. Soc.*, **105**, 2538 (1983).
11. M. S. Morales-Rios, J. Espiñeira and P. Joseph-Nathan, *Magn. Reson. Chem.*, **25**, 377 (1987).
12. E. Wenkert, C.-J. Chang, A. O. Clouse and D. W. Cochran, *J. Chem. Soc., Chem. Commun.*, 961 (1970).
13. Y. Schun and G. A. Cordell, *J. Nat. Prod.*, **48**, 969 (1985).
14. M. D. Johnston, Jr., M. Salazar, D. K. Kruger, R. N. Castle and G. E. Martin, *Magn. Reson. Chem.*, **26**, in press (1988).

# Index

# *INDEX*